OPERATIONAL AMPLIFIERS

OPERATIONAL AMPLIFIERS
Third edition

G. B. Clayton
BSc, FInstP
Principal Lecturer in Physics
Liverpool Polytechnic, England

and

B. W. G. Newby
BSc (Eng), CEng, MIEE
Senior Lecturer in Electronics
Brooklands Technical College,
Weybridge, England

 NEWNES

Newnes
An imprint of Butterworth-Heinemann Ltd
Linacre House, Jordan Hill, Oxford OX2 8DP

 PART OF REED INTERNATIONAL BOOKS

OXFORD   LONDON   BOSTON
MUNICH   NEW DELHI   SINGAPORE   SYDNEY
TOKYO   TORONTO   WELLINGTON

First published by Newnes-Butterworths 1971
Reprinted 1974, 1975, 1977
Second edition 1979
Reprinted 1980
Reprinted by Butterworths 1981, 1982, 1983, 1985, 1986
Third edition 1992

British Library Cataloguing in Publication Data
Clayton, G. B.
  Operational Amplifiers. – 3Rev.ed
  I. Title II. Newby, B. W. G.
  621.381535

ISBN 0 7506 0640 1

Printed and bound in Great Britain

CONTENTS

PREFACE TO THE THIRD EDITION

In writing this third edition of *Operational Amplifiers*, an attempt has been made not only to up-date the information contained in the second edition but at the same time to broaden its coverage. This new edition is intended to serve as a useful text book and reference work for students following National and Higher National Certificate and Diploma courses, undergraduates and practising engineers.

The first seven chapters of the second edition of this book have been repeated largely unchanged other than for minor up-dating of some of the circuit diagrams. While some of the operational amplifier type numbers mentioned in these early chapters are somewhat dated, they are still in service and most can still be purchased new. Rather than try to up-date each mention of the older type, a new Appendix A.5 has been included in this third edition to give references to and basic data about many of the latest devices available.

Chapter 8 has been given the new title of *Operational Amplifiers in Instrumentation Applications* and modified by the inclusion of additional information about the interfacing of operational amplifiers and transducers while the original small section on 'active filters' has been extracted, much expanded, and rewritten as a new Chapter 9.

The additional information included in the new Chapter 9 starts with a simple discussion of passive filters, at a level suitable for National Certificate student technicians, and progresses through the reasons for using active filters, their differing types and characteristics, to simple design considerations. The discussion does not pretend to be more than an in-depth introduction to active filters and does not require the reader to have an extensive mathematical knowledge.

Chapter 10, *Practical Considerations*, contains all the original material from the second edition together with substantial additions of new information about some of the technologies and developments in the operational amplifier field in the 1990s. Also included is a new section comparing the performance data and the characteristics of the present-day range of operational amplifiers.

There is a new Appendix A.4. This is intended for those readers who are to be involved with the 'front end' of instrumentation systems using transducers as mensurative sensors followed by operational amplifiers in signal conditioning circuits. Details are included of a range of commonly used transducers, all of which are supplied by RS Components Ltd and who have kindly given their permission for the reproduction of the appropriate technical specifications and photographs.

Finally, in order to give the book a more practical aspect and the reader a first-hand experience of the type of information available from the manufacturers about their operational amplifiers, a selection of data sheets and other information is included at Appendix A.5. Motorola Ltd, National Semiconductor (UK) Ltd and Texas Instruments Ltd are thanked for their kind co-operation in this matter.

Bruce W. G. Newby

CHAPTER ONE

FUNDAMENTALS

1.1 Introduction

The term 'operational amplifier' was originally introduced by workers
in the analogue computer field to denote an amplifier circuit which
performed various mathematical operations such as integration, differentiation, summation and subtraction. In such circuits the required
response is obtained by the application of negative feedback to a high
gain d.c. amplifier by means of components connected between amplifier
input and output terminals in a particular manner referred to as
'operational feedback'. The term operational amplifier is now more
loosely used to designate any high performance d.c. amplifier suitable
for use with this type of feedback. Operational amplifiers are still widely
used for analogue computation and, although analogue techniques are
being superseded by digital methods in computer engineering, the range
of applications for operational amplifiers in instrumentation and control
system engineering is rapidly expanding.

The increase in the use of operational amplifiers in analogue systems
is to a large extent attributable to the availability of high performance
ready built amplifiers in discrete component modular form and more
recently in the form of inexpensive integrated circuits. The use of such
ready built amplifiers makes it possible to adopt a new approach to
many electronic design problems. The approach involves the selection
of a suitable amplifier and the connection of a few discrete components
to it to form a complete sub-system. In many cases this is more economical than designing with individual components and it frees the
engineer from the tedious and often time consuming task of d.c.
amplifier design; it seems likely that it will displace discrete component
designs in a wide variety of fields. Amplification is probably the most
important basic capability of electronic circuits, for with an amplifier
it is possible to synthesise most other important circuits. The range of
possible applications in which the operational amplifier approach may
be usefully adopted is enormous and is only really limited by the
ingenuity of the user.

It is not essential that the user of operational amplifiers be familiar
with the intricacies of their internal circuit details, but he must under-

1

stand the function of the external terminals provided by the manufacturer and the terms used to specify the amplifier's performance if he is to be able to select the best amplifier for a particular application.

1.2 The ideal operational amplifier. Operational feedback

As far as input and output terminals are concerned operational amplifiers with three varieties of single ended and differential connections are available. Amplifier input and output terminals are said to be single ended if one of the pair of terminals is earthed (or common), otherwise they are said to be earth free or floating. In the case of floating amplifier input or output terminals, the signal of importance is the difference in the voltages at the two terminals irrespective of any common voltage they may both have with respect to earth. The conventional circuit symbols for the three basic varieties of amplifier are illustrated in Figure 1.1. Negative and positive signs are used to distinguish between the phase-inverting and non-phase-inverting input terminals of the differential amplifier. If the input terminal marked with a positive sign is earthed and a signal is applied to the input

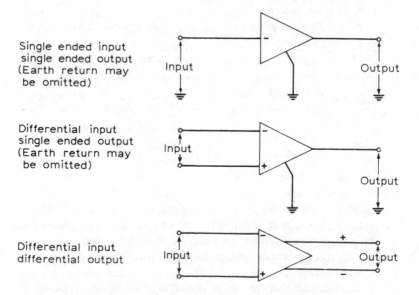

Single ended input
single ended output
(Earth return may
be omitted)

Differential input
single ended output
(Earth return may
be omitted)

Differential input
differential output

Figure 1.1 Circuit symbols for operational amplifiers

2

terminal marked with a negative sign it will appear phase inverted at the output of the amplifier. The converse holds if the terminal marked with a negative sign is earthed.

In a first consideration of an operational feedback it is found convenient to assume that the amplifier has certain ideal characteristics. The effects of departures from the ideal, exhibited by real operational amplifiers, may then be expressed later in terms of the errors to which they give rise. Differential input single ended output operational amplifiers are the ones most commonly used and we will refer specifically to this type of amplifier in the treatment which follows. A differential input amplifier allows a greater flexibility in the choice of feedback configuration than a single ended input amplifier. The output of the ideal differential input amplifier depends only on the difference between the voltages applied to the two input terminals. Any common voltage that the two input terminals may have with respect to earth is called a common mode input voltage. The output of the ideal differential amplifier is unaffected by any common mode input voltage. The ideal amplifier is further assumed to have the following characteristics.

Infinite Gain as we shall show makes the performance entirely dependent on input and feedback networks.

Infinite Input Impedance ensures that no current flows into the amplifier input terminals.

Infinite Bandwidth is a bandwidth extending from zero to infinity, ensuring a response to d.c. signals, zero response time and no phase change with frequency.

Zero Output Impedance ensures that the amplifier is unaffected by the load.

Zero Voltage and Current Offset ensures that when the input signal voltage is zero the output signal will also be zero regardless of the input source resistance.

There are two basic ways of applying operational feedback to a differential amplifier: Figure 1.2a shows the inverting configuration, the non-inverting configuration being illustrated in Figure 1.2b. In both circuits the signal fed back is proportional to the output voltage, feedback takes place via the resistor R_2 connected between the output and the phase-inverting input terminal of the amplifier. Phase inversion through the amplifier ensures that the feedback is negative. In the in-

3

verting circuit the feedback signal is effectively applied in shunt with the external input signal, in the non-inverter input signal and feedback are effectively in series.

The action of both circuits may be understood if a small voltage e_ϵ is assumed to exist between the differential input terminals of the amplifier. The signal fed back will be in opposition to e_ϵ and with the amplifier gain tending to infinity the voltage e_ϵ will be forced towards zero. This is an extremely important point and is worth restating in an alternative form. Thus, when the amplifier output is fed back to the inverting input terminal the output voltage will always take on that value required to drive the signal between the amplifier input terminals to zero. In the case of an amplifier having large but finite gain the error voltage e_ϵ is small but non-zero, and the effect will be discussed in the next chapter.

Figure 1.2 The two basic operational feedback circuits

A second basic aspect of the ideal circuit follows from the assumed infinite input impedance of the amplifier. No current can flow into the amplifier so that any current arriving at the point X, as a result of an applied input signal, must of necessity flow through the feedback path R_2. If instead of the single resistor R_1 connected to the inverting input terminal there are several alternative signal paths the sum of these several currents arriving at X must flow through the feedback path. It is for this reason that the phase-inverting input terminal of an operational amplifier (point X) is sometimes referred to as the amplifier summing point. The two basic aspects of ideal performance are called the summing point restraints; they are so important that they are repeated again.

4

1. When negative feedback is applied to the ideal amplifier the differential input voltage approaches zero.
2. No current flows into either input terminal of the ideal amplifier.

The two statements form the basis of all simplified analyses of operational feedback circuits; we use them to derive closed loop gain expressions for the circuits of Figure 1.2.

Inverter closed loop gain. (Refer to Figure 1.2a)

The first summing point restraint,

$$e_\epsilon = 0$$

gives
$$I_i = \frac{e_i}{R_1} \quad \text{and} \quad I_f = -\frac{e_o}{R_2}$$

The second restraint gives $I_i = I_f$

Thus,
$$-\frac{e_o}{R_2} = \frac{e_i}{R_1}$$

and the closed loop gain,

$$A_{VCL} = \frac{e_o}{e_i} = -\frac{R_2}{R_2} \tag{1.1}$$

Non-inverter closed loop gain. (Refer to Figure 1.2b)

Since no current flows into either input of the amplifier the voltage at the inverting input terminal,

$$e_B = \frac{e_o}{R_1 + R_2} R_1$$

But
$$e_\epsilon = 0$$

Therefore
$$e_i = e_B$$

or
$$e_i = \frac{e_o}{R_1 + R_2} R_1$$

and the closed loop gain

$$A_{VCL} = \frac{e_o}{e_i} = 1 + \frac{R_2}{R_1} \tag{1.2}$$

5

Apart from the difference in the sign of the closed loop gain the main difference between the two circuits lies in the effective input resistance which they present to the signal source e_i.

The effective input resistance of the ideal inverter measured at the amplifier summing point is zero. Feedback prevents the voltage at this point from changing; the point acts as a virtual earth. Note that any current supplied to this point does not actually flow to earth but flows through the feedback path R_2. The input current I_i in Figure 1.2a is thus determined by the resistor R_1 and the input resistance presented to the signal source is R_1.

The ideal non-inverting circuit Figure 1.2b takes no current from the signal source and thus presents an essentially infinite input impedance. It is worth noting that even if the differential input impedance of the amplifier itself is not infinite, as assumed, an infinite amplifier gain is a sufficient condition to ensure that the effective input impedance of the non-inverting circuit is infinite.

Infinite amplifier gain causes the voltage fed back to the inverting input terminal to be exactly equal in magnitude to the applied input voltage. The feedback voltage is effectively in series with the input voltage, it opposes the input voltage and so prevents any current from being driven into the amplifier. The inverter and non-inverter illustrate the difference between negative feedback applied in shunt and in series with the external signal. In general shunt negative feedback lowers input impedance, series negative feedback increases input impedance.

The comparative simplicity of the closed loop gain expressions should not make the reader forgetful of their significance. Closed loop gains in the ideal case depend only on the values of series and feedback components, not on the amplifier itself. Real amplifiers introduce departures from the ideal, and these are conveniently treated as errors. Using currently available operational amplifiers, errors can be made very small and one of the main features of the operational amplifier approach to instrumentation is the accuracy with which it is possible to set gain and impedance values.

1.3 More examples of the ideal operational amplifier at work

The ideal operational amplifier serves as a valuable starting point for a preliminary analysis of operational amplifier applications and in this section we present a few more examples illustrating the usefulness of

the ideal operational amplifier concept. Once the significance of the
summing point restraints are firmly understood, ideal circuit analysis
involves little more than the intelligent use of Ohm's law. Remember
that the ideal differential input operational amplifier with negative
feedback applied to it has an output voltage which takes on just that
value required to force the differential input voltage to zero and to
cause all currents arriving at the inverting input terminal to flow
through the feedback path.

1.3.1 THE IDEAL OPERATIONAL AMPLIFIER ACTS AS A CURRENT TO VOLTAGE CONVERTER

An ideal operational amplifier can act as a current to voltage converter.
In the circuit of Figure 1.3, the ideal amplifier maintains its inverting
input terminal at earth potential and forces any input current to flow

Feedback forces
the potential of
this point to
zero

$I_f = I_{in}$

R_f

I_{in}

$e_o = -I_{in}R_f$

Figure 1.3 An ideal operational amplifier acts as a current to voltage converter

through the feedback resistance. Thus $I_{in} = I_f$ and $e_o = -I_{in}R_f$. Notice
that the circuit provides the basis for an ideal current measurement; it
introduces zero voltage drop into the measurement circuit, and the
effective input impedance of the circuit measured directly at the inverting input terminal is zero.

1.3.2 THE IDEAL OPERATIONAL AMPLIFIER ADDS VOLTAGES OR CURRENTS INDEPENDENTLY

The principle involved in the current to voltage converter circuit of
Figure 1.3 may be extended. In the ideal operational amplifier circuit
of Figure 1.4 (a) the operational amplifier forces the sum of the several
currents arriving at the inverting input to flow through the feedback

7

path (there is no where else for them to go). The inverting input terminal is forced to be at earth potential (a virtual earth) and the output voltage is thus:

$$e_o = - [I_1 + I_2 + I_3] R_f$$

In Figure 1.4 (b) a number of input voltages are connected to resistors which meet at the inverting input terminal. The ideal operational amplifier maintains the inverting input at earth potential, thus each input

Figure 1.4 An ideal operational amplifier adds voltages and currents independently

current is independently determined by each applied input voltage and series input resistor. The sum of the input currents is forced to flow through R_f and the output voltage must take on a value which is equal to the sum of the input currents multiplied by R_f.

$$e_o = - R_f \left[\frac{e_1}{R_1} + \frac{e_2}{R_2} + \frac{e_3}{R_3} \right]$$

Figure 1.5 An ideal operational amplifier can act as a voltage to current converter

Figure 1.6 An ideal operational amplifier can act as an ideal unity gain buffer stage

8

1.3.3 THE IDEAL OPERATIONAL AMPLIFIER CAN ACT AS A VOLTAGE TO CURRENT CONVERTER

In maintaining its differential input voltage at zero the amplifier shown in the circuit of Figure 1.5 forces a current $I = e_{in}/R$ to flow through the load in the feedback path. The value of this current is independent of the nature or size of the load.

1.3.4 THE IDEAL OPERATIONAL AMPLIFIER CAN ACT AS A PERFECT BUFFER

In the circuit of Figure 1.6 the amplifier output voltage must take on a value equal to the input voltage in order to force the differential input signal to zero. The ideal circuit has infinite input impedance, zero output impedance and unity gain, and acts as an ideal buffer stage.

Figure 1.7 An ideal operational amplifier can act as a subtractor

1.3.5 THE IDEAL OPERATIONAL AMPLIFIER CAN ACT DIFFERENTIALLY AS A SUBTRACTOR

The circuit shown in Figure 1.7 illustrates the way in which an operational amplifier can act differentially as a subtractor. The voltage at the inverting input terminal is (by superposition)

$$e_- = e_2 \frac{R_2}{R_1 + R_2} + e_o \frac{R_1}{R_1 + R_2}$$

The voltage at the non-inverting input is

$$e_+ = e_1 \frac{R_2}{R_1 + R_2}$$

The operational amplifier forces $e_- = e_+$

Thus

$$e_2 \frac{R_2}{R_1 + R_2} + e_o \frac{R_1}{R_1 + R_2} = e_1 \frac{R_2}{R_1 + R_2}$$

or

$$e_o = \frac{R_2}{R_1} [e_1 - e_2]$$

1.3.6 THE IDEAL OPERATIONAL AMPLIFIER CAN ACT AS AN INTEGRATOR

In the circuit of Figure 1.8 negative feedback is applied by the capacitor C connected between the output and the inverting input terminal. The amplifier output voltage acting via this capacitor maintains the inverting input terminal at earth potential and forces any current arriving at the inverting input terminal to flow as capacitor charging current. Thus:

$$I_{in} = \frac{e_{in}}{R} = C \frac{dV_c}{dt}$$

The output voltage is equal in magnitude but opposite in sign to the capacitor voltage. Therefore:

$$\frac{e_{in}}{R} = - C \frac{de_o}{dt}$$

and

$$e_o = - \frac{1}{CR} \int e_{in} \, dt$$

The output is proportional to the integral with respect to time of the input voltage.

1.3.7 LIMITATIONS OF THE IDEAL OPERATIONAL AMPLIFIER CONCEPT

Real operational amplifiers have characteristics which approach those of an ideal operational amplifier but do not quite attain them. They have an open loop gain which is very large (in the region 10^5 to 10^6) but the gain is not infinite. They have a large but finite input imped-ance. They draw small currents at their input terminals (bias currents), they require a small differential input voltage to give zero output voltage (the input offset voltage), and they do not completely reject common mode signals (finite c.m.r.r.). In our discussion of ideal opera-tional amplifier circuits no mention has been made of frequency response characteristics—real amplifiers have a frequency dependent gain which can have a marked effect on the performance of operational amplifier applications.

10

Figure 1.8 An ideal operational amplifier can act as an integrator

The above features of real operational amplifiers cause the performance of a real operational amplifier circuit to differ from that predicted by an analysis based upon the assumption of ideal amplifier performance. In many respects the differences between real and ideal behaviour are quite small but in some aspects of performance, particularly those involving frequency dependent performance parameters, the differences are marked. The next chapter presents detailed discussions about the parameters which are normally given on the data sheets of practical operational amplifiers, and shows how a knowledge of these parameter values can be used to predict the behaviour of practical circuits. In the next section a brief practical introduction to real amplifiers is given to prepare the reader for his study of Chapter 2.

1.4 Integrated circuit operational amplifiers

Inexpensive integrated circuit operational amplifiers are now freely available, they are easy to use and they permit the comparatively inexperienced electronicist to build working circuits rapidly. The newcomer to operational amplifiers is strongly advised[1] to connect up a few of the basic operational amplifier applications and practically evaluate their performance; this forms a useful learning and familiarisation exercise which is worth performing before delving more deeply into the finer aspects of operational amplifier performance.

A preliminary practical evaluation of operational amplifier applications is most conveniently carried out using a general purpose or internally compensated operational amplifier type. The significance of this terminology will be discussed in the next chapter.

There are several general purpose amplifier types to choose from; at the present time the most widely used is the type 741 amplifier. Most manufacturers produce their version of this amplifier type, for example Analog Devices, A.D.741 J,K,L and S series, National Semiconductors L.M.741, Motorola's MC1741, Signetics μA 741. It seems likely that the 741 will eventually be superseded by the newer improved general purpose designs that are now appearing on the market. These new amplifier designs incorporate f.e.t. input stages formed on the same integrated circuit chip with standard bipolar transistors, for example, National Semiconductors Bi Fet amplifiers, type L.F. 155, 156 and 157 series, and R.C.A.'s Bi-Mos amplifiers type C A 3140 series.

As a user of operational amplifiers one does not need a detailed knowledge of their internal circuitry, and fortunately most general purpose operational amplifiers are pin compatible. It is with the func-tion of the external pin connections that the operational amplifier user is primarily concerned. The eight pin dual-in-line plastic package and the metal can are the most commonly used operational amplifier circuit packages. These are shown in Figure 1.9. Operational amplifiers are normally used with dual power supplies; input and output voltages are measured with respect to the potential of the power supply common terminal which acts as the zero signal reference point. The use of dual supplies allows input and output voltages to swing both positive and negative with respect to the zero reference point. Figure 1.10 shows the circuit connections which are required to make a practical form of the inverting amplifier circuit previously encountered in Section 1.2. Amplifier pin connections which are not shown in Figure 1.10 should be left with no connections made to them—their function will be described later.

In his preliminary experimentation with operational amplifiers the newcomer will inevitably make wrong connections; fortunately most general purpose operational amplifiers will tolerate many wrong connections. Particular care however should be taken to ensure that the power supplies are connected to the correct pins, as incorrect power supply connections can permanently damage an amplifier. Input signals should not be applied to an amplifier before power supplies are switched on, as application of input signals with no power supplies connected can damage an amplifier.

Practical forms of the ideal amplifier circuits given in the previous sections should be tried out. The ideal circuits require simply the addition of appropriate power supply connections in order to make working circuits. A preliminary acquaintance with real amplifiers

Figure 1.9 I.C. operational amplifier packages

Figure 1.10 Connections for practical inverting amplifier circuit

obtained in this way will help the reader to understand more readily the discussion of amplifier characteristics which is to be presented in the next chapter.

REFERENCE

1. CLAYTON, G. B., *Experiments with Operational Amplifiers—Learning by Doing*, Macmillan (1975). *88 Practical Op Amp Circuits you can Build*, Tab Books (1977)

Figure 1.11 Circuits for exercise 1.2

Exercises

1.1 Give component values and sketch diagrams of operational amplifier circuits for the following applications. Assume ideal operational amplifier performance.

(a) An amplifier voltage gain −5 and input resistance 100 kΩ.

(b) An amplifier voltage gain −20 and input resistance 2 kΩ.

(c) An amplifier voltage gain +100 with ideally infinite input resistance.

(d) An integrator with the circuit performance equation $e_o = -100 \int e_{in} dt$ and an input resistance 100 kΩ.

(e) A circuit which when supplied by an input signal of 2 V will drive a constant current of 5 mA through a variable load resistor.

1.2 Find the value of the amplifier output voltage for each of the circuits given in Figure 1.11. In all cases assume that the operational amplifier behaves ideally.

CHAPTER TWO

UNDERSTANDING OPERATIONAL AMPLIFIER PERFORMANCE PARAMETERS

A wide variety of operational amplifier types are now commercially available. The selection of the best amplifier type for a particular application is becoming an increasingly difficult problem. This is particularly true for the new user of operational amplifiers, for when first delving into manufacturers' catalogues, he is faced with what seems to be at first sight a bewildering variety of specifications and different amplifier types. It will usually be found that the more demanding the performance requirements the more costly will be the amplifier neces- sary to fulfil these requirements; so the choice of amplifier then will probably ultimately be governed by economic considerations. The least expensive one which will meet the design specifications will be chosen. In order to make this choice the design objectives must be completely defined and the relationship between published amplifier parameters and their effects on overall circuit performance in any intended applica- tion must be well understood.

This chapter will introduce the reader to the various amplifier specifications normally included in a manufacturer's data sheet; the significance of the parameters will be discussed. An added difficulty exists here in that there is not as yet any standardisation of specifica- tion definitions amongst the various manufacturers so that, when consulting data sheets, it is important to understand under exactly what conditions a particular parameter is defined. Differences are not great, but they do exist. The important question of amplifier selection will be

Figure 2.1 Idealised transfer curve for an operational amplifier

returned to in a later chapter after the reader has been introduced to amplifier applications; he should then more clearly appreciate design objectives and the way in which amplifier parameters limit their achievement.

2.1 Amplifier output and input limitations

The output of a differential amplifier is related to its inputs by the somewhat idealised open loop transfer curve illustrated in Figure 2.1; with open loop gains greater than 10^4 being quite typical only a very small voltage between the two input terminals is needed to cause saturation of the amplifier output voltage.

The *maximum output voltage swing* $V_{o\,max}$ is the maximum change in output voltage, (positive and negative) measured with respect to earth, that can be achieved without clipping of the signal waveform caused by amplifier saturation. Values of $V_{o\,max}$ are quoted for the amplifier working into a specified load (sometimes at full rated output current) and with specified values for amplifier power supplies. For applications requiring a d.c. response and an output voltage capable of swinging above and below earth potential (as shown by the transfer curve) operational amplifiers are generally designed to operate from symmetrical positive and negative voltage supplies (V^+, V^-). Maximum values for supply voltages are normally specified and should not be exceeded. Values for $V_{o\,max}$ will be found to be dependent on the magnitude of the supply voltages actually used.

Maximum voltage between inputs. The voltage between the input terminals of an operational amplifier is maintained at a very small value under most operating conditions by the feedback circuit in which the amplifier is used. If the application is such that the voltage between the terminals may be appreciable, care must be taken to ensure that it does not exceed the maximum allowable value for the particular amplifier as otherwise permanent damage to the amplifier may be caused. Some amplifiers are internally protected against input overload conditions; where such internal protection is not provided parallel back to back diodes may be connected externally to the amplifier input terminals to provide the necessary protection.

Maximum common mode voltage. The voltage at both inputs of a differential amplifier may be raised above earth potential. The input common mode voltage is the voltage above earth at each input when both inputs are at the same voltage. The maximum common mode voltage, E_{cm}, is the maximum value of this voltage which can be applied without producing clipping or nonlinearity at the output.

If an amplifier is to be used under conditions in which excessive common mode voltage may cause damage, protection can be obtained by the use of a suitable pair of zener diodes. The circuit of Figure 2.2 illustrates a method of protecting both against excessive voltage between inputs and excessive common mode voltage; current limiting by resistance in series with the input terminals is assumed.

2.2 Gain terminology. Feedback principles

2.2.1 OPEN LOOP VOLTAGE GAIN

The open loop voltage gain, A_{OL}, of a differential input operational amplifier may be defined as the ratio,

$$\frac{\text{change of output voltage}}{\text{change of input voltage}}$$

the input voltage being that measured directly between the inverting and non-inverting input terminals of the amplifier. A_{OL} is normally specified for very slowly varying signals and can in principle be determined from the slope of the non-saturated portion of the input/output transfer curve (Figure 2.1). The magnitude of A_{OL} for a particular amplifier depends on the amplifier load and on the value of the power supplies. Values of A_{OL} are normally quoted for specified supply voltages and load.

Operational amplifiers are seldom used open loop, being normally connected in negative feedback circuits in order to define precise operation; the significance of open loop gain is that it determines the accuracy limits in such applications. An assessment of the quantitative effects of the open loop gain magnitude requires a study of the principles underlying feedback amplifier operation.

2.2.2 NEGATIVE FEEDBACK AMPLIFIERS

In a negative feedback amplifier circuit a signal is fed back from the output to the input terminal, and this signal is in opposition to the externally applied input signal. The signal which actually drives the input terminals of the amplifier is that which results from a subtraction

Silicon diodes

Zeners

Figure 2.2 Circuit protection with diodes

process. The larger the gain of the amplifier without feedback (the open loop gain) the smaller is the signal voltage applied between the amplifier input terminals. If the open loop gain of the amplifier is infinite (as assumed for an ideal operational amplifier) negative feedback forces the amplifier input signal to zero; however, with a large, but finite amplifier open loop gain, a small input signal must exist between the amplifier input terminals. It is convenient to think of this as an input error voltage, which arises because the real amplifier has a finite open loop gain.

A differential input amplifier allows the application of negative feedback with a variety of external circuit configurations; this together with the power of the feedback techniques themselves accounts for the great versatility of the differential input operational amplifier. Certain general terminology is commonly used to distinguish between different types of feedback configuration. A negative feedback circuit which causes the feedback signal to be proportional to the output voltage produced by an amplifier is called a *voltage feedback* circuit. A circuit in which the signal fed back is proportional to output current is called a *current feedback* circuit. It is important to note that the terminology relates to the way in which the feedback signal is derived, and not to what is actually being fed back. Negative voltage feedback acts on the output voltage; it works towards making the amplifier output behave like an ideal voltage generator, i.e. as a source of e.m.f. linearly related to the externally applied input signal (no distortion) and having zero output impedance (output voltage unaffected by loading). Current feedback, on the other hand, works towards making the amplifier output behave like an ideal current source, i.e. a source of current linearly related to the input signal and unaffected by changes in the external load to which the current is supplied. The extent to which negative feedback achieves these performance ideals will be examined in the analyses which follow shortly.

19

Two other terms are used to distinguish feedback circuits: series feedback and shunt feedback. These terms relate to the way in which the feedback signal is introduced at the amplifier input terminals. In a series feedback circuit the feedback signal is applied in series with the external input signal to the amplifier input terminals; in a shunt feedback circuit the feedback signal and the external input signal are applied in parallel.

2.2.3 SERIES VOLTAGE FEEDBACK (THE FOLLOWER CIRCUIT)

A differential input amplifier with series negative voltage feedback applied to it is shown in Figure 2.3. The amplifier is represented in terms of its Thevenin equivalent. The output behaves like a source of e.m.f. $-A_{OL}\,e_\epsilon$ in series with the amplifier output impedance Z_o. (Note the minus sign simply comes from the assumed positive direction of the differential input signal e_ϵ).

A signal e_f which is directly proportional to the output voltage e_o is fed back to the inverting input terminal of the amplifier (negative feedback)

$$e_f = \beta e_o$$

The constant of proportionality β is called the *voltage feedback fraction*; it is an important quantity when analysing the quantitative effects of feedback. If, in Figure 2.3, we assume $Z_{in} \gg R_1$ and neglect the shunting effect of Z_{in} on R_1 we may write

$$\beta = \frac{R_1}{R_1 + R_2} \tag{2.1}$$

The output voltage of the amplifier may be written as

$$e_o = -A_{OL}\,e_\epsilon - i_o Z_o \tag{2.2}$$

It is simply a use of the general equation for the output voltage produced by a loaded source of e.m.f., Output voltage = Open circuit voltage − Internal volts drop. e_ϵ is the difference between the externally applied input signal e_i and the feedback signal e_f. Note that e_f and e_i are effectively applied in series to the differential input terminals of the amplifier.

$$-e_\epsilon = e_i - e_f \tag{2.3}$$

Substitution for e_ϵ in equation 2.2 gives

$$e_o = A_{OL}\,(e_i - e_f) - i_o Z_o$$

Figure 2.3 Series voltage feedback

Substituting $e_f = \beta e_o$ and rearrangement give

$$e_o = \frac{A_{OL}}{1 + \beta A_{OL}} e_i - i_o \frac{Z_o}{1 + \beta A_{OL}} \qquad (2.4)$$

According to equation 2.4 the circuit behaves like an amplifier with open circuit gain $A_{OL}/(1 + \beta A_{OL})$ and output impedance $Z_o/(1 + \beta A_{OL})$. These are the closed loop parameters for the circuit. Thus:

$$A_{CL} = \frac{A_{OL}}{1 + \beta A_{OL}} = \frac{1}{\beta} \left[\frac{1}{1 + \dfrac{1}{\beta A_{OL}}} \right] \qquad (2.5)$$

and

$$Z_{oCL} = \frac{Z_o}{1 + \beta A_{OL}} \qquad (2.6)$$

Note that if βA_{OL} is very large the quantity

$$\left[\frac{1}{1 + \dfrac{1}{\beta A_{OL}}} \right]$$

is as near unity as makes no difference and the closed loop gain is determined almost entirely by the value of the feedback fraction. The closed loop output impedance is made very small (i.e. output voltage little affected by loading). The product of the feedback fraction and the open loop gain is the gain around the feedback loop and it is called the *loop gain*. Loop gain, βA_{OL} is a most important parameter in determining the quantitative effects of feedback.

Let us derive an expression for the input impedance of the circuit. Note that e_f is applied in opposition to e_i and tends to prevent e_i from driving a current into the circuit. Series negative feedback may thus be expected to increase effective input impedance. We write

$$e_i = e_f - e_\epsilon = \beta e_o - e_\epsilon$$

But

$$e_o = - A_{OL}\, e_\epsilon\, \frac{Z_L}{Z_o + Z_L} \quad (\text{assuming } R_2 \gg Z_L)$$

Substitution gives

$$e_i = - e_\epsilon \left[1 + \beta A_{OL}\, \frac{Z_L}{Z_o + Z_L} \right]$$

Now

$$Z_{inCL} = \frac{e_i}{i_{in}} = - \frac{e_\epsilon}{i_{in}} \left[1 + \beta A_{OL}\, \frac{Z_L}{Z_o + Z_L} \right]$$

But

$$- \frac{e_\epsilon}{i_{in}} = Z_{in}$$

Thus

$$Z_{inCL} = Z_{in} \left[1 + \beta A_{OL}\, \frac{Z_L}{Z_o + Z_L} \right] \tag{2.7}$$

Series voltage feedback increases input impedance to an extent determined by the loop gain βA_{OL}.

2.2.4 SHUNT VOLTAGE FEEDBACK (THE INVERTER)

In Figure 2.4 the externally applied input signal voltage e_s and the output voltage e_o are effectively applied in parallel to the differential input terminals of the amplifier. The signal e_ϵ which drives the differential input terminals is a superposition of the effects of e_s and e_o.

$$e_\epsilon = e_s\, \frac{R_2}{R_2 + R_1 + R_s} + e_o\, \frac{R_1 + R_s}{R_1 + R_s + R_2} \tag{2.8}$$

It is assumed that $Z_{in} \gg R_1 + R_s$ and $Z_o \ll R_2$

22

The feedback fraction $\quad \beta = \dfrac{R_1 + R_s}{R_1 + R_s + R_2}$

The output voltage may be written as

$$e_o = -A_{OL}\, e_\epsilon - i_o Z_o$$

Substitution for e_ϵ and rearrangement give

$$e_o = -\frac{R_2}{R_2 + R_1 + R_s}\, \frac{A_{OL}}{1 + \beta A_{OL}}\, e_s - i_o\, \frac{Z_o}{1 + \beta A_{OL}}$$

The closed loop signal gain of the circuit is thus

$$A_{CL} = -\frac{R_2}{R_2 + R_1 + R_s}\, \frac{A_{OL}}{1 + \beta A_{OL}} = -\frac{R_2}{R_1 + R_s}\left[\frac{1}{1 + \dfrac{1}{\beta A_{OL}}}\right] \quad (2.9)$$

and the closed loop output impedance is

$$Z_{oCL} = \frac{Z_o}{1 + \beta A_{OL}} \quad (2.10)$$

Compare equations 2.9 and 2.10 with equations 2.5 and 2.6. Again, notice the importance of the loop gain βA_{OL}. If the loop gain is sufficiently large the closed loop performance is determined by the value of the components used to fix the feedback fraction β. If $R_1 \ll R_s$ and the loop gain is large the closed loop signal gain approximates to $A_{CL} = -R_2/R_1$.

Figure 2.4 Shunt voltage feedback

The main difference between shunt and series negative feedback lies in the effect on the input impedance of the circuit. In Figure 2.4

$$i_{in} = i' + i_f$$

$$= \frac{e_\epsilon}{Z_{in}} + \frac{e_\epsilon - e_o}{R_2}$$

If $Z_L > Z_o$,

$$e_o \cong - A_{OL} e_\epsilon$$

and

$$i_{in} = e_\epsilon \left[\frac{1}{Z_{in}} + \frac{1 + A_{OL}}{R_2} \right]$$

Figure 2.5 Series current feedback

The effect of the shunt feedback is to reduce the effective input impedance measured at the differential input terminals of the amplifier. An additional impedance $R_2/(1 + A_{OL})$ is effectively put in shunt with Z_{in} and if A_{OL} is very large the impedance measured at the differential input terminals is very small. The input impedance of the circuit is then effectively equal to the value of the resistor R_1.

2.2.5 NEGATIVE CURRENT FEEDBACK

In the circuit of Figure 2.5 a feedback signal is developed by the output current i_o flowing through a current sensing resistor R_f. The object of

24

applying negative current feedback is to achieve a stable output current. The circuit is analysed in order to find an expression for this output current. We may write

$$i_o = - \frac{A_{OL} \, e_\epsilon}{Z_o + Z_L + R_f}$$

But $e_\epsilon = e_f - e_i = i_o R_f - e_i$. (Assume $Z_{in} \gg R_f$).

Substitution for e_ϵ and rearrangement give

$$i_o = - A_{OL} \frac{e_i}{Z_o + Z_L + (1 + A_{OL}) R_f} \qquad (2.11)$$

Note that the effect of the negative current feedback is to increase the effective output impedance of the circuit (the term $A_{OL} R_f$ is added). Equation 2.11 may be written in an alternative form as

$$i_o = - \frac{e_i}{R_f} \left[\frac{1}{1 + \dfrac{1}{\beta A_{OL}}} \right] \qquad (2.12)$$

where in this case the voltage feedback fraction $\beta = R_f/(R_f + Z_L + Z_o)$ has a value which depends upon the load impedance. You should again notice the importance of the loop gain and the factor $[1/(1 + 1/[\beta A_{OL}])]$ which is sometimes referred to as the gain error factor.

In the current feedback circuit if βA_{OL} is very large, the output current is almost entirely determined by the input voltage and the value of the resistor R_f which is used to set the feedback.

2.3 Summary of some of the effects of negative feedback

It is instructive to summarise the effects of negative feedback as shown by the above analysis. If you analyse any negative feedback amplifier circuit you will find that the following general principles hold.

1. Series negative feedback increases input impedance.
2. Shunt negative feedback decreases input impedance.
3. Negative voltage feedback makes for a stable distortion-free output voltage.
4. Negative current feedback makes for a stable distortion-free output current.

2.3.1 GAIN TERMINOLOGY

It is important to distinguish between the several 'gain' terms which are used when discussing operational amplifier feedback circuits.

Open loop gain

A_{OL} is defined as the ratio of a change of output voltage to the change in the input voltage which is applied directly to the amplifier input terminals.

The other gains are dependent upon both the amplifier and the circuit in which it is used and are controlled by the feedback fraction β.

The feedback fraction

β, is the fraction of the amplifier output voltage which is returned to the input. It is a function of the entire circuit from output back to input, including both designed and stray circuit elements and the input impedance of the amplifier.

Loop gain

βA_{OL}, is the total gain in the closed loop signal path through the amplifier and back to the amplifier input via the feedback network. The magnitude of the loop gain is of prime importance in determining how closely circuit performance approaches the ideal.

Closed loop gain

Is the gain for signal voltages connected directly to the input terminals of the amplifier. The closed loop gain for an ideal amplifier circuit is $1/\beta$ and for a practical circuit is

$$\frac{1}{\beta}\left[\frac{1}{1 + \dfrac{1}{\beta A_{OL}}}\right]$$

The quantity $[1/(1 + 1/[\beta A_{OL}])]$ is called the gain error factor. The amount by which this factor differs from unity represents the gain error (usually expressed as a percentage).

Signal gain

Is the closed loop transfer relationship between the output and any signal input to an operational amplifier circuit.

Care should be taken to avoid confusion between closed loop gain and signal gain. In some circuits (the follower for example) the two

26

gains are identical. Reference to the circuit shown in Figure 2.6 should clarify the distinction between the two types of gain. The signals e_4 and e_3 are applied directly to the amplifier input terminals; they appear at the output multiplied by the closed loop gain $1/\beta$. The signals e_1 and e_2 are applied to the inverting input terminal through resistors R_1 and R_2 respectively; there are thus two different signal gains: $-R_f/R_1$ for signal e_1 and $-R_f/R_2$ for signal e_2.

$$\beta = \frac{\dfrac{R_1\,R_2}{R_1 + R_2}}{\dfrac{R_1\,R_2}{R_1 + R_2} + R_f}$$

$$e_0 = -\left[e_1 \frac{R_f}{R_1} + e_2 \frac{R_f}{R_2} + e_4 \frac{1}{\beta} - e_3 \frac{1}{\beta} \right]\left[\frac{1}{1 + \dfrac{1}{\beta A_{OL}}} \right]$$

e_1 Signal gain

e_2 Signal gain

Closed loop gain

Closed loop gain

Gain error factor

Signal source impedances are assumed negligibly small and the effects of amplifier offsets and input and output impedances are neglected

Figure 2.6   Differences between signal gain and closed loop gain

In evaluating an expression for β due account must always be taken for all paths from the inverting input terminal to earth. A general expression for the feedback fraction β in any circuit may be written as

$$\beta = \frac{Z_p}{Z_p + Z_f} \qquad (2.13)$$

where Z_p = parallel sum of all impedances between the inverting input terminal and earth

and Z_f = feedback impedance: the impedance between the output terminal and the inverting input terminal.

27

The performance equation for an operational amplifier feedback circuit can always be put in the form

$$\begin{bmatrix} \text{Actual Closed loop} \\ \text{performance equation} \end{bmatrix} = \begin{bmatrix} \text{Ideal Closed loop} \\ \text{performance equation} \end{bmatrix} \times \begin{bmatrix} \dfrac{1}{1 + \dfrac{1}{\beta A_{OL}}} \end{bmatrix}$$

The ideal closed loop performance equation is that obtained assuming that the operational amplifier behaves ideally (Chapter 1); ideal performance depends entirely upon the values of components external to the amplifier. If the loop gain βA_{OL} is very large, actual closed loop performance is very close to the ideal. Note that it does not matter if the open loop gain of an amplifier is non-linear, for provided that βA_{OL} is very large the closed loop performance will be linear if linear components are used to fix feedback conditions.

2.4 Frequency response characteristics

In our treatment of feedback amplifiers presented thus far no specific reference has yet been made to dynamic response characteristics. Gain has been defined as the ratio change of output voltage to change of input voltage, but we have not yet discussed the way in which amplifier response is governed by the rate of change of the input signals applied to it. It is usual to distinguish between sinusoidal and transient response characteristics: sinusoidal response parameters describe the way in which an amplifier responds to sinusoidal signals, in particular they show how response depends upon signal frequency; transient response parameters characterise the way in which an amplifier reacts to a step or squarewave input signal. An added complication arises when dealing with dynamic response parameters, in that it is necessary to distinguish between small signal and large signal response parameters; differences arise because of dynamic saturation effects which occur with large signals.

This section is concerned with small signal sinusoidal response characteristics. An ideal operational amplifier is assumed to have an open loop gain which is independent of signal frequency but the gain of real amplifiers exhibits a marked frequency dependence. Both the magnitude and the phase of the open loop gain are frequency dependent; this frequency dependence has a marked effect on closed loop performance.

2.4.1 BODE PLOTS

Gain/frequency characteristics are often presented graphically. It is usual to plot gain magnitude in dB against frequency on a log scale. Gain in dB is determined from the relationship

$$\text{Voltage gain in dB} = 20 \log_{10} \left| \frac{V_o}{V_{in}} \right| \tag{2.14}$$

The reader who is unfamiliar with the use of dB should get practice in working out the dB equivalents of some voltage ratios (try Exercises 2.5 and 2.6).

$|V_o/V_{in}| = 10$ represents a voltage gain of $20 \log_{10} 10 = 20$ dB; $|V_o/V_{in}| = 100$ represents 40 dB; $|V_o/V_{in}| = 1000$ represents 60 dB; $|V_o/V_{in}| = \frac{1}{10}$ represents -20 dB; $|V_o/V_{in}| = \sqrt{2}$ represents 3 dB; $|V_o/V_{in}| = 1/\sqrt{2}$ represents -3 dB.

Note, since power is proportional to (Voltage)^2, a fall in gain of 3 dB represents a halving of the output power.

Gain/frequency plots are often given as a series of straight line approximations rather than as continuous curves. The straight lines are called Bode approximations and the graphs are called Bode diagrams. The significance of Bode plots should emerge from the study of specific examples which will now be given.

The open loop frequency response of many operational amplifiers is designed to follow an equation of the form

$$A_{OL(jf)} = \frac{A_{OL}}{1 + j\dfrac{f}{f_c}} \tag{2.15}$$

$A_{OL(jf)}$ is a complex quantity representing the magnitude and phase characteristics of the gain at frequency f.

A_{OL} represents the d.c. value of the gain.

f_c is a constant, sometimes called the break frequency.

Equation 2.15 describes what is sometimes called a first order lag response; its magnitude and phase characteristics are shown plotted in Figure 2.7. The magnitude of the response is

$$|A_{OL(jf)}| = \frac{A_{OL}}{\sqrt{\left[1^2 + \left(\dfrac{f}{f_c}\right)^2\right]}} \tag{2.16}$$

At low frequencies for which $f < f_c$, $|A_{(jf)}| \to A_{OL}$ and the straight line $|A_{(jf)}| = A_{OL}$ is the low frequency asymptote. At high frequencies for which $f > f_c$, the response is asymptotic to the line $|A_{(jf)}| = A_{OL} f_c/f$ which has a slope of -20 dB/decade change in frequency.

| | Approximation errors | |
f	Magnitude of error	Phase error
$0.1 f_c$	0.04 dB	+ 5.7°
$0.5 f_c$	1 dB	− 4.9°
$1 f_c$	3 dB	0
$2 f_c$	1 dB	+ 4.9°
$10 f_c$	0.04 dB	− 5.7°

Figure 2.7   First order low pass magnitude and phase response and Bode approximations

For each ten times increase in frequency the magnitude decreases by 1/10, or -20 dB. (Note that a slope of -20 dB/decade is sometimes expressed as -6 dB per octave; it goes down by 6 dB for each doubling of the frequency.) Gain attenuation with increase in frequency is referred to as the *roll off* in the frequency response.

The two straight lines intersect at the frequency $f = f_c$ and at this frequency $|A_{(jf_c)}| = A_{OL}/\sqrt{2}$; the response is thus 3 dB down when $f = f_c$. The frequency f_c is sometimes referred to as the 3 dB band-width limit.

The phase/frequency characteristic associated with equation 2.15 is determined by

$$\theta = - \tan^{-1} \frac{f}{f_c} \qquad (2.17)$$

For $f \ll f_c$ $\theta \to 0$, for $f = f_c$ $\theta = -45°$ and for $f \gg f_c$ $\theta \to -90°$.

The Bode phase approximation approximates the phase shift by the asymptotic limits of $0°$ and $-90°$ for frequencies a decade below and above f_c respectively. The asymptotes are connected by a line whose slope is $-45°$ per decade of frequency as shown in Figure 2.7. The errors involved in using the straight line approximation for the magnitude and phase behaviour of equation 2.15 are tabulated in Figure 2.7.

Operational amplifier data sheets normally give values of A_{OL} and the unity gain frequency f_1, which is the frequency at which the open loop gain has fallen to 0 dB because of open loop roll off. In the case of amplifiers which exhibit a first order frequency response with a 20 dB per decade roll off down to unity gain the frequency f_1 is related to the 3 dB bandwidth frequency f_c by the expression

$$f_c = \frac{f_1}{A_{OL}} \qquad (2.18)$$

Frequency response characteristics are readily plotted from a knowledge of A_{OL} and f_1. The Bode magnitude approximations are obtained by simply drawing two straight lines, one horizontal line at the value of A_{OL} and the second through f_1 with a slope of -20 dB/decade. The two intersect at the frequency f_c.

Bode diagrams are useful in evaluating the frequency response characteristics of cascaded gain stages. The gain of a multistage amplifier is obtained as the product of the gains of the individual stages, but since gain is represented logarithmically in Bode plots the overall response may be determined by linearly adding the Bode plots for the separate stages as shown in Figure 2.8.

Note that the final roll off and limiting phase shift depend upon the number of gain attenuating stages. Two stages give a final roll off of -40 dB/decade and a limiting phase shift of $180°$; three stages give -60 dB/decade and $270°$ phase shift.

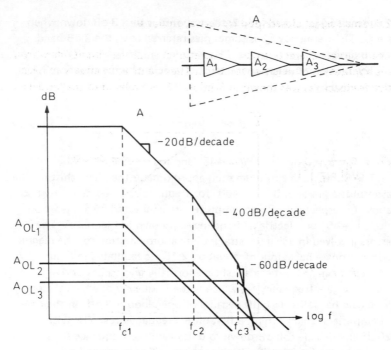

Figure 2.8 Frequency response of cascaded gain stages

Bode diagrams, as will be shown later, are particularly useful in assessing the stability and frequency response of feedback circuits and for this reason we give more examples of Bode diagrams for commonly encountered frequency response functions. The Bode magnitude and phase approximations for the function

$$T_{(jf)} = \frac{1 + j\dfrac{f}{f_{c2}}}{1 + j\dfrac{f}{f_{c1}}} \qquad (2.19)$$

are given in Figure 2.9. A response of this kind is produced by a so-called lag, lead network (see Exercise 2.8). Note that the response is obtained by adding the Bode approximation for $1/(1 + j\,[f/f_{c1}])$ (a response which exhibits lagging phase shift), to the Bode approximation of $1 + j\,(f/f_{c2})$ (a response which exhibits leading phase shift).

32

2.5 Small signal closed loop frequency response

The desirable characteristics of operational amplifier circuits stem from the application of negative feedback. The quantitative effects of negative feedback, as was shown in Section 2.3, are related to the loop gain

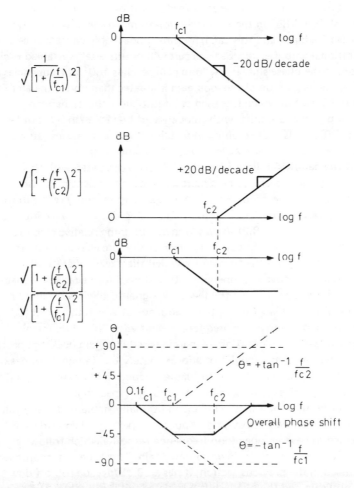

Figure 2.9 Bode magnitude and phase approximations for a lag lead type of response

βA_{OL}. Real operational amplifiers exhibit a frequency dependent A_{OL} and in some applications the feedback fraction β is also frequency dependent. Practical operational amplifier circuits therefore exhibit a frequency dependent loop gain and this has a marked effect on closed loop performance.

It should always be remembered that a frequency dependence implies both a magnitude change and a phase change with frequency and in a circuit designed to apply negative feedback it only needs an excess phase shift of $180°$ in the feedback loop to make the circuit apply positive feedback and this can have most undesirable consequences. An operational amplifier feedback circuit will produce self sustained oscillations if the phase shift in the loop gain reaches $180°$ at frequencies at which the magnitude of the loop gain is greater than unity; the amplifier and the circuit in which it is used must not allow this to happen.

Loop gain phase shifts with frequency of greater than $90°$ but less than $180°$, whilst not resulting in sustained oscillations, can cause a feedback circuit to have a closed loop frequency response which peaks up at the bandwidth limit before it rolls off. Associated with this closed loop gain peaking the circuit will have a transient response which exhibits overshoot and ringing. Transient response refers to the output changes produced in response to a step or squarewave input signal (see Section 3.8.2). A way of expressing the relative stability of a closed loop amplifier circuit is in terms of the so called phase margin. The phase margin is the amount by which the excess phase shift (phase shift over and above the inherent $180°$ required for negative feedback and obtained by returning the feedback signal to the inverting input terminal) is less than $180°$ at that frequency at which the magnitude of the loop gain is unity. A closed loop circuit with $90°$ phase margin shows no gain peaking; as phase margin is reduced gain peaking becomes noticeable for phase margins of approximately $60°$ (about 1 dB peaking) and becomes more marked with further reduction in phase margin ($20°$ phase margin gives approximately 9 dB of gain peaking).

In order that they should be unconditionally stable under any value of resistive feedback most general purpose operational amplifiers are designed to have an open loop frequency response which follows a first order characteristic down to unity gain. This type of response was discussed in the previous section; it has a 20 dB/decade roll off down to unity gain and the phase shift associated with gain attenuation never exceeds $90°$. The phase margin for any value of resistive feedback is therefore never less than $90°$.

The effect of open loop gain frequency dependence on closed loop gain frequency dependence is most conveniently demonstrated in

graphical form by sketching the appropriate Bode plots. We look for
the effect of A_{OL} on loop gain and then to the effect of loop gain on
the gain error factor. We may write

$$\left| \beta A_{OL(jf)} \right| = \left| \frac{A_{OL(jf)}}{\frac{1}{\beta_{(jf)}}} \right|$$

which when expressed in dB form gives

$$|\text{loop gain (in dB}| = |\text{open loop gain (in dB)}| - \left| \frac{1}{\beta} \right| \text{ (in dB)} \qquad (2.20)$$

The magnitude of the loop gain in dB at any frequency is equal to the
difference between the open loop gain magnitude in dB and $1/\beta$ in dB.

Figure 2.10 *Bode plots show frequency dependence of loop gain*

As an example of the graphical approach consider an operational
amplifier with a first order frequency response used with resistive feedback in the follower configuration. The circuit and its Bode plots are
illustrated in Figure 2.10. In order to display the frequency dependence
of the loop gain we merely superimpose the plot of $1/\beta$ (in dB) on the
open loop frequency response plot of the amplifier. If feedback is purely
resistive, as it is in the example under consideration, β is independent of

frequency and $1/\beta$ is a straight line parallel to the frequency axis. In a case such as this the frequency dependence of the loop gain is entirely due to the frequency dependence of the open loop gain.

Open loop gain attenuation with increase in frequency is accompanied by a corresponding decrease in loop gain and an increase in the gain error. Remember that gain error is related to the amount by which the gain error factor $[1/(1 + 1/\beta A_{OL})]$ differs from unity (see Section 2.3).

If it is required to compute the gain error at frequencies approaching or exceeding the open loop bandwidth f_c the phasor nature of the loop gain must not be forgotten. In illustration we evaluate the gain error for the circuit of Figure 2.10 at a frequency $f = 10^3$ Hz. At this frequency $|\beta A_{OL(jf)}| = 20$ dB $= 10$ and the phase shift in the loop gain is close to -90°. Thus

$$\left|\frac{1}{1 + \dfrac{1}{\beta A_{OL(jf)}}}\right| = \left|\frac{1}{1 + \dfrac{1}{-j10}}\right| = \frac{1}{\sqrt{(1 + 0.01)}} = 0.995$$

0.5% gain error.

Compare this with the value obtained by neglecting the phasor nature of the loop gain, that is

$$\frac{1}{1 + \dfrac{1}{10}} = 0.909$$

9% gain error!

At the frequency f_1' at which the open loop and $1/\beta$ magnitude plots intersect the magnitude of the loop gain is unity (0 dB); the two plots close at a rate of 20 dB per decade which is indicative of a 90° phase shift in the loop gain and a remaining 90° phase margin. The magnitude of the gain error at the frequency f_1 is

$$\left|\frac{1}{1 + \dfrac{1}{-j1}}\right| = \frac{1}{\sqrt{2}}$$

The closed loop gain magnitude is thus 3 dB down on its ideal value $1/\beta$ at the frequency f_1'. f_1' represents the closed loop bandwidth; at frequencies greater than f_1' the magnitude of the closed loop gain approaches the magnitude of the open loop gain. If $\beta A_{OL(o)} \gg 1$ the product: closed loop gain $\times$ closed loop bandwidth $= 1/\beta f_1' = f_1$ remains constant for different values of β. Negative feedback makes the closed loop bandwidth greater than the open loop bandwidth; the greater β the smaller the closed loop gain but the wider the closed loop bandwidth.

A second example of the graphical approach used to find closed

$$\beta \triangleq \frac{1}{1000}$$

$$\frac{1}{\beta} \equiv 1000 \equiv 60\,dB$$

$$e_1 \text{ Signal gain} = \frac{-R_f}{R_1}\left[\frac{1}{1+\dfrac{1}{\beta A_{OL\,(jf)}}}\right]$$

$$e_2 \text{ Signal gain} = \frac{-R_f}{R_2}\left[\frac{1}{1+\dfrac{1}{\beta A_{OL\,(jf)}}}\right]$$

Figure 2.11 Bode plots for inverting adder

loop signal bandwidth is illustrated in Figure 2.11. The circuit considered here is an inverting adder application. In this type of circuit the feedback fraction is influenced by the presence of the two input resistors R_1 and R_2.

$$\beta = \frac{R_p}{R_p + R_f} \text{ where } R_p = R_1 \| R_2 = \frac{R_1 R_2}{R_1 + R_2}$$

Substituting component values gives $\beta \cong 1/1000$ and $1/\beta \cong 1000$ or 60 dB. $1/\beta$ intersects the open loop frequency response at the frequency $f_1{}' = 1\,kHz$. This fixes the closed loop bandwidth at 1 kHz, but note that in this circuit the closed loop signal gain is not the same as the

37

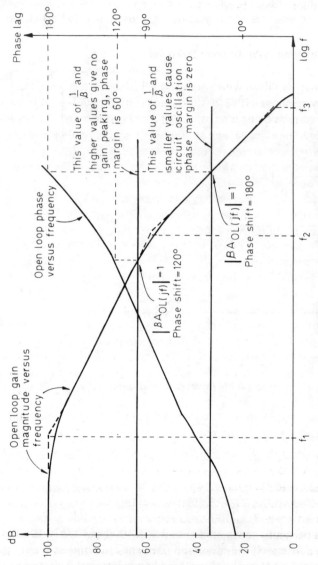

Figure 2.12 Gain magnitude and phase characteristics of amplifier with three pole response

Phase lag

180°

120°

90°

0°

log f

This value of $\frac{1}{\beta}$ and higher values give no gain peaking, phase margin is 60°

This value of $\frac{1}{\beta}$ and smaller values cause circuit oscillation phase margin is zero

Open loop phase versus frequency

$\left| BA_{OL(jf)} \right| = 1$
Phase shift = 120°

$\left| BA_{OL(jf)} \right| = 1$
Phase shift = 180°

Open loop gain magnitude versus frequency

dB

100

80

60

40

20

0

f_1

f_2

f_3

closed loop gain $1/\beta$ since there are two possible input signal paths. The ideal signal gain for the e_1 signal is $-R_f/R_1$ and is $-R_f/R_2$ for the e_2 signal. In this particular example $R_1 = R_2$, the two gains are equal and the closed loop signal bandwidth is 1 kHz.

2.6 Closed loop stability considerations

Operational amplifiers which are designed to exhibit an open loop frequency response with a 20 dB/decade roll off down to unity gain are referred to as being internally frequency compensated. A response of this kind, in principle, ensures that the amplifier will be closed loop stable under all conditions of resistive feedback. However, it is import- ant to be aware that the use of an internally frequency compensated amplifier does not always ensure closed loop stability. Capacitive load- ing at the output of an amplifier or stray capacitance between the inverting input terminal and earth can cause phase shifts leading to instability even in resistive feedback circuits (see Section 9.5); in dif- ferentiator applications in which the feedback fraction β is deliberately made frequency dependent, an internally compensated amplifier exhibits instability (see Section 6.7.1)

Some operational amplifiers exhibit a final roll off in their open loop frequency response of greater than 20 dB/decade; they are called fast roll off amplifiers or externally frequency compensated amplifiers. Fast roll off amplifiers are often used in applications requiring wide closed loop bandwidth in circuits with closed loop gain greater than unity. They require the external connection of frequency compensat- ing components (usually a capacitor) to make them closed loop stable.

The closed loop frequency response characteristics obtained with fast roll off amplifiers are conveniently discussed in terms of Bode plots, and are related to the effects of the magnitude/phase character- istics of the loop gain on the gain error factor. As an example, consider an operational amplifier with a response which exhibits three breaks connected in the follower configuration. The magnitude and phase characteristics of the open loop gain of such an amplifier are illustrated in Figure 2.12. The magnitude and phase characteristics of the loop gain for a particular feedback fraction are again obtained by super- imposing a plot of $1/\beta$ on the open loop frequency response plot. With resistive feedback β is frequency independent; the phase shift in the loop gain is determined by the phase shift in the open loop gain, which

Figure 2.13 Too much feedback gives gain peaking with uncompensated fast roll off amplifiers

can be found from inspection of the graphs in Figure 2.7. *Phase margin* is the amount by which this phase shift is less than 180° at the frequency at which the magnitude of the loop gain is unity (0 dB). Note that increasing β results in successively smaller phase margins; phase margins less than 60° cause the closed loop gain to peak up, and successively greater amounts of closed loop gain peaking occur with decrease in phase margin (Figure 2.13). With zero phase margin the circuit breaks out into sustained oscillations.

2.6.1 PHASE MARGIN DETERMINES CLOSED LOOP GAIN PEAKING

The gain peaking which occurs as a result of inadequate phase margin arises because the phase shift in the feedback loop at the higher frequencies causes the signal which is fed back to the input to have a component which is in phase with the externally applied input signal. If the phase shift in the loop gain reaches 180° at a frequency below that at which the loop gain magnitude has fallen to unity, the circuit oscillates. In assessing the extent of the gain peaking obtained as a result of inadequate phase margin we must look to the effect of the loop gain magnitude/phase behaviour on the gain error factor; we write.

$$\beta A_{OL(jf)} = \left| \beta A_{OL(jf)} \right| e^{-j\theta}$$

The value of the gain error factor may then be expressed as

$$\cfrac{1}{1 + \cfrac{1}{\beta A_{OL(jf)}}} = \cfrac{1}{1 + \cfrac{e^{j\theta}}{\left| \beta A_{OL(jf)} \right|}}$$

$$= \cfrac{1}{1 + \cfrac{1}{\left| \beta A_{OL(jf)} \right|} [\cos \theta + j \sin \theta]}$$

The magnitude of the gain error factor can then be written as

$$\left| \cfrac{1}{1 + \cfrac{1}{\beta A_{OL(jf)}}} \right| = \sqrt{\left(\cfrac{1}{1 + \cfrac{1}{\left| \beta A_{OL(jf)} \right|^2} + \cfrac{2 \cos \theta}{\left| \beta A_{OL(jf)} \right|}} \right)} \qquad (2.21)$$

Since the cosine of angles lying between 90° and 180° is negative we

Figure 2.14 Gain peaking versus phase margin for two commonly encountered situations

have the possibility of a gain error factor magnitude greater than unity for values of θ greater than 90°. It is this variation in the gain error factor which is responsible for closed loop gain peaking.

In many practical situations gain peaking arises as a result of phase shift with frequency controlled by a single first order function whose break frequency is remote (greater than a decade away) from other break frequencies. Two situations are illustrated in Figure 2.14. The break at f_c is remote from other break frequencies, and phase variations for frequencies in the neighbourhood of f_c are determined by this break at f_c. (The first break in the open loop frequency response has contributed a constant $90°$ phase shift at this frequency f_c.) The relationship between closed loop gain peaking and phase margin that is to be expected in a situation of this kind is shown graphically in Figure 2.14 (see Appendix A2). In order to assess the phase margin in a particular circuit configuration the $1/\beta$ graph (in dB) is superimposed on the open loop response; the intersection gives the frequency f_1' at which the magnitude of the loop gain is unity. The phase shift θ at this frequency is then determined from the phase/frequency variation in A_{OL}; the phase margin is $\theta_m = 180 - \theta$. The amount of gain peaking can be found from the graph. Note that the gain peaks in fact occur at frequencies slightly less than the frequency f_1' but as phase margin is reduced the gain peaks increase, and the frequencies at which they occur move closer to the frequency f_1'.

2.7 Frequency compensation (phase compensation)

Frequency compensation or phase compensation is the name given to the process of tailoring the loop gain magnitude/phase characteristics of a feedback amplifier circuit in order that it should have an adequate phase margin. Adequate phase margin ensures closed loop stability and freedom from closed loop gain peaking. General purpose operational amplifiers are normally internally frequency compensated so as to be unconditionally stable with all values of resistive feedback. The phase shift in their open loop gain is typically controlled to be $135°$ or less for all frequencies at which the open loop gain magnitude is greater than unity, assuring a minimum phase margin of $45°$ for all values of resistive feedback. Internal frequency compensation gives user convenience at the expense of the lost closed loop bandwidth and speed (slew rate—see later) which would otherwise be available when the amplifier is used at higher closed loop gains than unity.

Amplifiers without internal frequency compensation require the external connection of frequency compensating components to the amplifier. They allow the user to select a frequency compensating scheme appropriate to the particular closed loop application in which the amplifier is used. Closed loop bandwidth, slewing rate, full power response and noise performance (see later) are all affected by the frequency compensating method adopted. Compensation methods advocated for different amplifier types differ in detail because of internal circuit differences, but the general principles involved in frequency compensation are the same for all amplifiers. Once the techniques are understood stability problems should not present too serious a difficulty.

Operational amplifiers have a very large open loop gain, which is determined by the product of the gains of the separate internal gain stages which go to make up the amplifier. If gain is expressed in dB the open loop frequency response plot for an amplifier is obtained as a result of linearly adding the Bode plots for the individual gain stages. Each stage follows a first order characteristic (see 2.4.1) and the final rate of attenuation and phase shift in the overall response is governed by the number of internal voltage gain stages. Two stages give a final roll off of 40 dB/decade and a phase shift approaching 180°; three stages give 60°/decade final roll off and a phase shift approaching 270°.

In order that the open loop response of an operational amplifier shall exhibit a single 20 dB/decade roll off down to unity gain the frequency response of one of its internal stages must be made dominant. The break frequency associated with this dominant stage must be made sufficiently low to ensure that by its gain attenuation alone, it can get the loop gain magnitude down to unity at a frequency which is lower than that at which the other gain stages start to attenuate.

In most present day integrated circuit designs very high individual stage gains are obtained by using active loads[1] and they are thus able to achieve a sufficiently large overall gain using only two internal voltage gain stages. Operational amplifiers of this type are normally frequency compensated by means of a single capacitor connected as a feedback capacitor around the second inverting voltage gain stage in the amplifier. The technique requires only small values of frequency compensating capacitor (10 pF–30 pF). Capacitors of this size are small enough to be fabricated on the same integrated circuit chip as the rest of the amplifier circuitry; this is the system normally adopted in most general purpose internally compensated amplifiers.

The above frequency compensating technique is called Miller effect frequency compensation; a quantitative understanding of the technique requires some knowledge of an operational amplifier's internal circuitry. Even the reader who is familiar with discrete component transistor circuitry might find some difficulty when he first encounters the detailed internal circuit schematics of modern operational amplifiers, for example, the differential input gain stage in an operational amplifier may have as many as ten or more transistors associated with it. Most

44

of these transistors serve to set bias levels and to make the first stage give a single ended output. Readers interested in studying the details are referred elsewhere[1]. Fortunately an in-depth understanding of internal circuit operation is not needed in order to understand frequency compensation—a simplified model of the internal circuit suffices, and the equivalent circuit given in Figure 2.15 is such a model.

Figure 2.15 Equivalent circuit for frequency compensation

The differential input stage used in many operational amplifiers has a very high gain and a very high output impedance; it provides what is essentially a current drive to the second gain stage. In Figure 2.15 its action is represented by the current generator $g_m e_\epsilon$. The second gain stage is inverting; a capacitor connected between input and output of the high gain inverting stage gives that stage the frequency response characteristics of an integrator. Its output voltage is proportional to the

45

integral of the input current (see Section 6.4.4). Assuming ideal integrator action for the second stage its output and the output of the complete amplifier has a frequency roll off which can be approximated as

$$V_o = \frac{g_m e_\epsilon}{j\omega C_f}$$

and the open loop gain roll off is approximated by

$$A_{OL(jf)} = -\frac{g_m}{j2\pi f C_f} \tag{2.22}$$

Equation 2.22 must be made to dominate the overall frequency response. Unity gain frequency compensation requires that the value of C_f be chosen so that the equation brings the open loop gain down to unity at a frequency lower than the break frequency of other gain attenuating stages.

Setting $|A_{OL(jf)}| = 1$ in equation 2.23 gives the relationship between the unity gain frequency and the required frequency compensating capacitor as

$$f_1 = \frac{g_m}{2\pi C_1} \tag{2.23}$$

In many general purpose operational amplifier designs the current drive supplied by the first gain stage in the amplifier has a frequency dependence determined by the behaviour of pnp transistors in the first stage with a break frequency typically in the range 3 to 4 MHz. Dependent upon the unity gain phase margin required f_1 must be made to have the same order of magnitude, and most general purpose monolithic amplifier designs have C_f chosen to make f_1 typically 1 MHz.

Unity gain frequency compensation, although satisfactory, is not strictly necessary when an operational amplifier is used in a feedback configuration in which the closed loop gain $1/\beta$ is greater than unity. Externally compensated amplifiers in which the frequency compensating capacitor is user connected permit the designer to apply just sufficient compensation to achieve a desired phase margin. Use of the minimum frequency compensating capacitor consistent with achieving adequate phase margin gives a wider closed loop bandwidth than would be obtained if the amplifier were unity gain frequency compensated. In applications which are concerned only with slowly varying input signals, a wide closed loop bandwidth is of course not required;

*Figure 2.16 Open loop frequency response of amplifier with different values of
compensation capacitor*

in such cases it is often advantageous to restrict closed loop bandwidth (in order to reduce noise) by using much greater frequency
compensation than is required for closed loop stability. These points
are illustrated in Figure 2.16 which shows Bode approximations for the
open loop frequency reponse of a general purpose amplifier, using
values of the frequency compensating capacitor C_f larger than and
smaller than the value C_1 required for unity gain frequency compensation of the amplifier.

Adequate phase margin ($60°$ in the case considered in Figure 2.16) requires that the minimum value of C_f be chosen so that with the value of β used in the particular circuit configuration the magnitude of the loop gain βA_{OL} is reduced to unity at the frequency f_1. Use of equation 2.22 gives

$$\left| \beta A_{OL(jf)} \right| = \beta \, \frac{g_m}{2\pi f C_f}$$

We require

$$\left| \beta A_{OL(jf)} \right| = 1 \text{ at } f = f_1$$

Substituting for f_1 from equation 2.23 gives the required minimum value of C_f as

$$C_f \geqslant \beta C_1 \tag{2.24}$$

Remember that C_1 is the value of the frequency compensating capacitor required for unity gain frequency compensation.

An example of a practical amplifier which uses the above frequency compensating scheme is provided by the general purpose operational amplifiers of the 101/210/301 series. Using the standard frequency compensating technique these amplifiers require a frequency compensating capacity of value 30 pF for unity gain frequency compensation, but they can be compensated with 3 pF when used at a closed loop gain of 20 dB.

There are certain conditions however in which the use of minimum values of C_f can lead to instability problems: in circuits with large resistor values, appreciable stray capacitance to earth at the inverting input terminal or those in which the amplifier is expected to drive a capacitive load (see Section 9.5).

2.7.1 FREQUENCY COMPENSATION AND SLEW RATE CONSIDERATIONS

There is a limit to the rate at which the output voltage of an operational amplifier can change; slewing rate is the parameter used to characterise this effect. Slewing rate is usually expressed in volts per microsecond and is defined as the maximum rate of change of output voltage produced in response to a large input step. The basic mechanism governing slew rate is capacitor charging, the rate of change of voltage at any point in a circuit being limited by the maximum current available to charge the capacitance at that point. In many operational amplifier applications it is the charging of the frequency compensating capacitor which sets the output slew rate.

In amplifiers of the type discussed in the previous section the frequency compensating capacitor is charged by the output current supplied by the first gain stage in the amplifier. The limitation on the charging rate is thus determined by the first stage output current capabilities, thus:

$$\text{Slew rate } = \frac{dV_o}{dt_{max}} \cong \frac{I_o}{C} \qquad (2.25)$$

where I_o is the first stage operating current.

Equation 2.25 suggests that increased slewing rate may be achieved by simply increasing the first stage operating current, but this is not the case for operational amplifiers using bipolar transistor input stages. In normal bipolar transistor amplifier stages increase in operating current causes a corresponding increase in the transconductance of the stage:

$$\text{Transconductance } g_m = \frac{I_o}{2kT/q} \qquad (2.26)$$

k is Boltzmann's constant
T is temperature in $^\circ$K
q is electronic charge

Increase in transconductance which accompanies any increase in operating current requires a corresponding increase in C_f in order to set a particular value for f_1. Combining equations 2.25 and 2.26 with equation 2.23 gives

$$\text{Slew rate } = \frac{dV_o}{dt_{max}} = \frac{2kT}{q} 2\pi f_1 \qquad (2.27)$$

Slew rate is seen to be independent of input stage current level. Our approximate treatment explains why most internally compensated general purpose bipolar input operational amplifiers have slew rates of the order of 1 V/μs.

General purpose bipolar input operational amplifiers which are externally frequency compensated have this same slew rate limitation when compensated down to unity gain. When they are frequency compensated for closed loop gains greater than unity the smaller value of the frequency compensating capacitor which is required gives an increased slewing rate.

Operational amplifiers which use an f.e.t. input stage do not have the above limitation on slew rate because unlike bipolar transistors f.e.t.'s do not have their transconductance directly dependent upon operating current. General purpose f.e.t. input operational amplifiers normally feature a higher slew rate than bipolar input operational amplifiers. High slew rate bipolar input operational amplifiers are available; they feature specialised input stage circuitry which provides increased current output without at the same time giving an increase in the transconductance of the stage[1].

2.7.2 FEED FORWARD FREQUENCY COMPENSATION

Some amplifier types are suitable for use with so called feed forward frequency compensation, which, when it is applicable, can provide a significant increase in bandwidth and slewing rate over standard lag compensation techniques. In most operational amplifiers the first voltage gain stage provides the greatest single contribution to the overall gain of the amplifier but its frequency response is normally rather limited. In feed forward frequency compensation the high gain low bandwidth first stage is bypassed at the higher signal frequencies and these are fed directly to the wider bandwidth second stage of the amplifier. Using the technique the phase shift at the higher frequencies is primarily due to the wide band stage, and the phase shift due to the high gain low bandwidth stage is eliminated. The principle underlying the scheme is illustrated in Figure 2.17. The overall gain due to both stages may be expressed as

$$A_{OL(j\omega)} = \left[A_{1(j\omega)} + \frac{R}{R + \dfrac{1}{j\omega C}} \right] A_{2(j\omega)}$$

$$= \left[A_{1(j\omega)} + \frac{1}{1 + \dfrac{1}{j\omega CR}} \right] A_{2(j\omega)}$$

C is chosen so that when $f > 1/2\pi CR$ the gain of the 1st stage has fallen to below unity, making the overall gain approximately that of the second gain stage

The second stage 20 dB/decade roll off takes the overall gain down to unity. Bode plots showing the uncompensated response and the response with feed forward compensation are illustrated in Figure 2.17. Feed forward frequency compensation is only applicable to inverting feedback configurations.

50

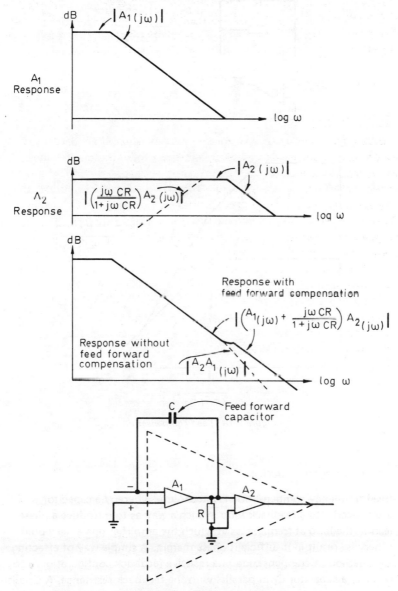

Figure 2.17 Principle of feed forward frequency compensation

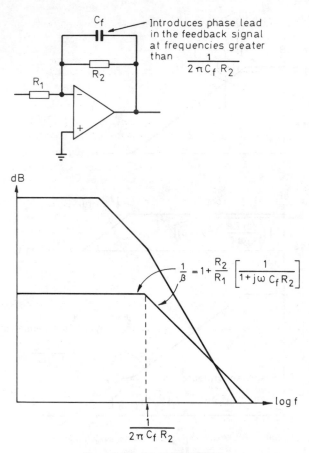

Figure 2.18   Lead compensation

2.7.3 LEAD COMPENSATION

Lead frequency compensation is a technique whereby a capacitor is introduced into a feedback loop in such a way as to introduce a phase lead in the loop at frequencies at which the amplifier phase lag would otherwise result in insufficient phase margin. A simple way of effecting this method of compensation in a resistive feedback configuration is to connect a capacitor C_f in parallel with the feedback resistance. A circuit using this method of lead compensation is shown, together with its associated Bode plots, in Figure 2.18. We write

Attenuation introduced
by $C_1 = \dfrac{1}{1 + j\omega C_1 R_0}$

Externally connected
frequency compensating
capacitor

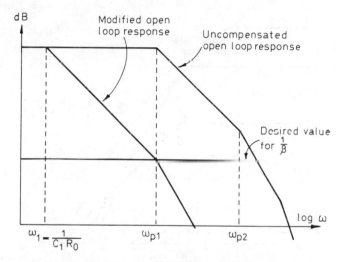

Modified open
loop response

Uncompensated
open loop response

Desired value
for $\dfrac{1}{\beta}$

$\omega_1 = \dfrac{1}{C_1 R_0}$ ω_{p1} ω_{p2} $\log \omega$

dB

Figure 2.19 Simple lag compensation with single capacitor

$$\frac{1}{\beta} = 1 + \frac{R_2}{R_1}\left[\frac{1}{1 + j\omega C_f R_2}\right]$$

$$= \left[1 + \frac{R_2}{R_1}\right]\frac{1 + j\omega C_f R_1 \| R_2}{1 + j\omega C_f R_2}$$

At frequencies greater than $1/2\pi C_f R_2$ the capacitor introduces a phase
lead in the feedback fraction which approaches $90°$. If C_f is chosen so
that the frequency $1/2\pi C_f R_2$ is a decade below the frequency at which
the $1/\beta$ and open loop response plots intersect a phase margin of
approximately $90°$ is obtained. Use of a lead capacitor in parallel with

53

Attenuation introduced by $C_1 R_1$

$$F_{(j\omega)} = \frac{1 + j\omega\, C_1 R_0}{1 + j\omega\, C_1 (R_1 + R_0)}$$

Figure 2.20 Frequency compensation with resistance capacitance shunt

a feedback resistor is a convenient way of getting extra phase margin; it is a technique which can be used to overcome the effect of stray capacitance between the inverting input terminal of an amplifier and earth (see Section 9.5).

2.7.4 OTHER FREQUENCY COMPENSATING TECHNIQUES

Techniques other than those described in the above sections are sometimes used for frequency compensation but whatever technique is used the same basic principle is involved. Frequency compensation involves attenuating the loop gain magnitude down to unity without, at the same time, introducing an excessive phase shift leading to closed loop instability.

Frequency compensation is sometimes affected by simply shunting a signal point in the feedback loop with a capacitor (Figure 2.19). Assuming the output resistance at the signal point is R_o the added capacitor introduces a 20 dB/decade rate of attenuation which starts at the break frequency $1/2\pi C_1 R_o$; the maximum phase shift associated with a CR lag network is $90°$. The capacitor value must be chosen so that the loop gain magnitude is attenuated down to unity at a frequency lower than other break frequencies of attenuating stages.

Shunting a signal point with a capacitor resistor combination (a lag lead network) is an alternative technique which allows wider closed loop bandwidths (Figure 2.20). At frequencies above $1/2\pi C_1 R_1$ (the break back frequency) a network of this kind produces an attenuation $R_1/(R_1 + R_o)$ but the phase shift returns to zero.

2.8 Transient response characteristics

Previous sections have been concerned with factors influencing the small signal frequency response characteristics of operational amplifier feedback circuits. Attention is now directed to the factors influencing their behaviour in time, namely their transient behaviour in response to large and small input step or squarewave signals. An understanding of the significance of transient behaviour and the terminology used to describe it is greatly helped by actually performing transient tests. The experimental procedure required is not difficult and the new user of operational amplifiers is urged to perform the test measurements suggested in Section 3.8.2, concurrently with his reading of this present section. Frequency compensating component magnitude, load capacity, input capacity and any stray feedback capacity all influence closed loop transient behaviour.

2.8.1 SMALL SIGNAL TRANSIENT RESPONSE

Small signal characteristics are those obtained when there are no saturation effects (no output, slew rate limiting) and the amplifier and all associated circuit elements are operating in their linear range. In small signal operation circuit relationships are independent of the level of the output voltage and current and of their previous history.

55

The small signal transient behaviour of an operational amplifier feedback circuit is closely related to its small signal sinusoidal frequency response. In our previous discussion of small signal closed loop frequency response we distinguished between two different closed loop situations.

Sinusoidal response

Step response

$$V_O = V_f \left[1 - \exp \frac{-t}{T_c} \right] \qquad \text{where } T_c = \frac{1}{2\pi \beta f_1}$$

Step response

Figure 2.21   Small signal sinusoidal and transient response for unity gain frequency compensated amplifier with resistor feedback

Firstly, closed loop configurations with $90°$ phase margin which showed no gain peaking; secondly, closed loop configurations which showed gain peaking because of inadequate phase margin. Systems which give no closed loop gain peaking give an output waveform which varies exponentially with time when subjected to a step input signal. Systems which exhibit gain peaking give a more rapid transient output response but the output overshoots its final value and rings (oscillates) before settling. The

56

greater the peaking in the closed loop sinusoidal response the more the transient response overshoots and rings. Quantitative relationships governing the two types of transient response are now given.

Unity gain frequency compensated amplifier with resistive feedback
Figure 2.21 illustrates the considerations governing the behaviour of a unity gain frequency compensated amplifier used with resistive feedback. In response to an input step signal the output follows an exponential governed by the relationship

$$V_o = V_f \left[1 - \exp \left(-\frac{t}{T_c} \right) \right] \tag{2.28}$$

The time constant $T_c = T_1/\beta$ where $T_1 = -1/2\pi f_1$. Notice that T_c increases for increasing values of closed loop gain (decrease in β) and decreases for increasing values of the unity gain frequency f_1.

Rise time is a parameter which is frequently used to characterise the response of an amplifier to an input step. Rise time is defined as the time taken for the output to rise between 10% and 90% of its final value. Neglecting the time for the initial 10% rise an approximate expression for rise time can be obtained by substituting $V_o = 0.9 V_f$ in equation 2.28. Thus

$$0.9 V_f = V_f \left[1 - \exp \left(-\frac{T_r}{T_c} \right) \right]$$

giving

$$\frac{T_r}{T_c} = \log_e 10$$

or

$$T_r = \frac{\log_e 10}{2\pi f_{3dB}}$$

$$T_r \cong \frac{1}{3 f_{3dB}} \tag{2.29}$$

$f_{3dB} = \beta f_1$ is the closed loop small signal 3 dB bandwidth limit.

Closed loop configurations with a lightly damped transient response
The most commonly encountered closed loop configurations which exhibit a lightly damped transient response are those in which loop gain variation with change in frequency is governed by two breaks, and in which the break frequencies are remote from each other by at least a decade (see Figure 2.22). Systems of this kind have a response which

57

Sinusoidal response

Input

Output

Step response

Figure 2.22   Lightly damped closed loop response

is characteristic of a so called second order system. The response equation is obtained by substituting the frequency dependent loop gain expression into the gain error factor (see Appendix A2). Second order system response characteristics are well established and readers wishing to study the mathematical relationships in greater detail are referred elsewhere[2]. We simply quote the relevant equations which relate the second order system parameters to the circuit parameters of the closed loop amplifier configurations of Figure 2.22.

A second order system is characterised by parameters called the damping factor ζ and natural frequency f_o. A step input V_{st} applied to a second order system gives rise to an output response which for $\zeta < 1$ has a time variation governed by the relationship

$$V_{o(t)} = A_{CL(o)} V_{st} \left\{ 1 - \frac{\exp(-\zeta\omega_o t)}{\sqrt{(1 - \zeta^2)}} \sin\left[\omega_o\sqrt{(1 - \zeta^2)}t + \cos^{-1}\zeta\right] \right\}$$

where $\omega_o = 2\pi f_o$. (2.30)

Treating $A_{CL(o)} V_{st}$ as a scaling factor and plotting time in units of $\omega_o t$ the normalised step response for different values of the damping

Figure 2.23 Second order step response

factor is plotted in Figure 2.23. The step response shows an increasing overshoot and ringing as the value of the damping factor is successively reduced below unity. The damping factor is related to the break frequencies governing the frequency dependence of the loop gain of the operational amplifier feedback configuration by the expression

$$\zeta = \frac{\sqrt{f_{c2}}}{2\sqrt{(|\beta A_{(o)}| f_{c_1})}} \tag{2.31}$$

The amount of gain peaking to be expected in the small signal closed loop frequency response is related to the damping factor by (Appendix 2)

$$P_{(dB \text{ of peaking})} = 20 \log_{10} \frac{1}{2\zeta\sqrt{(1 - \zeta^2)}} \qquad (2.32)$$

$$\zeta < \frac{1}{\sqrt{2}}$$

Note that there is no peaking in the sinusoidal response, for $\zeta > 1/\sqrt{2}$.

The value of f_o is related to the circuit parameters by

$$f_o = \sqrt{(|\beta A_{OL(o)}| f_{c_1} f_{c_2})} \qquad (2.33)$$

The frequency of the damped frequency oscillation is related to the natural frequency f_o by

$$f_d = f_o \sqrt{(1 - \zeta^2)} \qquad (2.34)$$

The times at which the peaks in the ringing occur, found by differentiating equation 2.30 with respect to time and equating to zero, are

$$t = \frac{N\pi}{\omega_o \sqrt{(1 - \zeta^2)}} \qquad (2.35)$$

$N = 1$ gives first max
$N = 2$ gives first min
$N = 3$ gives second max etc.

Overshoot

In the case of a lightly damped response ($\zeta < 1$) the amount by which the first ringing peak exceeds the final value is referred to as overshoot. Expressed as a percentage of the final value

$$\text{Overshoot \%} = 100 \exp\left(-\frac{\zeta\pi}{\sqrt{(1 - \zeta^2)}}\right) \qquad (2.36)$$

Small signal settling time

Overshoot represents the maximum output transient error following initial rise in response to a step input. The time taken by the output to settle within a certain accuracy (settling time) following a transient is often of greater interest. A conservative estimate of the small signal settling time for a second order system can be made by finding the smallest value of N which satisfies

$$100 \exp\left(- \frac{\zeta N\pi}{\sqrt{(1 - \zeta^2)}}\right) \leqslant x \text{ \%}$$

where x % represents a specified accuracy.

This value of N is then substituted into equation 2.35 to give the estimate of settling time.

Small signal settling time is clearly directly related to the value of the damping factor. For fast settling to high accuracy nothing is to be gained by using damping factors less than unity, for although light damping gives a faster initial rise any ringing prolongs settling time. It is for this reason that designers of fast settling operational amplifiers strive to have the open loop frequency characteristic strongly dominated by a single 20 dB per decade roll off down to unity gain in the open loop frequency response.

2.8.2 LARGE SIGNAL TIME RESPONSE CHARACTERISTICS

In applications in which signal amplitudes or their time rates of change are large enough to drive circuit operating conditions into non-linear regions the small signal transient response characteristics discussed in the previous section no longer provide an adequate description of circuit behaviour. In this section some of the effects accounting for the difference between small and large signal characteristics are considered.

Slewing rate

Within an operational amplifier there are inherent semiconductor and circuit capacitances as well as those added for frequency compensation; at the output there is load capacitance. The rate of change of voltage in each point in the circuit is limited by the available current to charge the capacitance at that point:

$$\frac{dV}{dt_{max}} = \frac{I_{max}}{C}$$

This mechanism sets an upper limit to the rate at which the output voltage of an operational amplifier can change. Slewing rate, usually expressed in V/μs is the parameter which is used to characterise the effect. As discussed in Section 2.7.1 it is often the charging of the frequency compensating capacitor which determines the output slew rate, but there are applications in which the charging of some other circuit capacitance sets the limit, for example, large capacitive loads.

Slewing rate is the performance parameter which determines the maximum frequency at which an amplifier can give a full scale sinusoidal output signal, and is one of the important factors in determining large signal settling time. Slewing rate determines the maximum operating frequency in such applications as precise rectifiers (see later).

Overload recovery

An amplifier when in a saturated overload condition takes a finite time to recover to a linear operating mode. Overload recovery defines the time required for the output voltage to recover to within its rated value from a saturated condition. Saturation takes place both when an amplifier's output voltage exceeds its rated value and during non-linear slewing with the output within rated limits. Saturation imposes charge changes away from normal operating values on the circuit capacitances; these must be discharged back to equilibrium values before the amplifier can operate normally. Thus, in an amplifier circuit required to give a full scale output step there is a period of recovery which is comparable to the period of slewing, but it may be substantially greater if many internal stages are involved. Fast slew rate therefore, is not by itself a good indicator of a fast settling amplifier. Some amplifiers with extremely large slew rates have excessive recovery time and greater overall settling time than other amplifiers having more modest slew rates.

Large signal settling time

Settling time is defined as the time elapsed from the application of a perfect step input to the time when the amplifier has entered and remained within a specified error band symmetrical about the final value. Large signal settling time is usually specified for the condition of unity gain and a full scale output step. The main contributions to settling time are slew rate and overload recovery (see Section 3.9.4) but there are other non-linear effects which can adversely affect settling time to high accuracy. Dielectric absorption in capacitors can be an important factor in circuits requiring settling to 0.01 % accuracy. Amplifiers which exhibit thermal feedback effects (see Section 3.6) can have settling times prolonged because of the effect.

2.9 Full power response

The inability of an amplifier output voltage to slew faster than a limiting rate can lead to distortion of sinusoidal signals with frequencies in the amplifier pass band, even though their amplitude is below the maximum rated output voltage for the amplifier. Some manufacturers specify the effect by amplifier full power response f_{p1} defined as the maximum frequency, measured at unity closed loop gain, for which full output can be obtained at rating load without distortion. An approximate relationship between slewing rate and full power response is readily derived if it is remembered that in the case of a sinusoidal signal the maximum rate of change occurs as the signal passes through zero. Consider a sinusoidal output signal with amplitude equal to the rated output voltage E_o and frequency f_p:

$$e_o = E_o \sin 2\pi f_p t$$

$$\frac{de_o}{dt} = 2\pi f_p E_o \cos 2\pi f_p t$$

$$\text{Slew rate} = \left| \frac{de_o}{dt} \right|_{max} = 2\pi f_p E_o \qquad (2.37)$$

If the output amplitude is reduced, distortion due to slew rate does not occur until the frequency is increased above f_p. Amplifier data sheets sometimes give graphs which relate maximum sinusoidal output voltage obtainable without distortion to frequency; they show what is called the power bandwidth of the amplifier.

2.10 Offsets, bias current and drift

The provision of a d.c. response for an operational amplifier introduces design problems that are not encountered in amplifiers intended to operate only on a.c. signals. An operational amplifier is normally required to give zero output voltage (referred to earth) when the voltage between its input terminals is zero, and with a constant d.c. input signal its output should remain constant. Parameters are defined which indicate how far real amplifiers depart from this ideal behaviour.

A practical amplifier with its input terminals shorted together is found to give a non-zero output voltage or 'offset', indeed in the case of a high gain amplifier the output offset under these conditions will normally cause the output voltage to be at one of its saturated levels. It is therefore usual to specify amplifier offsets by referring them to the input of the amplifier.

The *input offset voltage, V_{io},* is defined as that input voltage which would have to be applied in order to cause the amplifier output voltage to be zero. It is specified at a particular temperature.

All operational amplifiers require some small relatively constant current at their input terminals, called an *input bias current*. In the case of a differential amplifier the input bias current, I_b, is defined as the average value (half the sum) of the currents at the two input terminals with the amplifier output voltage at zero. It too is specified at a particular temperature. Ideally the currents taken by the two input terminals should be the same under these conditions but in practice some degree of mismatch always exists.

The *input offset current, I_{io},* is defined as the difference in the input bias currents to the two input terminals; it is specified at a particular temperature. With equal source impedances connected to the two input terminals, it is only this mismatch, or difference current, which causes an offset error. The effects of bias and offset currents tend to overshadow the effects of input offset voltage when the input source impedances are high.

Provision is normally made for balancing out the effects of initial amplifier offsets by means of a suitable potentiometer. After this adjustment has been made the output voltage of an amplifier is still found to change, even though the applied input signal is zero or at a constant d.c. value. This slow change in the output voltage of an amplifier is referred to as *drift*. Drift problems do not arise in a.c. amplifiers because in these amplifiers any d.c. change in voltage level is effectively blocked off from the output by the capacitors used to couple signals between stages.

A specification for the drift in the output voltage of an operational amplifier would in itself give little criterion for the selection of an amplifier for drift performance. The observed output drift is dominated by drift in the early stages of the amplifier, for this is magnified many times by subsequent stages before appearing at the amplifier output. It is usual to characterise drift performance by referring the drift to the input; the various contributions to drift are specified by their effects on amplifier input offsets.

2.10.1 TEMPERATURE DRIFT

In solid state operational amplifiers drift with temperature normally represents the largest single source of drift and is the one which gives rise to the biggest errors in many applications. It arises because of the temperature dependence of the characteristics of both active (transistors, etc.) and passive (resistors, etc.) components. Temperature drift may be specified by the temperature coefficients of bias current and input offsets. The coefficients, $\Delta I_b/\Delta T$, $\Delta I_{io}/\Delta T$ and $\Delta V_{io}/\Delta T$ are usually defined as the average slope over a specified temperature range being obtained by the subtraction of the offset values at the end points of the range divided by the temperature range. The drift to be expected for a defined temperature change from ambient is found by multiplying the specified drift rate by the temperature excursion. Drift of bias current, input offset current and input offset voltage are in general a non-linear function of temperature, with the drift rates normally being greater at the extremes of temperature than at ambient temperature. In drift critical applications it is therefore important to know exactly how the drift coefficients published in amplifier data sheets are specified; manufacturers are normally very willing to answer queries. A further point which should be borne in mind is that published temperature drift specifications assume that the whole of the amplifier is at the same temperature; amplifiers used in environments where thermal gradients exist are likely to give offsets greater than that predicted by the drift coefficients. This point is more important with discrete component amplifiers than with monolithic circuits.

2.10.2 SUPPLY VOLTAGE SENSITIVITY

Changes in the magnitude of amplifier power supplies cause changes in amplifier output voltage. The effect is usually specified by the effect of supply voltages on input bias current and input offsets. Supply voltage coefficients, $\Delta V_{io}/\Delta V$, $\Delta I_b/\Delta V$ and $\Delta I_{io}/\Delta V$ are included in data sheets. In the case of amplifiers using twin power supplies the positive and negative supply voltage coefficients will not normally be the same, but in any case with well regulated power supplies the drift due to power supply changes will normally be negligible compared with temperature drift.

2.10.3 DRIFT WITH TIME

Operational amplifiers maintained at a constant temperature and used with constant power supplies are still found to have some remaining drift over any appreciable time period. This drift with time is due to long term ageing of amplifier components. Not all amplifier data sheets include a time drift characterisation, but when they do the effect is normally specified in terms of the change in bias current and offsets to be expected in some specified period of time. No agreed standard time period for this specification is in existence, it may be one hour, one day,

Offsets represented by equivalent input error generators

Output offset

$$= E_{os} \frac{1}{\beta}$$

$$\frac{1}{\beta} = 1 + \frac{R_f}{R_p}$$

$$E_{os} = \pm V_{io} + I_b^- R_s^- - I_b^+ R_s^+$$

$$R_s^- = R_p \,//\, R_f$$

$$R_s^+ = R_3$$

Input error generators combined as single input offset error voltage

Figure 2.24 Evaluating offset errors

or even one month. It should be realised however that published time drifts do not accumulate linearly. Chopper stabilised amplifiers exhibit the best long term voltage stability, a voltage drift (drift of V_{io}) of 1 μV per day might be a typical specification, but the voltage drift of this

66

amplifier for 30 days would not normally exceed 5 μV. Accumulative drift can sometimes be extrapolated by multiplying the specified drift per day by the square root of the number of days.

2.10.4 EVALUATING ERRORS DUE TO INPUT OFFSET VOLTAGE AND BIAS CURRENT

In applications requiring a response down to zero frequency amplifier input offset voltage and bias current and their drift coefficients are usually the limiting performance parameters which determine the obtainable processing accuracy. We set out a general method for evaluating offset errors.

The effect of amplifier input offset voltage and bias current is conveniently represented by error generators at the input of an otherwise ideal operational amplifier (see Figure 2.24). Further simplification is provided by combining the effects of the separate error generators into a single input offset voltage error generator. The effects of bias current are expressed in terms of the equivalent voltages connected directly to the input terminals of the amplifier. Thus I_b^- applies a voltage $-I_b^- R_{source}^-$ to the inverting input terminal and I_b^+ applies a voltage $-I_b^+ R_{source}^+$ to the non-inverting terminal. R_{source}^- and R_{source}^+ represent the effective source resistances connected at the inverting and non-inverting input terminals respectively. They represent the parallel combinations of all resistive paths to ground, including in the case of R_{source}^- the path through any feedback resistor and the amplifier output resistance to ground.

V_{io} is directly applied to the input terminal so we may represent the total equivalent input offset voltage as

$$E_{os} = \pm V_{io} + I_b^- R_{source}^- - I_b^+ R_{source}^+ \qquad (2.38)$$

Drift in the total equivalent input offset voltage is obtained by substituting values of the drift coefficients of I_b and V_{io}.

Graphs showing the dependence of E_{os} drift on source resistance are given in some amplifier data sheets. E_{os} appears at the output multiplied by the 'noise gain' $1/\beta$; the resultant error may be referred to any signal input by simply dividing by the signal gain associated with that input. A numerical example should serve to clarify the evaluation of offset error:

An amplifier with $I_b = 100$ nA, $I_{io} = 10$ nA and $V_{io} = 1$ mV is to be used in the inverting summing circuit shown in Figure 2.25. Find the minimum signals which can be amplified at the two input signal points with less than 1 % error due to offset.

Figure 2.25 Circuit for example of offset error evaluation

In the circuit under consideration, the non-inverting input terminal is connected directly to earth, making R_{source}^+ zero. The effective source resistance through which bias current must flow to the inverting input terminal is

$$R_{source}^- = R_1 \| R_2 \| R_f$$

$$= 8.3 \text{ k}\Omega$$

According to equation 2.38

$$E_{os} = 10^{-3} + 10^{-7} \cdot 8.3 \times 10^3 \qquad \text{(worst case)}$$

$$= 1.83 \text{ mV}$$

In this circuit

$$\frac{1}{\beta} = 1 + \frac{R_f}{R_1 \| R_2} \cong 12$$

The output offset error is thus $1/\beta \, E_{os} = 22$ mV. Referring this error to the e_1 input the equivalent input error is $22/10 = 2.2$ mV. Referring the output error to the e_2 input the equivalent input error is $22/1 = 22$ mV. The smallest input signal at the e_1 input for less than 1% error is thus 220 mV and at the e_2 input is 2.2 V.

Error due to bias current can be reduced by connecting a resistor equal in magnitude to R_{source}^{-} between the non-inverting input and earth. This makes

$$E_{os} = \pm V_{io} + I_{io} R_{source}$$

Accuracy is still relatively low but can be improved if the initial offset is balanced out by using one of the offset balancing methods discussed in Section 9.6. An evaluation of subsequent offset error would then require knowledge of the temperature coefficient of I_b and V_{io} and an estimate of the possible ambient temperature variations.

Values $\qquad V_{io} = \dfrac{\Delta V_{io}}{\Delta T} \delta T, I_B = \dfrac{\Delta I_B}{\Delta T} \delta T$

should then be substituted in equation 2.38 in order to find the equivalent input error due to temperature drift.

2.11 Common mode rejection

An ideal differential amplifier responds only to the difference in the voltages applied to its input terminals and produces no output for a common mode input voltage. In practical amplifiers, because of slightly different gains between the inverting and non-inverting inputs, common mode input voltages are not entirely subtracted at the output. The gain of an amplifier for common mode input voltages is known as the common mode response and the ratio of the gain with the signal applied differentially to the common mode response is called the common mode rejection ratio, c.m.r.r. It is often expressed in dB by taking 20 times log (base 10) of the ratio.

Common mode rejection presents no problem in the case of amplifiers used in the inverting configuration, for with one input earthed, the input common mode voltage e_{cm} must be zero. In the case of the non-inverting circuit feedback causes the voltage at the inverting input to follow that at the non-inverting input, the input common mode voltage thus varies directly with the input signal. With finite c.m.r.r. an output signal is produced in response to this common mode input signal, thus an error is introduced which affects the overall circuit accuracy. The error is conveniently represented in terms of an equivalent input common mode error voltage, $e_{\epsilon_{cm}}$, where $e_{\epsilon_{cm}}$ = common mode output/ differential gain. If the amplifier is considered to have $e_{\epsilon_{cm}}$ applied to its non-inverting input terminal along with the input signal it may then be treated as though it completely rejected the actual input common mode signal e_{cm}. The relationship between input common mode error voltage and input common mode voltage is readily obtained as

$$\frac{e_{\epsilon_{cm}}}{e_{cm}} = \frac{1}{\text{c.m.r.r.}} \tag{2.39}$$

It is instructive to consider a numerical example. Consider an amplifier with c.m.r.r. 1 000, (60 dB) used in the non-inverting configuration with say an input signal of 1 V, the input common mode voltage e_{cm} would also be 1 V. The input common mode error voltage is seen to be 1 mV and this represents a 0.1 % measuring error. The amplifier is illustrated in Figure 2.26. If common mode error voltage varied linearly with common mode voltage (e_{cm}), errors such as this would not be important since they could be compensated for by adjustment of closed loop gain (adjustment of R_2/R_1).

Figure 2.26 Representation of common mode error

It is not always possible to compensate for common mode errors because the c.m.r.r. for some amplifiers shows a dependence on the magnitude of the input common mode signal and the common mode error voltage is a nonlinear function of common mode voltage. There is also an added complication of a temperature dependence. Since linearity of common mode error voltage with common mode voltage is really more important than the actual value of the c.m.r.r., a graph illustrating this relationship is valuable if an amplifier is to be used in an application which is critically dependent on commmon mode performance. Figure 2.27 illustrates an example, and although this type of graph is not normally included in amplifier data sheets it is not difficult to obtain experimentally, the method being explained in Chapter 3. Specified values of c.m.r.r. where nonlinearities exist are usually average values, assuming a measurement of $e_{\epsilon_{cm}}$ at the end points corresponding to the maximum common mode voltage E_{cm}. Published common mode specifications generally apply to d.c. input signals; c.m.r.r. is usually found to decrease at the higher frequencies.

Figure 2.27 *Common mode error voltage as a function of common mode input voltage*

2.12 Amplifier impedances

The ideal operational amplifier of Section 1.2, when open loop, was assumed to have an infinite input impedance and zero output impedance; real amplifiers show departures from this ideal.

The *input impedance* of an amplifier is the effective impedance seen looking into the amplifier input terminals. In the case of differential amplifiers there is in addition to the differential input impedance an effective impedance from each input to earth or power supply common called the *common mode input impedance*. The parallel sum of the impedances from each input to earth is sometimes defined as the common mode input impedance, thus

$$Z_{cm} = \frac{Z_{cm}^{+} \, Z_{cm}^{-}}{Z_{cm}^{+} + Z_{cm}^{-}}$$

In other cases Z_{cm}^{+} and Z_{cm}^{-} are specified separately. Z_{cm} is normally some ten to a hundred times greater than Z_i, the impedances are largely resistive, and reactive components may normally be represented by shunt capacitances of a few picofarad. Provided that the input impedance of an amplifier is larger than the resistors connected to its input terminals finite input impedance has no significant role in determining closed loop performance.

The open loop *output impedance* of practical amplifiers is not zero. It can vary from a few ohms to as much as several thousand ohms with the majority of amplifiers having output impedances in the range 100–500 Ω. The effect of the open loop output impedance is to form a voltage divider with the effective amplifier load (external load plus feedback components) thus attenuating the open loop gain. Manufacturers normally specify open loop gain at a rated load thus effectively allowing for Z_o; open loop gain will however vary slightly with change in amplifier load because of non-zero output impedance.

2.13 Noise in operational amplifier circuits[3,4]

The output of an amplifier is always found to contain signals which are unrelated to the input signals and which cannot be predicted from an accurately known closed loop transfer function. These unwanted signals are called noise. Errors such as drift error can, theoretically at least, be reduced to negligible proportions (by say using a temperature controlled environment) but there always remains a noise error which limits the attainable accuracy and resolution. The operational amplifier user needs to concern himself with noise errors whenever he is required to process low level signals with high accuracy.

There are two basically different types of noise in a circuit: interference noise, which is noise picked up from outside the circuit, and inherent noise which is noise arising within the circuit itself. Sources of interference noise are many and varied; they include electromagnetic or electrostatic pickup from power sources at mains frequency, radio broadcast stations, electrical arcing at mechanical switch contacts, voltage spikes from switching in reactive circuits, and mechanical vibrations of circuit components or leads. Fortunately the circuit designer can usually minimise interference noise by suitable shielding and guarding and elimination of earth loops (see Section 9.4) and by proper attention to mechanical design. Inherent noise however, is a function of a particular amplifier and the circuit in which it is used; the only way in which the designer can influence inherent noise is through his choice of amplifier and circuit components.

The choice of amplifier type for a particular application can result
in several orders of magnitude variation in the noise and if the designer
is to make a satisfactory choice he must start with a sound knowledge
of the methods used to characterise random noise. He must understand
the methods used to specify the noise performance of operational
amplifiers and he must be able to relate these specifications to the noise
performance to be expected when an amplifier is used in a specific
application. The treatment which follows does not attempt to cover the
underlying mechanisms governing noise generation, but is intended
simply to give the reader a basic working knowledge of noise characterisation sufficient to enable him to make an order of magnitude noise
evaluation in an operational amplifier circuit.

2.13.1 CHARACTERISATION OF RANDOM NOISE SOURCES

The noise that is inherent in an operational amplifier circuit (or any
circuit for that matter) can be thought of as a combination of the effects
of several separate noise sources. These inherent noise sources are
essentially random. They give an electrical signal whose waveform has
no defined shape, amplitude or frequency, which may be thought of as
a superposition of signals at all possible frequencies, with amplitude and
phase varying in a completely random fashion. Readers who are unfamiliar with the concept of random noise sources may well find that
some practical noise measurements (see Section 3.2) performed concurrently with their reading of this section are a great help in giving a
feel for the significance of random noise.

R.M.S. value of a noise source
It is a characteristic of most forms of random noise source that averaged
over a sufficiently long time interval their r.m.s. value in a specified
bandwidth remains substantially constant. R.M.S. value in a specified
bandwidth is thus a useful and meaningful way of characterising a random noise source. The general defining equation is

$$N_{r.m.s} = \sqrt{\left(\frac{1}{T} \int_0^T n_i^2 \, dt \right)} \qquad (2.40)$$

where n_i is the instantaneous noise amplitude (current or voltage), and
$N_{r.m.s.}$ is the value of the noise which would be indicated by a true
r.m.s. meter connected to the noise source. In order to be meaningful a

practical measurement of the r.m.s. value of a noise source must have the bandwidth clearly defined, for the wider the bandwidth the greater is the r.m.s. value of the noise measured from a particular noise source.

Combining noise sources

The combined effect of several random noise sources is found by root sum of the squares addition of the r.m.s. values of the separate noise sources. Thus, if E_1, E_2, E_3 are the r.m.s. voltage values of three separate voltage noise generators their combined effect when connected in series is equivalent to a single noise voltage generator of r.m.s. value

$$E = \sqrt{(E_1^2 + E_2^2 + E_3^2)} \qquad (2.41)$$

Equation 2.41 can be used as a means of expressing the combined effect of separate noise sources. It can also be used to combine the noise contributions from different parts of the frequency spectrum of a single noise source. It should be noted that the root sum of the squares addition has the effect of emphasising the larger quantities; for example, if one quantity is three times greater than another the error incurred by ignoring the smaller quantity is only about 5 %: $\sqrt{(1^2 + 3^2)} = 3.16 \cong 3 \times 1.05$.

Peak to peak noise

In some applications it is peak to peak noise which really sets the limit to a system performance. Peak to peak noise is the difference between the largest positive and negative peak excursions to be expected during some arbitrary time interval. Random noise is for all practical purposes Gaussian in amplitude distribution, the highest noise amplitudes having the smallest (yet not zero) probabilities of occurring. Peak to peak noise is thus difficult to measure repeatedly, but a useful rule of thumb for converting from an r.m.s. noise value to a peak to peak value is to multiply the r.m.s. value by a factor of 6. Rigorously this gives an amplitude which is exceeded less than 0.25 % of the time by a random noise signal of the given r.m.s. amplitude.

Noise density spectrum

The noise generated by any random noise source exists in all parts of the frequency spectrum, and in a noise measurement the noise contributed by a source varies with the range of frequencies over which the observation is made. A noise density spectrum shows the way in which the noise produced by a given source is distributed over the frequency spectrum; noise density n is shown as a function of frequency, usually on log-log axes. The noise density for a particular source is defined in such a way that the r.m.s. value of the noise contributed by the source in the range of frequencies $f_1 \rightarrow f_2$ is determined by the relationship

$$N_{r.m.s.(f_1-f_2)} = \sqrt{\left(\int_{f_1}^{f_2} n^2 \, df \right)} \qquad (2.42)$$

n is the noise density (current or voltage) at frequency f
n^2 is expressed in A^2 or V^2 per Hz. (Note $n^2 = d/df \, N_{r.m.s.}^2$
n is normally expressed in $pA/\sqrt{Hz}$ for a noise current source and $nV/\sqrt{Hz}$ for a noise voltage source.

A graph of n^2 against frequency shows the way in which the noise power produced by a given source is distributed through the frequency spectrum and the area under such a graph, say between frequency limits f_1 and f_2, is proportional to the average noise power in this frequency band. Examples of noise spectra are given in Figures 2.28, 2.30 and 2.31.

In the spectral regions of interest in operational amplifier applications, the noise sources encountered often have spectral distribution belonging to one of two types: in the one n is constant as a function of frequency and in the other n varies inversely with the square root of frequency.

White noise
A noise source for which n is constant with change in frequency is called a white noise source; the noise from a white noise source is distributed uniformly throughout the frequency spectrum. Thermal agitation of electrons in a resistor causes random voltages to appear across it. The spectrum of this noise voltage is characterised by a noise density which is constant as a function of frequency, and resistance noise (Johnson noise) is thus an example of white noise. The noise voltage associated with a resistor has a noise density:

$$\text{Resistance noise } e = \sqrt{(4 \, kTR)} \text{ V per } \sqrt{Hz} \qquad (2.43)$$

k = Boltzmann's constant = $1.37 \times 10^{-23} \, J/^\circ K$
T = the temperature in $^\circ K$
R = the resistor value in Ω.

Substitution of equation 2.43 into equation 2.42 gives an expression for the r.m.s. noise voltage generated by a resistor R in the range of frequencies f_1 to f_2 as:

Resistance noise $E_{r.m.s.f_1-f_2} = \sqrt{[4kTR\,(f_2 - f_1)]}\,V$ (2.44)

At room temperatures with rather more convenient units this becomes

Resistance noise $E_{r.m.s.} = 0.13\sqrt{[R\,(f_2 - f_1)]}\,\mu V$ (2.45)

(R in MΩ, $f_2 - f_1$ in Hz).

1/f or 'pink' noise

A noise source which has noise density varying inversely with the square root of frequency is referred to as a $1/f$ noise source; noise power density is proportional to $1/f$, and $1/f$ noise is sometimes called 'pink' noise. The noise density for a pink noise source is determined by an equation of the form

$$\text{Pink noise } n = K\sqrt{\frac{1}{f}}$$ (2.46)

where K is the value of n at $f = 1$ Hz.

A graph of n against frequency for a pink noise source when shown as a log-log plot is a straight line of slope -10 dB/decade; n^2 against frequency gives a straight line of slope -20 dB/decade.

The contribution which a pink noise source makes to the r.m.s. value of the noise in a frequency range f_1 to f_2 may be found by substituting equation 2.46 into the general equation 2.42, thus:

$$N_{r.m.s.(f_1-f_2)} = K\sqrt{\left(\int_{f_1}^{f_2}\frac{df}{f}\right)} = K\sqrt{\left(\log_e\frac{f_2}{f_1}\right)}$$ (2.47)

Note that the r.m.s. noise contributed by a pink noise source in a particular bandwidth depends upon the ratio of the frequencies defining that bandwidth. Every frequency decade of noise from a pink noise source has the same r.m.s. value as every other decade.

Evaluation of r.m.s. noise from a noise density spectrum

The contribution that a particular noise source makes to the r.m.s. noise in any specified bandwidth can in principle be found by evaluating the integral in equation 2.42. Equations 2.44 and 2.47 are the results of such evaluations for the particular cases of a white noise source and a pink noise source. Note that in order to evaluate the integral the equation defining the noise density as a function of frequency must of course be known.

In the spectral regions of interest the noise generators used to represent the effect of internally generated amplifier noise often exhibit a noise density spectrum of the form shown in Figure 2.28. A spectrum of this kind can be thought of as consisting of two components, a white noise component which is the predominant noise component at high frequences and a $1/f$ component which predominates at low frequencies. The r.m.s. value of the noise contributed by the source in any bandwidth can be found by a root sum of the squares addition of the r.m.s. contributions of the two separate components in that bandwidth.

Figure 2.28   Noise density spectrum for amplifier generated noise

1/f component
At the frequency f_0 the noise density of the $1/f$ component is equal to that of the white noise component n_w; f_0 lies at the point of intersection of the high and low frequency asymptote to the curve. It is called the $1/f$ corner frequency. Substituting values in equation 2.46 gives the value of K for the pink noise component as

$$K = n_w \sqrt{f_0}$$

Substituting this value of K into equation 2.47 gives the contribution of the pink noise component to the r.m.s. noise in the frequency range f_1 to f_2 as

$$N_{(pink)} = n_w \sqrt{\left(f_0 \log_e \frac{f_2}{f_1} \right)}$$

White noise component

The contribution of the white noise component to the r.m.s. noise in the frequency range $f_1 \rightarrow f_2$ is

$$N_{(\text{white})} = n_\text{w} \sqrt{(f_2 - f_1)}$$

Total r.m.s. noise contributed by both components is

$$N_{(f_1 - f_2)} = \sqrt{[N^2_{(\text{pink})} + N^2_{(\text{white})}]}$$

$$N_{(f_1 - f_2)} = n_\text{w} \sqrt{\left[f_0 \log_\text{e} \frac{f_2}{f_1} + (f_2 - f_1) \right]} \qquad (2.48)$$

2.13.2 OPERATIONAL AMPLIFIER NOISE SPECIFICATIONS

Following on the previous section in which general methods of characterising noise were discussed, this section now introduces the reader to the types of noise information he is likely to find included in operational amplifier data sheets. Noise data need careful inspection for noise specifications are probably the least standardised of all operational amplifier performance parameters. There is a lack of standardisation

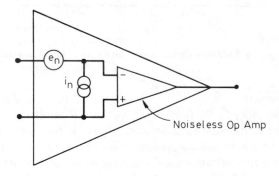

Figure 2.29 Amplifier model for internally generated noise

between different manufacturers and to make matters even more complicated for the user any one manufacturer cannot be relied upon to provide noise information in a standardised manner.

The noise which is present at the output of an amplifier is a combination of the amplified noise present at its input and noise generated internally inside the amplifier. Noise produced internally by an operational amplifier is conveniently modelled as shown in Figure 2.29, by

Broadband noise for various bandwidths

Total noise referred to input - μVrms

$V_S = \pm 15V$
$T_A = 25°C$

10.100kHz

10.10kHz

10.1kHz

(a) Source resistance - Ω

Input noise voltage as a function of frequency

Mean square voltage - V^2/Hz

$V_S = \pm 15V$
$T_A = 25°C$

(b) Frequency (Hz)

Input noise current as a function of frequency

Mean square noise current - A^2/Hz

$V_S = \pm 15V$
$T_A = 25°C$

(c) Frequency (Hz)

Figure 2.30 Graphical noise data.
Amplifier type $\mu A741$ typical data

a noiseless amplifier with a noise voltage and a noise current generator at its input terminal. Noise generated by resistors at the input are similarly represented by equivalent noise generators connected to the amplifier input terminals. The several input contributions to the noise are combined as a single resultant input referred noise source and in closed loop amplifier applications this total input referred noise appears at the output multiplied by the closed loop noise gain $1/\beta$.

The technique for noise evaluation as just described is similar to the technique for evaluating offset and drift errors given in Section 2.10.4. The main difference between a drift error evaluation and the evaluation of noise errors is the dependence of noise on bandwidth. In making a noise assessment of an operational amplifier application the designer must use published noise data in order to estimate the amount of noise present in the noise bandwidth as set by the particular closed loop configuration in which the amplifier is used.

Operational amplifier noise data will be found presented in both graphical and numerical form. Some of the different types of noise information to be found in operational amplifier data sheets are now discussed.

Graphical data

Three main types of graphical noise data are to be found: total input referred noise in various bandwidths as a function of source resistance, amplifier equivalent input noise voltage density as a function of frequency, and amplifier equivalent input noise current density as a function of frequency. Examples of these graphs are shown in Figure 2.30; they represent the typical noise characteristics of a 741 type operational amplifier. It is important to note carefully the units in which the graphical information is presented; note that the ordinate of the graph giving the spectral distribution of the input noise voltage is in V^2/Hz, Figure 2.30(b), and that the spectral distribution of the input noise current is in A^2/Hz, Figure 2.30(c). It is the square of the noise density function (see equation 2.42) which appears as the ordinate in both figures 2.30(b) and (c).

The information provided by the graphs of Figures 2.30(b) and (c) is sometimes presented in a different way; an example of an alternative presentation is shown in Figure 2.31. The graphs show the spectral distribution of the amplifier input noise voltage and input noise current, but note that it is the noise density function which appears as the ordinate, noise voltage density expressed in $nV/\sqrt{Hz}$, and noise current density in $pA/\sqrt{Hz}$. The graphs have log–log scales as before but some data sheets will be found to give noise distributions plotted in a linear/log fashion.

Numerical noise data

Graphical data present a detailed characterisation of the noise performance of an amplifier but it is convenient sometimes to have a simple numerical indication of an amplifier's noise performance included in a list of amplifier specifications.

Figure 2.31   Noise density spectra. Operational amplifier type SSS 725 (Precision Monolithics)

Peak to peak noise values

Some manufacturers specify typical peak to peak input voltage and current noise in a low frequency band (say 0.01 to 1 Hz). A peak to peak specification of this kind is particularly useful in assessing accuracy limits (as limited by noise) in applications in which the signals of interest are essentially d.c. or very slowly varying quantities. Wide bandwidth noise will of course be present in the amplifier output but it can be removed by following the amplifier with a suitable low pass filter.

R.M.S. values of noise can be used as a means of estimating peak to peak values. The rule of thumb multiplication factor of 6 mentioned previously is used.

Wide band noise
A numerical indication of the r.m.s. value of the amplifier generated input noise voltage in a specified bandwidth is sometimes given in amplifier data sheets, but it should be noted that the bandwidth for which this specification is given is not the same for all amplifiers. Typical noise bandwidths used in defining this parameter are 10 Hz to 10 kHz, 500 Hz to 5 kHz, and 5 Hz to 50 kHz. It is important to note the noise bandwidth when attempting to compare the noise specification of different amplifiers.

2.13.3 EVALUATING NOISE ERRORS USING NOISE SPECIFICATIONS

The problem facing the circuit designer is to assess accuracy and resolution limits as determined by noise. Clearly if the noise level at the output is comparable to the signal level the signal is obscured by the noise. In wide band applications signal to noise ratio is a useful figure of merit in describing how well the signal 'stands out' from the noise. The signal to noise ratio at the output is defined as

$$\text{Signal to noise ratio, s.n.r.} = \frac{\text{signal power out}}{\text{noise power out}} = \frac{P_s}{P_n} \qquad (2.49)$$

The ratio is sometimes expressed in dB by the relationship

$$\text{s.n.r.}_{\text{dB}} = 10 \log_{10} \frac{P_s}{P_n} \qquad (2.50)$$

In d.c. and low frequency applications accuracy limits as determined by noise can be related to peak to peak noise, noise error being expressed as a percentage from the relationship

$$\text{Noise error} = \frac{\text{peak to peak value of output noise}}{\text{peak to peak value of output signal}} \times 100\% \qquad (2.51)$$

Signal gain $= \dfrac{R_2}{R_1} \left[\dfrac{1}{1 + \dfrac{1}{\beta A_{OL(jf)}}} \right]$

Noise gain $= \dfrac{1}{\beta} \left[\dfrac{1}{1 + \dfrac{1}{\beta A_{OL(jf)}}} \right]$

E_v = r.m.s. value of amplifier input noise voltage

E_i = IR_s, where I is the r.m.s. value of the input noise current of the amplifier

E_r = r.m.s. value of Johnson noise produced by resistors

Figure 2.32 Noise evaluation example 1

An estimate of the amount of noise to be expected in a given application is made by employing the noise specifications for the amplifier in the circuit, but the actual computation technique best adapted to a particular noise evaluation is very much dependent upon the nature of the circuit conditions and upon the type of noise data available for the amplifier. Rigorous noise evaluations are time consuming and can be of dubious practical value if they are based upon 'typical' noise data. In many applications the effect of a single noise source can be dominant, and the ability to identify the most significant noise contributions facilitates a rapid order of magnitude noise assessment. Examples of

83

noise evaluations are now given. A study of these examples will, it is hoped, help the reader make a noise assessment in his own operational amplifier application.

As a starting point in any noise evaluation the signal gain and noise gain in the application should be found. Bode plots giving their frequency dependence help to show up the spectral regions where significant noise contributions are to be expected. An evaluation of the noise performance of an operational amplifier used in a basic resistive feedback configuration is taken as a first example, Figure 2.32. Input noise sources appear at the amplifier output multiplied by the noise gain; in this example the noise gain is 100, its 3 dB bandwidth limit is 10^4 Hz and it rolls off at 20 dB/decade beyond this frequency.

Using total noise specifications
Total noise specifications of the type shown in Figure 2.30(a) relate the total r.m.s. noise in a specified bandwidth to the value of the source resistance. They can often be used directly in applications in which noise bandwidth is clearly defined and in which noise gain rolls off at 20 dB/ decade. Some minor difficulties in the interpretation of total noise specifications can arise because the source resistance condition under which they are defined and measured is not always clearly stated. In general source resistance can be present at both input terminals of an amplifier. The source resistance at the inverting input terminal is the parallel combination of input resistors and feedback resistor, and an equal source resistance is often connected between the non-inverting input terminal of the amplifier and earth for bias current compensation. Total noise specifications in general refer to the conditions in which equal value source resistors are connected into each input, and it is this value of resistance which is specified as the source resistance. In follower applications used to buffer high input signal sources the internal resistance of the signal source itself constitutes the source resistance at the non-inverting input terminal of the amplifier, and it is normally much greater than the source resistance at the inverting input. In cases in which there are unequal value source resistors at the two input terminals the larger of the two source resistors should be regarded as the effective source resistance when using total noise specifications.

The source resistance in the circuit application under consideration is $\cong 1$ kΩ. If the amplifier in the circuit is assumed to have the noise characteristics given in Figure 2.30(a) (a 741) inspection of this graph shows that the total input referred noise in the bandwidth 10 Hz–

10 kHz is 2.5 μV r.m.s. We are in fact concerned with the noise in the whole bandwidth 0–10 kHz but the contribution to the noise in this whole bandwidth at frequencies less than 10 Hz is not very significant. The r.m.s. value of the noise at the output in this application is $2.5 \times 100 \times 10^{-6} = 0.25$ mV r.m.s. Since in this example signal gain and noise gain are the same, the signal to noise ratio can be expressed in terms of the input signal and the r.m.s. value of the total input referred noise.

Using equivalent input noise generators
In cases in which total input referred wide bandnoise characteristics are not available, or not applicable, the several input noise contributions in an application must be separately identified. In the application under consideration in Figure 2.32, the noise at the output may be considered as a combination of the effects of several independent noise sources at the input. The equivalent input noise voltage sources are

1. Amplifier equivalent input noise voltage generator with spectral density e_v.
2. Noise voltages generated by amplifier input current noise flowing through input source resistors. If there are resistors at both in put terminals there will be two such generators with densities respectively

$$e_{I_1} = i R_s^+$$

$$e_{I_2} = i R_s^-$$

The two may be combined in terms of a single generator and effective source resistance R_s as

$$e_I = i R_s$$

where $\quad R_s = \sqrt{[(R_1 // R_2)^2 + R_3^2]}$

i is the spectral density function of the amplifier generated input current noise.
3. Voltage noise generated by the source resistances themselves. The spectral density function for resistance generated noise being

$$e_R = 0.13 \sqrt{R} \ \mu V / \sqrt{Hz}$$

with R in MΩ.

We assume that the amplifier used in the application is a type 741 with input noise generators having spectral distributions shown in Figures 2.30(b) and (c). Sketching in the high and low frequency asymptotes (see Figure 2.28) to these noise distribution curves gives

Amplifier input voltages noise

$$\text{White noise component density } e_\omega = 20 \text{ nV } \sqrt{Hz}$$

$$\frac{1}{f} \text{ corner frequency } = 50 \text{ Hz}$$

Noise voltage due to amplifier noise current

$$\text{White noise component } = i_\omega R_s$$
$$= 5.5 \times 10^{-13} . \sqrt{2} . 10^3 \text{ V}/\sqrt{Hz}$$
$$= 0.78 \text{ nV}/\sqrt{Hz}$$

$$\frac{1}{f} \text{ corner frequency } = 1.4 \times 10^3 \text{ Hz}$$

Resistance noise

$$\text{Noise density } e_R = 0.13 \sqrt{(R_1 \| R_2 + R_3)} \text{ } \mu V/\sqrt{Hz}$$

(R is in MΩ)

$$e_R \cong 5.8 \text{ nV}/\sqrt{Hz}$$

The only significant contribution to the r.m.s. noise in the bandwidth 10 Hz–10 kHz is that contributed by the amplifier input noise voltage. Substitution in equation 2.48 gives

$$E_{\text{r.m.s.(10 Hz-10 kHz)}} = 2 \times 10^{-8} \sqrt{(50 \log_e 10^3 + 9990)}$$
$$= 2 \times 10^{-8} \sqrt{(345 + 9990)}$$
$$= 2.03 \text{ } \mu V$$

Note that it is in fact the white noise component of the amplifier input voltage noise which makes the only significant contribution. Readers should evaluate the other r.m.s. contributions due to the amplifier current noise and resistance noise in order to satisfy themselves that they give a negligible contribution to the r.m.s. value of the input referred noise. Find also the effect of extending the low frequency bandwidth down to say 0.1 Hz. Remember that root sum of the squares addition must be used when combining noise sources and that this emphasises the effect of the larger contributions.

Filter skirt

Equation 2.48 is based upon the assumption of infinitely sharp bandwidth limits. In fact, in practical noise measurements made, say, with a first order low pass filter characteristic there is a significant amount of noise passed in the frequency band beyond the nominal cut off frequency of the filter. For example, equation 2.45 when used to give an expression for the r.m.s. white noise in the whole frequency band below f_2 must be multiplied by 1.26 to account for the white noise passed at frequencies higher than f_2 by a first order low pass filter. In effect the frequency $f_2 \frac{\pi}{2}$ should be used to express the noise bandwidth for white noise defined by a first order low pass filter with cut off frequency f_2. The r.m.s. noise from a white noise source of density e_w passed by a first order low pass filter with cut off at frequency f_2 is

$$E = e_w \sqrt{\left(f_2 \; \frac{\pi}{2} \right)} \qquad (2.52)$$

If allowance is made for the noise passed at frequencies beyond 10 kHz by the 20 dB/decade roll off in the noise gain in the application of Figure 2.32, the total input referred noise becomes

$$E_{r.m.s.} = 2 \times 10^{-8} \sqrt{\left(10^4 \; \frac{\pi}{2} \right)}$$

$$10 \text{ Hz–}10 \text{ kHz} = 2.5 \, \mu V$$

This figure agrees, as it should, with the noise evaluation made earlier using total input referred noise specifications.

Noise evaluation—Example 2

The circuit of Figure 2.33 is given as a second example of a noise evaluation. It is simply the circuit of Figure 2.32 with a capacitor $C_f = 0.15 \, \mu F$ connected in parallel with the feedback resistor so as to limit both signal and noise bandwidth. The important thing to realise is that although capacitor C_f makes the signal gain roll off continuously at 20 dB/decade the value of the noise gain $1/\beta$ reaches a minimum value of unity. In all circuits using passive components to fix the feedback fraction β the value of $1/\beta$ can never be less than unity. An expression for the noise gain and a Bode plot showing its frequency dependence is given in Figure 2.33.

Total noise specifications cannot be used in this example; a noise

$$\text{Noise gain} = \left[1 + \frac{R_2}{R_1}\right] \frac{1 + j\omega\, CR_1 /\!/ R_2}{1 + j\omega\, CR_2} \left[\frac{1}{1 + \frac{1}{\beta A_{OL(jf)}}}\right]$$

Figure 2.33 Noise evaluation example 2

evaluation requires a knowledge of the spectral distributions of the input noise sources. These spectral distributions are multiplied by the frequency dependent noise gain and appear at the output as output spectral distributions. If input noise density graphs and noise gain are shown on compatible log/log scales the multiplication process can be achieved by a simple addition of graphs shown in Figure 2.33. Only the amplifier generated voltage noise is considered, because with the comparatively low value source resistance used in this example the noise contributions due to current noise and source resistance noise are again negligible. In a more general treatment with larger source resistances the output noise spectral distributions due to all input noise sources would need consideration in order to identify the significant contributions.

Having obtained output noise spectral distributions the problem is to evaluate the total value of the r.m.s. output noise. A rigorous evaluation involves an integration process based upon the use of equation 2.42; this is a time consuming process and it is usually sufficient to make an approximate evaluation. Approximations are made by identifying the spectral regions in which a significant contribution to the total r.m.s. output noise is made. This may be accomplished by lowering a line of slope -10 dB/decade until it touches a noise density graph. A -10 dB/decade slope represents the spectral distributions of a pink noise source; all frequency decades of the noise from a pink noise source have the same r.m.s. content. Because of root sum of the squares addition of r.m.s. noise sources it follows that only those parts of noise distribution curves which lie less than 10 dB below $(-10$ dB $= 1/3)$, the pink noise tangent, make a significant contribution to the total r.m.s. noise.

In the example under consideration the most significant output noise contributions are the pink noise component of the amplifier input noise voltage (multiplied by 100) at frequencies less than 10 Hz and the white noise component of the amplifier input voltage (multiplied by unity) at frequencies below 1 MHz and multiplied by a 20 dB/decade roll off at frequencies greater than 1 MHz. The K value of the pink noise component of the output noise voltage is

$$K = 20 \times 100 \sqrt{50} \cong 14 \times 10^3 \text{ nV}$$

The contribution of the pink noise component to the total output noise may be approximated by 3 decades of this pink noise (frequency 0.1 to 10 Hz). Pink noise component of output noise

$$\cong 14 \times 10^3 \sqrt{(\log_e 10^3)} \text{ nV} = 37 \text{ } \mu\text{V}$$

Figure 2.34 The general noise evaluation problem

The white noise component and 20 dB/decade roll off give a contribution to the output noise

$$E_w \cong 20 \sqrt{(10^6 \frac{\pi}{2})} \text{ nV} = 25 \,\mu\text{V}$$

An approximation for the r.m.s. value of the total output noise is

$$E_{out} = \sqrt{(37^2 + 25^2)} = 44 \,\mu\text{V}$$

The s.n.r. to be expected at the output is thus

$$\text{s.n.r.} = \frac{(E_s \times 100)^2}{(44 \times 10^{-6})^2}$$

where E_s is the r.m.s. value of the signal input.

The operational amplifier used in this example is assumed to have an ample unity gain phase margin. An inadequate phase margin would cause the noise gain to peak up at the intersection frequency of $1/\beta$ with the open loop response (see Section 2.6.1). Note, that in cases in which there is an inadequate phase margin the noise gain can peak up even though the signal gain rolls off smoothly, and in such a situation the noise at frequencies outside the signal pass band can become the dominant factor. In applications in which there are significant noise contributions at frequencies outside the signal pass band, signal to noise ratio can be improved by adding a low pass filter to the amplifier output.

Summary of noise considerations

If the noise bandwidth in an application is well defined and noise gain rolls off at 20 dB/decade total noise specifications, if available, give the most rapid method of noise assessment in an application. If total noise specifications are not available, or the frequency dependence of the noise gain is such that they are not applicable, the contributions of the several input noise sources to the total output noise must be evaluated. Consider the general case shown in Figure 2.34. The input noise voltage sources expressed in terms of their density functions are:

1. Amplifier generated input voltage noise e_v.
2. Noise voltage sources due to amplifier input current noise flowing through input source impedances; these are

$$iz_1 \| z_f \text{ and } iz_2$$

where i is the density function of the equivalent amplifier generated current noise.

3. Thermal noise voltage sources due to input source resistors; these are

$$e_{R_2} = 0.13 \sqrt{R_2} \text{ and } e_{R_1} = 0.13 \sqrt{(R_1 \| R_f)}$$

R_1, R_2 and R_f are the resistive components of impedances z_1, z_2 and z_f.

The r.m.s. contribution of the several input noise sources in a specific bandwidth can be evaluated by use of the general equation 2.42, or if the spectra of the noise sources have only pink and white noise components equation 2.48 can be used. The r.m.s. value of the total input referred noise is found by root sum of the squares addition of the r.m.s. values of the separate noise sources; when signal gain and noise gain are the same the total input referred noise can be compared with the signal in order to estimate the value of the signal to noise ratio.

In cases in which the frequency response of the noise gain is such that significant contributions to the output noise are made at frequencies outside the signal bandwidth, output noise density distributions must be found. Output density functions are obtained by multiplying input density functions by the noise gain; the multiplication process is conveniently performed by the addition of compatible log/log graphs (for example see Figure 2.33). A pink noise tangent lowered to contact with the output density functions allows the significant spectral contributions to be located, and an approximate evaluation of the total output noise can be made. Signal to noise ratio is assessed from the relationship

$$s.n.r. = \frac{(E_s \times \text{signal gain})^2}{(\text{r.m.s. value of total output noise})^2}$$

Note that in summing applications there may be several different signal gains associated with several input points to an operational amplifier circuit; in such cases each input would have a different signal to noise ratio associated with it.

In wideband applications (say 10 kHz upwards) with source resistance value of 100 kΩ upwards the Johnson noise of input sources resistors begins to dominate the noise performance, and in such cases there is little to choose between different amplifier types from the point of view of noise performance. In low source resistance, low frequency narrow band applications, bipolar input operational amplifiers normally give a smaller total noise than f.e.t. input amplifiers, but as source resistance value is increased f.e.t. amplifiers generally give a smaller total output noise. The input noise current associated with an f.e.t. input amplifier is very much smaller than that of a bipolar input operational amplifier.

Noise factor

Although signal to noise ratio at the output of an amplifier is the important parameter which shows how well the signal is distinguished from the noise, it does not give a direct indication of the noisiness of the amplifier itself. In any bandwidth the noise present at the output of an operational amplifier is a combination of the noise produced in the signal source, which appears amplified at the output, together with noise generated internally inside the amplifier. In radio frequency and audio frequency communication systems a quantity called the noise factor is commonly used to express the noisiness of receivers or amplifiers. Noise factor is defined by the relationship

$$\text{Noise factor} = \frac{\text{s.n.r. at input}}{\text{s.n.r. at output}} \tag{2.53}$$

The noise factor for an ideal amplifier which generates no noise would be unity.

The noise factor for an operational amplifier in a specified noise bandwidth can be expressed by the relationship

$$\text{Noise factor} = 1 + \frac{E_a^2 + I_a^2 R_s^2}{4kT\,R_s B}$$

E_a is the r.m.s. value of the amplifier equivalent input noise voltage in the bandwidth B.

I_a is the r.m.s. value of the amplifier equivalent input noise current In the bandwidth B.

R_s is the value of the source resistance.

The noise factor is seen to depend upon the magnitude of the source resistance value R_s. The optimum value of source resistance for minimum noise factor may be obtained by differentiation and equating to zero; this gives

$$R_{s(opt)} = \frac{E_a}{I_a}$$

Note, that both E_a and I_a depend upon bandwidth and so also does the value of the optimum source resistance for minimum noise factor.

The concept of noise factor is not, in fact, all that useful in operational amplifier circuits since with a specified amplifier the use of the optimum source resistance for minimum noise factor does not necessarily give the best signal to noise ratio at the output. In operational amplifier circuits the value of source resistance is usually determined by constraints other than noise, and it is the designer's task to choose an amplifier and circuit configuration to fulfil all operating requirements whilst at the same time maximising signal to noise ratio at the output.

REFERENCES

1. GREBENE, A. B., *Analog Integrated Circuit Design*, Van Nostrand (1972)
2. DI STEFFANO, J. J., STUBBERD, A. R. and WILLIAMS, I. J., *Feedback and Control Systems*, McGraw Hill (1967)
3. SMITH, L. and SHEINGOLD, D. H., Analog Dialogue 3 No 1 (March 1969)
4. SODERQUIST, D., *Minimisation of Noise in Operational Amplifier Applications*, Precision Monolithics Application Note A.N. 15

Exercises

2.1 An operational amplifier is to be used in the inverting feedback configuration with a closed loop signal gain of 100 and an input resistance of 10 kΩ.

(a) Assuming ideal amplifier performance what values of input and feedback resistor should be used?

(b) If the operational amplifier is assumed ideal except for a finite loop gain of 10^4, by how much will the signal gain differ from 100?

(c) If the open loop gain of the amplifier changes by 5% what effect will this have on the closed loop signal gain?

2.2 The amplifier used in the circuit of Figure 2.3 has an open loop gain 5×10^4 and differential input resistance 100 kΩ, resistor $R_1 = 1$ kΩ, $R_2 = 4$ kΩ. Find the closed loop gain and the effective input resistance of the circuit. Assume that the common mode input impedance of the amplifier is infinite and that its output resistance is negligible.

2.3 An amplifier with the characteristics given in exercise 2.2 is used in the circuit of Figure 2.5. The current sensing resistor R_f has a value 1 kΩ and the circuit is supplied by an input signal of 1 V. What current is supplied to the load? What is the percentage change in load current if the load resistance is changed from 10 kΩ to 10 Ω? What is the effective output resistance through which current is supplied to the load?

2.4 Write expressions for the feedback fraction β and the closed loop gain $1/\beta$ for the circuits given in Figures 1.3, 1.4(b), 1.6, 1.7, 1.8.

2.5 Express the following voltage ratios in dB to the nearest whole dB.

(a) 1, (b) 2, (c) 3, (d) 10, (e) 100, (f) 1000, (g) 10^6.

2.6 Without using log tables, using only the results of Exercise 2.5, calculate the dB equivalents of the following voltages ratios to the nearest whole dB.

(a) 6, (b) 15, (c) 3.33, (d) 333, (e) 9, (f) 0.01, (g) 0.05, (h) $\sqrt{2}$, (i) $1/\sqrt{2}$.

(Hint: 3.33 = 10/3 Thus (3.33 expressed in dB) = (10 expressed in dB) − (3 expressed in dB)

Figure 2.35 Circuits for exercise 2.8

2.7 Sketch the Bode magnitude and phase approximations for the function $A_{(jf)} = A_1/[1 - j(f_c/f)]$. Assume that $A_{(jf)}$ represents the gain of an amplifier and that $A_1 = 100$, $f_c = 10$ Hz. If a sinusoidal signal of amplitude 1 mV is applied to the amplifier find the amplitude of the output signal and its phase relative to the input signal if the frequency is (a) 1 Hz, (b) 10 Hz, (c) 100 Hz.

2.8 Derive expressions for the frequency response $v_o/v_{i(j\omega)}$ for the circuits shown in Figure 2.35. Sketch the Bode magnitude and phase approximations for these functions.

2.9 An amplifier consists of three cascaded stages, each exhibiting a first order low pass frequency response and having $|A_1| = 40$ dB, $f_{c_1} = 10$ kHz, $|A_2| = 20$ dB, $f_{c_2} = 100$ kHz, $|A_3| = 10$ dB, $f_{c_3} = 500$ kHz. Sketch the Bode approximations for the overall response and for the response of the individual stages (see Figure 2.8).

2.10 An operational amplifier has an open loop frequency response which exhibits a 20 dB/decade roll off down to unity gain. Its open loop gain at zero frequency is 100 dB and its unity gain frequency is 1 MHz. Sketch the open loop frequency response on a dB/log f plot:

(a) The amplifier is connected as a non-inverting feedback amplifier (a follower) with closed loop gain (i) 2, (ii) 10, (iii) 50. Find the small signal closed loop bandwidth in each case and sketch the appropriate Bode plots.

(b) The amplifier is connected as an inverting adder, as in Figure 1.4(b), so as to form the weighted sum of three separate signals. Input resistors $R_1 = 30$ kΩ, $R_2 = 40$ kΩ, $R_3 = 60$ kΩ, and a feedback resistor $R_f = 120$ kΩ, are used. Find

(i) the ideal performance equation, (ii) the value of $1/\beta$ for the circuit, (iii) the small signal closed loop bandwidth, (iv) by how much the ideal performance equation is in error at a frequency 20 kHz (see Section 2.5).

2.11 An internally frequency compensated operational amplifier has a unity gain frequency 2×10^4 Hz. Find the small signal closed loop signal bandwidth and output rise time in response to a small input step signal:

(a) when the amplifier is connected in the circuit of Figure 2.10 with (i) $R_1 = 1.2$ kΩ, $R_2 = 2.2$ kΩ; (ii) $R_1 = 1$ kΩ, $R_2 = 39$ kΩ;

(b) when the amplifier is connected in the circuit of Figure 2.11 with (i) $R_1 = 10$ kΩ, $R_2 = 5$ kΩ, $R_f = 100$ kΩ; (ii) $R_1 = 20$ kΩ, $R_2 = 50$ kΩ, $R_f = 500$ kΩ (see Sections 2.5 and 2.8.1).

2.12 An operational amplifier has two internal voltage gain stages; it is frequency compensated by a feedback capacitor connected around the second gain stage. Stable closed loop operation at unity gain with no closed loop gain peaking is found to require a frequency compensating capacitor of minimum value 50 pF and this gives the amplifier a unity gain frequency of 8×10^5 Hz and a slew rate of 0.4 V/μs.

(a) Find the minimum value of frequency compensating capacitor required (with no gain peaking) if the amplifier is to be used (i) as a follower with closed loop gain 10, (ii) as an inverting adder with two signal inputs each at gain 2. Find the closed loop bandwidth and slew rate that you expect in each case.

(b) It is required to restrict the closed loop bandwidth by using a greater than normal value of frequency compensating capacitor. What value is required for a closed loop bandwidth of 1 kHz at a closed loop gain of 40 dB? What slew rate do you expect if this capacitor value is connected? (see Section 2.7).

2.13 An operational amplifier has a slew rate of 0.5 V/μs. What is the maximum frequency for which the amplifier will give an undistorted sinusoidal output signal of (a) 20 V peak to peak; (b) 10 V peak to peak? (see section 2.9).

2.14 An operational amplifier employing the frequency compensating technique discussed in Section 2.7 has a frequency compensating capacitor of value 10 pF connected to it; and when it is used as a unity gain follower it has a closed loop frequency response which exhibits 5 dB of gain peaking. What damping factor and overshoot do you expect in the small signal step response of the follower? What minimum value of frequency compensating capacitor would be required for no gain peaking, and what would be the overshoot in the small signal step response if this value of capacitor were connected? (Hint: use equations 2.32, 2.36, A.4. Note that $f_{c_1} \propto 1/C_f$ and for no gain peaking $\zeta \geqslant 1/\sqrt{2}$.

2.15 The open loop gain of an operational amplifier is 100 dB at zero frequency and its open loop frequency response exhibits two breaks at frequencies $f_{c_1} = 100$ Hz, $f_{c_2} = 4 \times 10^6$ Hz. The amplifier is connected as a unity gain follower. At what frequency is the magnitude of the loop gain unity? What is the phase margin in the circuit? By how much does the closed loop gain peak and at what frequency does the gain peak occur? Estimate the settling time to 0.1% if a small input step signal is applied. Find the minimum closed loop gain for which the amplifier will exhibit (a) no closed loop gain peaking; (b) a critically damped response with no overshoot in the transient response.
Find the closed loop 3 dB bandwidth for each of these values of closed loop gain (see Section 2.8.9, Appendix A.2).

2.16 Draw up a table of values and sketch a graph relating phase margin to percentage overshoot in small signal step response, for a feedback configuration in which the phase variation of the closed loop gain is determined as a result of two break frequencies which are separated by more than a decade (use equation A.9 and equation 2.36).

2.17 An operational amplifier has the following offset and temperature drift specifications:
$V_{io} = 2$ mV; $\Delta V_{io}/\Delta T = 10 \, \mu V/^\circ C$; $I_B = 500$ nA; $\Delta I_B/\Delta T = 1$ nA/$^\circ$C; $I_{io} = 50$ nA; $\Delta I_{io}/\Delta T = 0.1$ nA/$^\circ$C.
The amplifier is connected as a simple inverter with $R_1 = 10$ kΩ, $R_f = 1$ mΩ, and is supplied by a signal source of negligible resistance. Find:
(a) The output offset voltage;
(b) The change in output offset voltage to be expected from a temperature change of 10 $^\circ$C;
(c) Assuming initial offset balanced, the smallest input signal that can be amplified with less than 1% error, due to a 10 $^\circ$C temperature change;
(d) The value of a resistor R_c which should be connected between the non-inverting input terminal and earth to reduce the offset error due to amplifier bias current.

Repeat parts (a), (b) and (c), assuming that the resistor R_c is connected in the circuit. In all cases assume worst case errors (see Section 2.10.4).

2.18 An operational amplifier with the offset and temperature drift specifications given in Exercise 2.17 is to be used as a follower with a feedback resistor of 10 kΩ and a resistor of 1 kΩ connected between the inverting input terminal and earth. The circuit is supplied by a signal source of internal resistance 100 kΩ. Find:
(a) the output offset error with no offset balance;
(b) the smallest input signal that can be amplified with no more than 1% error if initial offsets are balanced and the temperature changes by 10°C (see Section 2.10.1).

2.19 An inverting adder has three input points with input resistors 200 kΩ, 100 kΩ, 50 kΩ, and a feedback resistor of value 200 kΩ. The operational amplifier used in the circuit has the offset and temperature drift specifications given in Exercise 2.17. If initial offsets are balanced and the temperature then changes by 10°C find the smallest input signals at the three input points which can be processed with less than 1% error. Repeat the problem assuming that a bias current compensating resistor is connected between the non-inverting input terminal of the amplifier and earth (see Section 2.10.1).

2.20 A differential input operational amplifier, assumed ideal except for finite open loop gain and finite c.m.r.r., has an open loop gain of 5 $\times$ 10^4. When both amplifier input terminals are connected together and a signal of 1 V with respect to earth is applied to them the output voltage of the amplifier is found to be 5 V. Find the c.m.r.r. of the amplifier and the measurement error due to common mode signals (expressed as a percentage), when the amplifier is used as a non-inverting feedback amplifier.

2.21 A random noise voltage source has a noise density function which varies inversely with frequency; the r.m.s. value of the noise voltage produced by the source is 2 μV in the frequency range 20 Hz to 100 Hz. Find the r.m.s. noise voltage produced by the source in the frequency range
(a) 1 Hz to 10 Hz, (b) 10 Hz to 100 Hz, (c) 1 Hz to 1 kHz (use equation 2.47).

2.22 The input connected noise voltage and noise current generators which are used to represent the noise generated by an operational amplifier have noise density spectra consisting of white noise and $1/f$ components. The noise voltage generator has a white noise component with density 20 nV/$\sqrt{Hz}$ and $1/f$ corner frequency 50 Hz; the current generator has a white noise component with density 0.3 pA/$\sqrt{Hz}$ and $1/f$ corner frequency 1 kHz. Sketch the noise density spectra. Find:

(a) the r.m.s. value of the noise voltage generator;

(b) the r.m.s. value of the noise current generator, in the frequency ranges (i) 0.1–10 Hz, (ii) 1–100 Hz, (iii) 1 Hz–1 kHz, (iv) 1 Hz–10 kHz;

(c) the r.m.s. value of the total input referred noise voltage in the above frequency ranges for source resistances 1 kΩ, 10 kΩ, and 100 kΩ (use equation 2.48; see also Section 2.13.3 'Using Equivalent Input Noise Generators').

2.23 An operational amplifier with noise characteristics specified by the noise density spectra given in Exercise 2.22 is used as a simple inverter as in Figure 1.2(a). The amplifier is internally frequency compensated and has unity gain frequency 3×10^0 Hz. Estimate the r.m.s. value and peak to peak value of the random output noise voltage for input and feedback resistor values (a) $R_1 = 1$ kΩ, $R_f = 10$ kΩ; (b) $R_1 = 100$ kΩ, $R_f = 1$ MΩ (see Section 2.13.3).

CHAPTER THREE

AMPLIFIER TESTING. MEASUREMENT OF PARAMETERS

In this chapter some general test procedures are given; they are applicable to the measurement of the amplifier parameters discussed in Chapter 2. Most specified amplifier parameters are open loop parameters; a knowledge of open loop parameter values provides a basis for predicting the behaviour of an amplifier in closed loop applications. It should not be forgotten that the ultimate test of an amplifier must always be its ability to meet a desired performance specification in a particular application. Overall performance specifications (closed loop specifications) depend both upon the amplifier open loop characteristics and upon the nature of the external components which are connected to the amplifier in order to define a particular mode of operation.

Setting up test circuits provides the new user of operational amplifiers with an opportunity to gain familiarity with amplifier characteristics, but test measurements can be time consuming, so before embarking on an extensive series of tests you should make sure that you really need the information that the tests will provide. Device data sheets give performance parameters, but they do not give exact values. Considerable spread in parameter values is often found between individual units of the same amplifier type. 'Typical values' of performance parameters, as published in data sheets, are not of much value to the designer wishing to implement an application in which some aspect of the amplifier behaviour is of a critical performance limiting nature. He must assume a worst case value of critical performance specifications if he is to be sure that all units will meet a desired performance requirement. Alternatively he must set up a test circuit to measure the critical performance parameter and use the circuit as a means of selecting better than average units from a batch of amplifiers of the same type.

Open loop characteristics are the ones which are normally specified for an amplifier, but the very large open loop gain of most modern amplifiers (and thermal feedback effects, see Section 3.6) makes it impracticable to use open loop test circuits. The majority of integrated circuit amplifiers are not stable under open loop conditions. The test circuits given in this chapter are therefore closed loop circuits; open loop parameter values are inferred from measurements made on the closed loop circuits. Component values given in circuit diagrams are meant to serve as a guide, and are applicable to general purpose moderate performance amplifiers. Very high input impedance low current amplifiers may require changes in component values or possibly different test methods. Circuit diagrams do not include details of power supplies or frequency compensating components. In a practical test circuit the amplifiers must be connected to appropriate power supplies: the supplies should be r.f. bypassed to earth with bypass capacitors connected at, or as near as possible to, the amplifier socket. Externally compensated amplifiers should be connected with the recommended frequency compensating components.

Figure 3.1   Test circuit for offset voltage, bias current and difference current measurement

3.1   Measurement of bias current, input difference current, input offset voltage and their drift coefficients

A closed loop test circuit for the measurement of I_b, I_{io} and V_{io} is shown in Figure 3.1. The device under test is made to amplify its own input offset errors by the closed loop gain of the test circuit, and voltage measurements are made at the low impedance output terminal of the amplifier. The following procedure should be adopted.

1. Measure V_{io}, the input offset voltage.

Close switches S_1 and S_2 and measure V_o. Deduce the value of V_{io} from

$$V_o = \left[1 + \frac{R_2}{R_1}\right] V_{io} \qquad (3.1)$$

Assume $I_b R_1 \ll V_{io}$

2. Measure I_b^-, the bias current taken by the inverting input terminal. Close S_2, open S_1 and measure V_o. Deduce I_b^- from

$$V_o = \left[1 + \frac{R_2}{R_1}\right]\left[V_{io} + I_b^- R_s\right] \qquad (3.2)$$

The value of V_{io} found from measurement 1 should be substituted in equation 3.2. The output voltage in measurement 2 may be greater or less than that in measurement 1 depending upon the sign of V_{io}.

3. Measure I_b^+, the bias current taken by the non-inverting input terminal.

Close S_1 and open S_2, and measure V_o. Deduce I_b^+ from

$$V_o = \left[1 + \frac{R_2}{R_1}\right]\left[V_{io} - I_b^+ R_s\right] \qquad (3.3)$$

4. Measure I_{io}, the difference between the two bias currents. Open S_1 and S_2, and measure V_o. Deduce I_{io} from

$$V_o = \left[1 + \frac{R_2}{R_1}\right]\left[V_{io} + I_{io} R_s\right] \qquad (3.4)$$

Note that there is normally an initial warm up drift of V_{io} and I_b following the application of power supply voltages. Measurements should be taken after this initial warm up has taken place. Warm up time varies between different amplifier types; it may be from 2 to 5 minutes for monolithic amplifiers but may be greater than 20 minutes for some modular types of amplifier. Initial offsets are normally specified at $25°C$ ambient temperature.

3.1.1 TEMPERATURE DRIFT OF I_b, I_{io}, V_{io}

The temperature coefficients of I_b, I_{io} and V_{io} can be measured by using the test circuit of Figure 3.1 and repeating the measurements described in the previous section with the amplifier held at various temperatures within its operating range. Care should be taken to ensure that the amplifier has reached thermal equilibrium before taking measurements. Thermal gradients within the temperature chamber in which the amplifier is housed can cause large errors in the measurements, particularly with modular amplifiers. Input offset voltage is particularly sensitive to any thermal gradient. Thermal gradient can be reduced by surrounding the amplifier by a thermal insulating material or by using heat sinks. In monolithic amplifiers the small chip size reduces amplifier thermal gradients due to external thermal gradient.

3.1.2 SUPPLY VOLTAGE SENSITIVITY

V_{io}, I_b and I_{io} change with change in power supply voltages. Static or d.c. supply voltage coefficients ($\Delta V_{io}/\Delta V_s$, $\Delta I_b/\Delta V_s$, $\Delta I_{io}/\Delta V_s$) can be measured with the test circuit of Figure 3.1. Positive and negative supplies are each changed separately by say 1 V and the measurement procedure previously outlined repeated. Note that the positive and negative supply voltage coefficients are normally quite different.

3.2 Amplifier noise

Noise is generated internally in an amplifier. It appears at the output terminals of an amplifier as random fluctuations in output voltage. Noise errors are similar to the errors due to amplifier bias current and to input offset voltage in the way in which they are specified, measured and analysed. Amplifier generated noise is specified in terms of equivalent input noise generators: a noise current generator i_n and a noise voltage generator e_n.

The test circuit used for the measurement of i_n and e_n is basically the same as in Figure 3.1, but in the equations for the circuit and in the equivalent circuit for the amplifier under test, I_b is replaced by i_n and V_{io} is replaced by e_n. The main difference in both measuring and specifying noise as opposed to d.c. offset is that bandwidth must be considered. Two sets of noise measurements are usually taken: e.g. low frequency noise in a bandwidth from 0.01 to 1 Hz or 0.1 to 10 Hz, and wide band noise in a bandwidth of 5 Hz to 50 kHz. The noise measurement bandwidth is set by connecting a sharp cut off band pass filter to the output of the test circuit in Figure 3.1.

Low frequency noise measurements are made by recording the noise output voltage variations on a strip chart recorder; the low frequency noise is then specified in terms of a peak to peak variation. The wide band noise is specified as an r.m.s. value and should be measured by a true r.m.s. reading voltmeter connected to the amplifier output (after the filter).

These are practical points which require particular attention in amplifier noise measurements. The noise measurement bandwidths must be set only by the output band pass filter. Any capacitors in the test circuit must not limit the bandwidth of the noise gain $1/\beta$ in the measurement bandwidth. This point makes wide band measurements of low noise currents particularly difficult; low noise currents require large values of the resistor R_s in the test circuit. Stray capacitance inevitably interacts with large values of R_s and thereby affects the frequency dependence of $1/\beta$. Another practical point is that the test circuit must be adequately shielded otherwise mains frequency or radio frequency pick-up may be mistaken for amplifier generated noise.

An example of a low frequency peak to peak noise measurement test circuit is shown in Figure 3.2. The bandwidth of the noise measurement is set by a combination of high pass and low pass active filters (see Section 8.6) connected to the output of the device under test. The output noise voltage fluctuations recorded by a low frequency storage oscilloscope are shown in Figure 3.2. In the upper trace, which was obtained with switch S_1 closed, the oscilloscope vertical amplifier was set at 20 mV/division; this corresponds to 2 μV/division when referred to the input of the amplifier under test. The noise measured at the output of the test circuit is the input noise voltage multiplied by the gain in the circuit. The amplifier under test is in a closed loop configuration with noise gain 100, and a further gain 100 is provided by the high pass filter.

The lower trace in Figure 3.2 recorded with the oscilloscope vertical amplifier set at 1 V/division was obtained with switch S_1 open; it represents the noise voltage generated by the input noise current of the amplifier under test in flowing through the 1 MΩ sampling resistor R_s. The trace corresponds to $1/(R_s \times 10^4) = 100$ pA/division when referred to the input noise current. Note that with the 1 MΩ sampling resistor in circuit input noise in the bandwidth 0.1 to 10 Hz is dominated by the effect of the amplifier input noise current.

In the time period recorded (10 $\times$ 5 = 50 s) the oscilloscope traces indicate a peak to peak amplifier input noise voltage of approximately 2.4 μV and a peak to peak amplifier input noise current of approximately

160 pA. Both measurements are made in the bandwidth 0.1 to 10 Hz, the lower bandwidth limit being defined by a first order high pass characteristic and the upper limit by a second order low pass characteristic.

The device used as the test amplifier for the recordings was a type μA 741 C. Observed noise is in fair agreement with estimated values of peak to peak noise (see Section 2.13.3) based upon typical noise data published for the 741 (Figure 2.30).

Upper trace recorded at 20 mV/division.
Lower trace at 1V/division.
Horizontal 5 s/division.

Figure 3.2 Low frequency peak to peak noise measurement

Examination of Figure 2.30(b) gives $e_\omega^2 \cong 4 \times 10^{-16}$ V^2/Hz and $e_\omega \cong 2 \times 10^{-8}$ V/$\sqrt{\text{Hz}}$ as the density of the white noise component of the amplifier input noise voltage, and $f_o \cong 50$ Hz as the $1/f$ corner frequency. Substitution in equation 2.48 gives

$$E_{(0.1 \text{ to } 10 \text{ Hz})} = 2 \times 10^{-8} \sqrt{(50 \log_e 100 + 10)}$$

$$\cong 0.3 \ \mu\text{V r.m.s.}$$

Use of the rule of thumb multiplication factor of 6 gives the peak to peak input noise voltage as 1.8 μV (observed value 2.4 μV).

Examination of Figure 2.30(c) gives $i_\omega \cong \sqrt{(3 \times 10^{-25})}$ A/$\sqrt{\text{Hz}}$ as the density of the white noise component of the amplifier input noise current and $f_o \cong 1.4$ kHz as the $1/f$ corner frequency. Substitution in equation 2.48 gives

$$I_{(0.1 \text{ to } 10 \text{ Hz})} = \sqrt{(3 \times 10)^{-25}} \cdot \sqrt{(1.4 \times 10^3 \log_e 100 + 10)}$$

$$= 44 \text{ pA r.m.s.}$$

Figure 3.3 Differential input impedance test circuit

Use of the rule of thumb multiplication factor gives the estimated peak to peak input noise current as 264 pA (observed value 160 pA peak to peak).

3.3 Measurement of input impedances

An operational amplifier has two impedances at its input terminals which are of interest: its differential input impedance and its common mode input impedance. Differential input impedance is the impedance measured directly between the inverting and non-inverting input terminals of the amplifier. Common mode input impedance is the impedance which exists between each input terminal and earth, (or power supply common). Input impedances are largely resistive and it is the differential input resistance R_d and common mode input resistance R_{cm} of an amplifier which are normally specified and measured. Input capacitances which are effectively in parallel with input resistances normally have values of a few pF.

In principle the measurement of amplifier differential input resistance is straightforward; it can be measured by comparison with a known external resistor. The main practical difficulties lie in the elimination of noise pick up; this is particularly troublesome when attempting to measure the input resistance of high input impedance amplifiers.

A test circuit which can be used for the measurement of differential

Increase R_s to reduce V_0 by
10% then $R_{cm} = 9R_s$

Figure 3.4 Common mode impedance test circuit

input resistance is shown in Figure 3.3. The amplifier under test is connected in the inverting feedback configuration with the addition of the potential divider R_1, R_2 at its input terminals. The function of the potential divider is to allow a magnified value of the input error signal to be measured at the point X in the circuit. The test procedure simply consists of finding a value for the resistor R_s such that switching R_s into circuit by opening switch S_1 doubles the size of the signal measured at X. This occurs when $R_s = R_d$, R_s and R_d form a potential divider and the signal at X increases to hold the amplifier input error signal constant. A low frequency ($f = 1$ to 5 Hz) sine wave is used as the test signal; its amplitude is set at say 10 V ($V_s \pm 15$ V).

Figure 3.4 shows a test circuit for the measurement of the common mode input resistance at the non-inverting input terminal of an amplifier. A low frequency test signal is used (e.g. frequency 1 to 5 Hz, amplitude 10 V). The external series resistor is increased in order to get a 10% reduction in the amplitude of the signal measured at the amplifier output terminal. The common mode input resistance must then be equal to 9 times this external series resistor: $R_{cm} = 9 R_s$.

Note that the differential input resistance of the amplifier has no significant shunting effect on R_{cm} since its effective value is multiplied by the open loop gain of the amplifier (see Section 2.2.3).

3.4 Maximum common mode voltage

When an amplifier is used in the follower configuration the externally applied input signal is a common mode signal to the amplifier. Maximum common mode input can be found by using the test circuit in

Figure 3.5 Maximum common mode voltage test

Figure 3.5 and increasing the amplitude of the sinusoidal input signal until distortion is evident on the output waveform. The test frequency must be well below the full power response frequency f_p.

3.5 Measurement of open loop voltage gain and output dynamic range

Open loop gain is defined as the ratio of change in output voltage to the change in the input voltage between the differential input terminals of the amplifier required to produce the output change. It is normally specified at d.c. with an output voltage swing of ± 10 V across the rated load and with the amplifier worked with ± 15 V power supplies.

The test circuit shown in Figure 3.6 can be used for the measurement

108

of open loop gain. In this test circuit the amplifier is connected in the inverting feedback configuration with a resistive divider R_3, R_4 used to make the measured input voltage 100 times greater than the actual input error voltage. Before taking measurements the input test signal is set at zero and the offset balance potentiometer adjusted for zero output. A low frequency sinusoidal signal with frequency approximately 5 Hz is then applied and its amplitude is adjusted for an output voltage swing of ± 10 V. Gain is measured by observing the peak to peak output swing at the test points X and Y in the circuit using a double beam oscilloscope.

$$A_{OL} = \frac{100\,p - p \text{ signal at } X}{p - p \text{ signal at } Y}$$

Care is required in avoiding noise pickup; if the amplifier has $A_{OL} = 10^5$

Open loop gain $= 100 \dfrac{\text{p-p signal at X}}{\text{p-p signal at Y}}$

Figure 3.6   Open loop gain test circuit

the input error signal is only ± 100 μV for ± 10 V output, and the Y axis signal is only ± 10 mV. Leads to the summing junction should be as short as possible and all wiring associated with the error signal should be shielded.

The circuit of Figure 3.6 can be used to measure values of open loop gain over a range of frequencies. Decrease in open loop gain with increase in frequency causes the input error voltage to increase in order to maintain a constant output voltage. At the higher frequencies when open loop gain has fallen significantly it is possible to measure the input error voltage directly and the resistive divider R_3, R_4 should then not be used. Also at high frequencies the amplitude of the output voltage should be reduced to avoid exceeding the slewing rate of the amplifier, and output voltage should be adjusted so that

$$V_o \text{ (peak)} < \frac{\text{slew rate}}{2Hf_i}$$

where f_i is the test frequency.

Dynamic output range

The dynamic output range of the amplifier can be measured with the circuit of Figure 3.6 by simply increasing the amplitude of the input test signal until slipping is observed at the output. The output can then be measured directly from the oscilloscope trace. Values of open loop gain and output range are both dependent to some extent upon supply voltage and load impedance. The effect can be examined by repeating the above measurements using different values of power supply voltage and with different values of load impedance other than the rated load.

3.6 Anomalies in d.c. open loop gain values revealed by open loop transfer curves

Some amplifiers, whilst meeting their large signal d.c. open loop gain specifications, may be found to have a much smaller d.c. gain for small output voltage changes. Also the change in input signal required to produce an output change from 0 to +10 V might be quite different from that required to produce the output change 0 to −10 V. This means that the gain error in a d.c. application may be greater than the user would be led to expect from a specified value of d.c. open loop gain.

Anomalies of this kind are shown up from an examination of d.c. (that is very low frequency) input/output transfer curves for the device. The input/output transfer curve for many monolithic operational amplifiers is very different from the near linear transfer curve often assumed by circuit designers. The mechanism which describes severe d.c. transfer characteristic non-linearity is commonly referred to as thermal feedback; examples of the effect of thermal feedback are shown in Figures 3.9, 3.10 and 3.11.

The monolithic chip in an integrated circuit operational amplifier has high thermal conductivity and low thermal mass. Internal power dissipation is greatest at the location of the output stage transistors and heat produced here is readily conducted to the input stage of the amplifier. Any thermal gradient in the chip at the input stage upsets the precisely matched base emitter or gate source voltages of the input stage transistors and results in a change in input offset voltage. For slowly varying output voltage changes the position of the hottest spot on the chip changes according to whether the output is positive or negative, and this, dependent upon the chip geometry, changes the nature of the thermal gradient at the input stage and produces a corresponding change in V_{io}. Induced input offset voltage changes due to thermal feedback can thus vary with both the polarity and the magnitude of the output load current. They can override externally applied summing point voltages and can give the appearance of infinite open loop gain or even an apparent change in the sign of the open loop gain.

Figure 3.7 Effect of thermal feedback represented by input error voltage

Amplifiers which, because of thermal feedback effects, exhibit an apparent negative gain characteristic, are open loop unstable but behave quite acceptably when used in a negative feedback\configuration. An understanding may be gained by representing thermal feedback in terms of an equivalent input feedback voltage e_{th} (Figure 3.7); e_{th} varies in a non-linear manner with the d.c. value of V_o (strictly with I_o) and its sign may be positive or negative. The error voltage at the amplifier summing point in a closed loop application may be written as

$$e_\epsilon = e_\epsilon^1 + e_{th}$$

where $e_\epsilon^1 = -V_o/A_{OL}^1$ is the error voltage between the input terminals of the equivalent amplifier with open loop gain A_{OL}^1 and with no thermal feedback effects; e_ϵ is the actual gain error voltage.

$e_\epsilon = -V_o/A_{OL}$ and A_{OL} is the gain of the amplifier as measured from the observed transfer characteristic.

Dependent upon the sign and magnitude of e_{th}, A_{OL} may be greater than, less than or of opposite sign to A_{OL}^1. An amplifier with A_{OL} negative is closed loop stable because the externally applied negative feedback overrides the positive feedback due to the thermal effect and the overall feedback is therefore negative.

The effect of thermal feedback is to make the input gain error voltage vary nonlinearly with V_o and possibly change sign but this is of little significance in a closed loop application provided that

$$|e_\epsilon| < \frac{V_o}{A_{OL} \text{ spec}}$$

The only way to make absolutely sure that this condition is satisfied in a device which exhibits thermal feedback is actually to observe the open loop transfer curve of the device. The extent of thermal feedback effects is very much dependent upon chip geometry; devices which are nominally the same type but made by different manufacturers can show very different thermal feedback characteristics.

Thermal feedback effects disappear at the higher frequencies when localised temperature changes are smoothed out and some steady temperature gradient is established in the chip due to the internal dissipation.

3.7 Input/output transfer curves allow the measurement of amplifier d.c. parameters

The test circuit used in the previous section for open loop gain measurement (Figure 3.6) provides a simple method for the display of open loop transfer characteristics. The output of the amplifier under test is used to provide vertical deflection and the amplifier error voltage produces the horizontal deflection for a display. A very low frequency test signal should be used (0.01 to 0.1 Hz) and this requires the use of a storage oscilloscope or X Y Recorder to provide the display function.

In circumstances in which operational amplifiers are to be used

Figure 3.8 Curve tracing test circuit for open loop transfer curve and common mode transfer curve

under non-standard conditions, device data sheets often do not provide adequate information. Curve tracing allows a rapid characterisation under non-standard conditions and if many amplifiers are to be tested it is probably worthwhile designing a semi-permanent test fixture for the purpose. The circuit given in Figure 3.8 will be found a useful starting point for such a design. It provides a display of open loop transfer curve, common mode transfer curve (allowing the measurement of c.m.r.r. and common mode range) and it allows the measurement of V_{io} and I_b. In Figure 3.8 the amplifier under test is included within a

113

feedback loop together with an external second amplifier which is used to boost the loop gain and allow a greater voltage sensitivity. A third amplifier connected as an integrator is used to allow automatic centring of the display without the necessity for trim potentiometer adjustment. It provides an automatic offset balance. The circuit requires a low frequency sinusoidal or triangular test signal of frequency in the range 0.01 to 1 Hz with amplitude adjustable up to a maximum value which is sufficient to sweep through the common mode range of the amplifier under test. A waveform generator circuit of the type discussed in Chapter 7 could well be used and built in as part of a permanent test fixture.

3.7.1 DISPLAY OF OPEN LOOP TRANSFER CURVES

The switches in Figure 3.8 are set in the position appropriate for the display of the open loop transfer curve. The test signal amplitude is first set at zero and switch S_1 is closed and opened again. This has the effect of nulling the initial offset voltage of the device under test and centring the display. Any voltage present at the output of A_2 is integrated by A_3 and applied to the non-inverting input of the device under test so as to force its output voltage to zero. The output of amplifier A_2 is forced to zero to within a limit determined by the input offset voltage of A_3. If a precise initial offset balance is desired an offset voltage potentiometer should be connected to A_3 in order to balance its initial offset voltage. The voltage at the output of the integrator A_3 is held constant when switch S_1 is opened; it applies a fixed offset to balance the initial offset of the device under test.

The test signal amplitude is now turned up. It is applied together with the output of the amplifier under test to the non-inverting input terminal of the external feedback amplifier A_2. The output voltage of the amplifier under test varies in such a way as to maintain the non-inverting input terminal of A_2 at earth. The output of A_2, attenuated by R_1, R_2, provides the input error voltage to the amplifier under test and at the same time gives the horizontal deflection for the display. Using the component values shown in the circuit R_2 = 100 kΩ, R_1 = 100 Ω. One volt at the output of A_2 corresponds to a mV differential input signal applied to the amplifier under test; horizontal scaling can of course be changed by varying say the resistor R_2. The output signal of the amplifier under test is used to provide vertical deflection for the display.

114

Figure 3.9   Input/output transfer curve for 741 amplifier with R_L = 10 kΩ and R_L = 2.2 kΩ

Figure 3.10   Input/output transfer curve for 741 amplifier with R_L = 2.2 kΩ V_s = ± 15 V, V_s = ± 10 V, V_s = ± 5 V

Some examples of the information provided by open loop transfer curves may be gleaned by examination of the curves given in Figures 3.9 to 3.12. Figure 3.9 shows the transfer curve for a 741 amplifier. In the upper trace obtained with R_L = 10 kΩ the amplifier behaves normally and gives a near linear transfer curve with slope corresponding to an open loop gain of approximately 1.3 × 10^5. In the lower trace R_L is decreased to 2.2 kΩ and the curve shows marked non-linearity.

Figure 3.11 Input/output transfer curve for 741 sample which does not meet its open loop gain specification

Figure 3.12 741 sample: effect of internal offset balance on input/output transfer curve

The curve shown in Figure 3.10 is for the same amplifier with load 2.2 kΩ but used with power supplies $V_s = \pm 15$ V, $V_s = \pm 10$ V and $V_s = \pm 5$ V.

Results for another sample are shown in Figure 3.11 for $R_L = 10$ kΩ and 2.2 kΩ. Note that this sample meets its open loop gain specification for $R_L = 10$ kΩ (albeit a 'negative gain') but with a load of 2.2 kΩ and a negative output voltage the slope corresponds to a gain of only

116

6.7×10^3. In d.c. applications with negative output signals the gain error would clearly be significantly greater than one would expect from published values of open loop gain.

A factor often overlooked by designers is the effect the internal offset balancing can have on the performance of certain amplifiers; this is readily shown up by the curve tracer. The curves shown in Figure 3.12 are for a 741 sample but many other amplifier types show similar behaviour. The linear transfer curve was obtained with no internal offset balance. The other two curves were obtained with the recommended 10 kΩ offset balance potentiometer connected to the amplifier and set to provide offset of $+3$ mV and -3 mV. The $+3$ mV offset produces a marked decrease in open loop gain. This gives a warning that you should not attempt to balance bias current induced offset voltage with an internal offset balance potentiometer unless you are prepared to accept the consequences!

3.7.2 MEASUREMENT OF V_{io}

Input offset voltage values can be readily determined with the curve tracer. The offset nulling amplifier A_3 is disconnected from the non-inverting input terminal of the amplifier under test and the transfer curves are displayed with reduced horizontal sensitivity by reducing the value of R_2. The point where the curve cuts the horizontal scale gives the value of V_{io}.

3.7.3 DISPLAY OF COMMON MODE TRANSFER CURVE AND MEASUREMENT OF C.M.R.R.

Display of common mode transfer curve (see Section 2.11) shows up non-linearities in common mode performance. The displays are readily obtained with the curve tracer using the circuit shown in Figure 3.8. The input offset voltage of the device under test is first nulled with amplifier A_3 by closing S_1 and then opening it again. The switches are then set in the position appropriate for common mode display. In this position the low frequency test signal drives the power supply common of the device under test. Note that the device under test has its own separate power supplies for this purpose. The test signal also drives the inverting input terminal of amplifier A_2 whilst the non-inverting input terminal is driven by the output signal of the device under test. Resistive dividers are used at the input of A_2 so that test signals will not exceed its common mode range.

Figure 3.13 Common mode transfer curves for 741 sample

Figure 3.14 Effect of internal offset balance on c.m.r.r. 741 sample

The action of the feedback loop is to maintain the output of the amplifier under test at zero voltage with respect to its power supply common. The output of A_2 supplies a signal to the inverting input terminal of the amplifier which is equal and opposite to the common mode error signal of the amplifier under test, and it provides horizontal deflection for the display. Vertical deflection is provided by the input test signal which is equal but of opposite sign to the common mode input voltage to the device under test.

Displays obtained for a 741 sample are shown in Figure 3.13; the curves were obtained for $V_s = \pm 10$ V and $V_s = \pm 5$ V. Note that the positive common mode range is greater than the negative common mode range. The slope of the curves $V_{cm}/e_{\epsilon_{cm}}$ gives the c.m.r.r. of the amplifier; the curve shows up any non-linearities in common mode behaviour.

Internal offset balancing can have a marked effect on c.m.r.r.; the curves shown in Figure 3.14 are for a 741 sample. With no internal offset balance the amplifier exhibits a very high c.m.r.r. and the common mode transfer curve exhibits significant non-linearity only towards the extremes of the common mode range. The other two curves were obtained with the usual 10 kΩ offset potentiometer connected to the amplifier to provide an offset balance of ± 2 mV. Note the marked decrease in c.m.r.r. and increase in nonlinearity of the transfer curves.

Figure 3.15 Bias currents of 741 sample

3.7.4 MEASUREMENT OF I_b

The curve tracer can be used to measure I_b and show any variation of I_b with common mode input voltage. The switches are set in the position for measurement of common mode transfer curve. The test signal is first set at zero and the initial offset voltage of the device under test nulled with the offset integrator A_3 (by closing and opening switch S_1). Bias current sampling resistors are switched in as source resistances, first at the inverting input of the device under test, then at the non-inverting input. The source resistance magnitude is chosen to make the

119

offset voltage due to bias current considerably larger than the common mode error voltage, allowing common mode error voltage to be neglected. The test signal amplitude is turned up so as to sweep through the common mode range; the test signal is displayed vertically against the

Figure 3.16   741 sample: V_{io} variations with supply voltages

output of amplifier A_2 horizontally. The curves in Figure 3.15 were obtained for a 741 sample used with $\pm$ 10 V power supplies; they give values $I_b^+ = 120$ nA, $I_b^- = 130$ nA, $I_{io} = 10$ nA. The horizontal deflecting signal is inverted for the display of I_b.

Figure 3.17   Test circuit for unity gain small signal response

3.7.5 MEASUREMENT OF POWER SUPPLY REJECTION

Offset voltage variations may be displayed as a function of power supply voltage. Switches are set in the position for display of open loop trans-

120

fer curve, the test signal is set at zero and the offset nulling integrator is used first to null the initial offset of the device under test to bring the display to centre screen. Positive and negative supplies to the device under test are varied separately, the power supply voltage is used to provide vertical deflection and the output of A_2 provides the horizontal deflection for the display. Note that an alternating signal is not used on this measurement. The output of the device under test is held at zero by the action of the test loop. Scaling is chosen so as to provide adequate horizontal deflection. The curves obtained for a 741 sample are shown in Figure 3.16; notice the much greater sensitivity to positive supply variations than to negative supply variations.

3.8 Dynamic response measurements

The tests described thus far have measured d.c. characteristics of the amplifier. In the remainder of this Chapter we discuss the measurement of dynamic response characteristics. In dynamic response measurements it becomes more difficult to separate behaviour which is due to the amplifier from that which is due to external circuit influences. This is particularly the case for wide band amplifiers where measured performance can be dominated by external factors if proper attention is not paid to external circuit arrangements (stray capacitance in parallel with feedback resistor, load capacitance, etc.).

3.8.1 UNITY GAIN SMALL SIGNAL RESPONSE

The unity gain small signal response of an amplifier is the frequency at which its open loop gain becomes unity. Small signal implies that the output signal amplitude is small enough to ensure that no slew rate limiting takes place. Output signal amplitude $< S/(2\pi f_i)$, where S is the slew rate and f_i is the input signal frequency.

Unity gain small signal response is an open loop parameter but its measurement is simplified if it is remembered that open loop and closed loop sinusoidal frequency response characteristics coincide at frequencies appreciably higher than the closed loop bandwidth. A test circuit of the type illustrated in Figure 3.17 can be used. The frequency of the input test sinusoid is simply increased until input and output signals have the same amplitude; this occurs when the open loop gain is unity.

121

3.8.2 EXAMINATION OF CLOSED LOOP STABILITY. TRANSIENT RESPONSE TESTING

An important aspect of the design of operational amplifier feedback circuits lies in ensuring that the circuit has an adequate phase margin, otherwise it might be unstable and exhibit undesirable gain peaking (see Section 2.5). Amplifier data sheets for externally compensated amplifiers give recommended values of frequency compensating components for a limited number of closed loop circuit configurations. The designer, in an effort to save time, often uses recommended values of frequency compensating components for closed loop configurations which are not actually given on the data sheet. In doing this he perhaps uses more compensation than is strictly necessary and therefore restricts closed loop bandwidth and decreases slewing rate. This does not matter in low frequency applications, indeed it might be advantageous to restrict bandwidth in order to reduce noise. It does matter however if the designer is seeking to exploit the full bandwidth and slew rate capabilities of the particular amplifier type in use.

Finding the optimum frequency compensation for use with a new design from purely theoretical considerations can be a time consuming process, but a practical evaluation of closed loop stability can be achieved quite quickly by observing the small signal transient response of the circuit. Transient response is measured by observing the output signal produced by the circuit in response to a step or square wave input signal. The size of this signal is made sufficiently small to ensure that no slew rate limiting is observed at the output. The pulse or square wave generator and the oscilloscope vertical amplifier, including any probe used, should have a rise time less than half that of the closed loop amplifier circuit under test. The square wave generator should produce a signal with negligible departures from flatness, either through overshoot or droop, and its waveform should not be degraded by inattention to proper signal connections and matching. It is a good idea to monitor both input and output signals to make quite sure that the observed transient response is really produced by the circuit under test, and is not due to anomalies in the input signal.

Typical output transients, which were obtained in a test, are shown in Figure 3.18. The top trace shows a marked overshoot and ringing indicating inadequate phase margin. The other two traces were obtained using successively bigger values for the frequency compensating capacitor, phase margin increases, overshoot and ringing decrease. This type of behaviour is typical of what is called a second order system. The

Figure 3.18 Small signal transient response of closed loop configuration with different values of frequency compensating capacitor

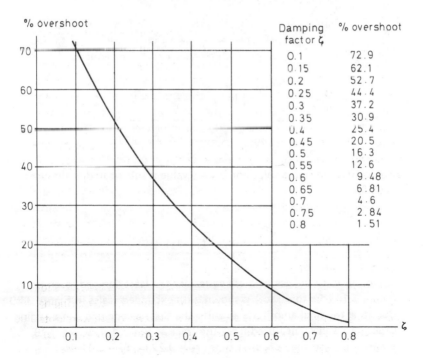

Damping factor or ζ	% overshoot
0.1	72.9
0.15	62.1
0.2	52.7
0.25	44.4
0.3	37.2
0.35	30.9
0.4	25.4
0.45	20.5
0.5	16.3
0.55	12.6
0.6	9.48
0.65	6.81
0.7	4.6
0.75	2.84
0.8	1.51

Figure 3.19 Overshoot/damping factor

mathematical equations governing the behaviour of such systems are well developed, and are characterised by a damping factor ζ and a natural frequency w_o of the system. The lower trace in Figure 3.18 was obtained with the frequency compensating capacitor value increased just sufficiently to prevent overshoot and yet maintain as fast a rise time as possible. This response is said to be critically damped; a critically damped response has a value of damping factor $\zeta = 1$.

Damping factors less than unity give a second order system a transient response which overshoots and rings. A measurement of the overshoot may be used as a method of evaluating the damping factor from the relationship

$$\text{Overshoot \%} = 100 \exp\left[-\frac{\zeta \pi}{\sqrt{(1 - \zeta^2)}}\right] \tag{3.5}$$

For convenience values of overshoot and damping factor, obtained by the use of equation 3.5, are tabulated and plotted graphically in Figure 3.19.

An alternative way of evaluating the damping factor and natural frequency for a lightly damped response is to measure the first three ringing peaks and the times at which they occur. The equation[1]

$$\zeta = \frac{\left| \log_{10} \dfrac{V_{p1} - V_{p3}}{V_{p2}} \right|}{\log_{10}^2 \dfrac{V_{p1} - V_{p3}}{V_{p2}} + 1.8615} \tag{3.6}$$

may be used to find a value for ζ. This value is substituted in the equation

$$w_o = \frac{2\pi}{(t_3 - t_1)\sqrt{(1 - \zeta^2)}} \tag{3.7}$$

in order to find w_o.

To illustrate the use of these equations the least damped response in Figure 3.18 (the top trace) is shown with expanded scales in Figure 3.20. Counting the final amplitude as unity the first, second and third ringing peaks have values measured from the trace as $V_{p1} = 1.4$, $V_{p2} = 0.78$, $V_{p3} = 1.13$, and they occur at times $t_1 = 0.22\ \mu s$, $t_2 = 0.41\ \mu s$, $t_3 = 0.79\ \mu s$.

Substitution of these values in equation 3.6 and 3.7 gives $\zeta = 0.32$, $w_o = 12 \times 10^6$ rad/s. Accuracy is of course limited by the inevitable errors associated with estimating amplitudes and times from the observed oscilloscope trace.

Figure 3.20 Transient response with extended scale for finding ζ and ω_0

In an operational amplifier closed loop configuration damping factor, phase margin and peaking in the systems gain/frequency response are closely related. A system which is lightly damped gives a marked overshoot and ringing in its transient response and exhibits a sinusoidal response with a pronounced gain peaking. Gain peaking and damping factor are related by the equation

$$P_{(dB)} = 20 \log_{10} \frac{1}{2\zeta \sqrt{(1 - \zeta^2)}} \qquad (3.8)$$

The gain peak occurs as the closed loop bandwidth limit frequency is approached. P is the amount by which the gain at the peak (expressed in dB) exceeds the midband gain (also expressed in dB)

$$P_{(dB)} = \text{Peak Gain}_{(dB)} - \text{Mid Band Gain}_{(dB)}$$

As an aid to evaluation values of ζ against dB of closed loop gain peaking are tabulated and plotted in Figure 3.21.

A transient test provides a rapid practical characterisation of the closed loop stability of a new design; gain peaking can be found using

ζ	Peaking (dB)
0.01	33.9803
0.02	27.9609
0.03	24.4412
0.04	21.9454
0.05	20.0111
0.06	18.4323
0.07	17.099
0.08	15.9457
0.09	14.9301
0.1	14.0232
0.15	10.5565
0.2	8.13619
0.3	4.84662
0.4	2.69544
0.5	1.2494
0.6	0.35458
0.7	0.00173

Figure 3.21   Gain peaking as a function of damping factor

equation 3.8 without the necessity of performing time consuming measurements of gain/frequency. In many cases it is sufficient merely to ensure an adequate stability phase margin without working out values of damping factor. This can be done by adjusting frequency compensating component values whilst experimentally observing the transient response of the system. The value of the frequency compensating components required for critical damping can be quickly found in this way. The small signal closed loop bandwidth in the critically damped case (lower trace of Figure 3.18) can be inferred from the equation

$$f_{(3\,dB)} \cong \frac{1}{3T_r} \tag{3.9}$$

T_r = small signal rise time (10% – 90%).

Often it is required to ensure maximum closed loop bandwidth with frequency compensation just sufficient to prevent any significant gain peaking. A critically damped response, $\zeta = 1$, gives no gain peaking, but it is possible by using slightly less than critical damping to get a wider closed loop bandwidth with almost imperceptible gain peaking. A damping factor $\zeta = 0.7$ gives only 0.00173 dB of gain peaking.

126

3.9 Large signal response measurements

Full power response, f_p, is the maximum frequency for which the rated output can be obtained for a sinusoidal signal at rated load without distortion due to slew rate limiting. It is normally specified at unity closed loop gain. Slew rate S is the maximum rate at which the output signal of the amplifier can change for a large step change in input. It is usually expressed in V/μs and is normally specified at unity closed loop gain.

The two parameters f_p and S are closely connected quantities and for many amplifiers the equation

$$S = \left| \frac{dV_o}{dt} \right|_{max} = 2\pi t_p E_o \tag{3.10}$$

where E_o = rated output voltage, gives a close approximation for the relationship between them. A knowledge of the one enables the other to be calculated.

Certain amplifiers, particularly fast response types, give slewing rate higher than that predictable from a knowledge of f_p and it may then be necessary to specify f_p and S separately. Also some amplifiers have unequal positive and negative going slew rates; in such cases it is normally the slower of the two which is specified.

3.9.1 FULL POWER RESPONSE MEASUREMENT

In order to measure the f_p of an amplifier it is connected with resistive feedback in a closed loop circuit. A sinusoidal test signal is applied as input signal, its frequency is first set at some low value and its amplitude adjusted so that the amplifier under test gives its rated output. Test signal frequency is then increased until the output signal shows evidence of distortion, either through oscilloscope observation or by the use of a distortion meter. There is no industry-wide accepted value for the distortion level which determines the f_p but a number like 1% to 3% is a reasonable figure.

Closed loop output distortion depends upon the amount of feedback or loop gain—therefore it depends upon the closed loop gain used in the measurement circuit. Specified f_p values are normally measured at unity gain where loop gain is the highest; at higher closed loop gains output distortion will be greater at this frequency. Small amounts of output distortion are difficult to separate from possible distortion in the input test signal. Rather than measuring output distortion a more sensitive method of detecting the onset of distortion as the full power response frequency is reached is to observe the error signal at the input summing point as suggested in the test circuit of Figure 3.22. Marked distortion of the error signal occurs when the f_p is reached.

Figure 3.22   Measurement of f_p

The full power response and slew rate of externally compensated amplifiers are very much dependent upon the type of frequency compensation applied and upon the values of the frequency compensating components used. Both of the parameters are best measured using a test circuit which most closely resembles the final circuit configuration in which the amplifier is to be used.

3.9.2 SLEWING RATE MEASUREMENT

Slewing rate values can be inferred from measured f_p values, but a more direct method of measuring slewing rate is to use a square wave or square pulse as the input test signal. If the input test signal and the output signal are viewed with a double beam oscilloscope and the input signal amplitude increased, slew rate limiting of the output wave will be readily detected as an inability of the output wave to follow the rapidly changing input signal. The frequency of the test signal is adjusted for convenient oscilloscope observation and slew rate may be measured

directly from the slope of the observed output waveform. An example
of the oscilloscope traces obtained in this type of test is shown in
Figure 3.23.

Figure 3.23   Large signal transient response of AD518 sample

3.9.3 OVERLOAD RECOVERY

Overload recovery defines the time taken to recover from a saturated
condition to rated output voltage. It can be measured at the same time
as slew rate by simply turning up the input squarewave amplitude so
that the output of the amplifier under test goes into saturation. The
waveform in Figures 3.23 and 3.24 should be compared; in Figure 3.24
the input signal amplitude was increased slightly so as to drive the amplifier output into saturation. The dead time before the amplifier output
responds to the input signal represents the overload recovery time.

Figure 3.24 Large signal transient response of AD518 sample—amplifier in saturation takes finite time to recover

Figure 3.25 Measurement of settling time—inverting configuration

3.9.4 SETTLING TIME

Settling time is the time taken for the output voltage of an amplifier to settle to within some specified percentage of its final value after the application of an input step signal. It is normally specified for a full scale output step. During the settling time the output of an amplifier is in error, and settling time is therefore an important consideration when operational amplifiers are used to process linearly fast changing signals (for example, in sample hold circuits).

130

The typical form of a large signal output response produced by a step input is shown by the oscilloscope waveforms in Figure 3.23. It is not possible to measure settling time accurately from a display of this kind because the specified error band is normally too small to observe directly on the output signal. Also, the squarewave produced by the test signal generator may not be flat to within the specified error band of the settling time measurement. Accurate settling time measurements require that the error be separated from the signal before measurement.

Subtraction of the error from the signal is readily accomplished in a unity gain inverting test circuit by summing input and output signals, using additional resistors connected to the output and input of the test circuit. The method is illustrated in Figure 3.25 which shows a unity gain test configuration used to measure settling time. The error signal attenuated by a factor of 2 appears at the junction of the added resistors and is displayed by the oscilloscope.

Measurement of settling time for the non-inverting unity gain configuration requires the use of an additional high speed differential amplifier to subtract the output signal of the test circuit from that of its input. Some high speed oscilloscopes include fast differential amplifiers in their vertical plug in range.

The results obtained with settling time test circuits can be greatly influenced by the external components used in the test circuit and by their physical arrangement. In order that results should not be misleading it is important to test settling time in a circuit configuration which closely resembles the final configuration in which the amplifier is to be used. Care should be taken to ensure that the instruments used in the test arrangements do not give rise to spurious errors. The oscilloscope amplifier used to observe the error signal is inevitably driven into saturation by the large initial error voltage. The oscilloscope should have an overshoot recovery time which is much less than the settling time to be measured. Overload recovery waveform of the oscilloscope should be 'clean': it should not distort and ripple whilst recovering from the large initial signal that is driven into it. The settling time of the oscilloscope and the settling time of any differential amplifier used for the non-inverting test circuit should be much less than the settling time of the amplifier under test.

REFERENCE

1. Motorola Application Note A.N. 460

131

CHAPTER FOUR

APPLICATIONS. BASIC SCALING CIRCUITS

4.1 Introduction

The application chapters of this book set out to provide a survey of the circuits which illustrate the commonly used modes of operation for an operational amplifier. A classification of applications is inevitably somewhat arbitrary; the subject has an underlying unity and the versatility of the operational amplifier approach is such that a particular application can often be equally well treated under a range of different headings.

Operational amplifiers are used extensively in chemical, physical, geophysical, and biomedical instrumentation. The operational amplifier approach to an instrumentation application involves breaking down the instrument or system function into a series of specific operations, each of which can be simulated by an operational amplifier circuit. This type of approach is very valuable in many research fields where existing commercial instrumentation is perhaps inadequate, and the ability to modify or even start from scratch to create a 'tailor-made' instrument is of tremendous importance.

The requirements of the instrumentation systems appropriate to the various disciplines vary considerably but the specific operations required in the different systems are common to many systems. It is with these specific operation circuits that the applications chapters are mainly concerned, and the designer should be able to pick out from the many circuits given the ones appropriate to his own particular system. A further collection of operational amplifier circuits will be found in Appendix A1.

The great thing about operational amplifier circuits is that they really work and even the beginner can expect to get accurate performance. The more experienced practitioner, fully conversant with the several factors controlling accuracy (Chapter 2) will be able, using operational amplifiers, to build instrumentation systems of extreme accuracy.

In this first chapter of the applications section we deal with linear scaling circuits, i.e. circuits which are designed to change signal magnitudes and impedance levels to linearly add and subtract signals and to change linearly from current to voltage or voltage to current. Modifications to these basic circuits will be found in many guises in subsequent chapters. An intriguing aspect of the operational amplifier approach to instrumentation lies in the many variations of a basic circuit that are possible.

Specific amplifier types are for convenience sometimes mentioned in the text but since new amplifier designs continue to emerge, in order to avoid obsolescence the circuits do not in general refer to particular amplifiers. Most applications will in fact function (in a fashion) with any operational amplifier type; the particular amplifier used in a circuit determines the errors and performance limits of the application. In order to make a working circuit from those given in the text all that is normally required is to add power supply connections to the amplifier (see Section 1.4). A general purpose operational amplifier will be found suitable for the majority of applications; only those applications involving very small signals or in which wide bandwidth and very fast changing signals are involved will normally require the use of more specialised (and more expensive) amplifiers.

In the applications given in this chapter, resistors are the external components which are most frequently connected to the operational amplifier in order to define a precise circuit operation. The circuit designs given do not in general give specific values for resistors and the reader wishing to implement a practical application must choose resistor values for himself. The newcomer to operational amplifier circuits is sometimes hesitant about this choice of component values; he knows what ratios of component values to use in order to set a specific scaling factor but he does not know what actual values to use. As a general guideline to resistor value selection the underlying principle should be to select values which are not significantly larger than those required to minimise signal loading. Large resistor values increase offset errors due to bias current (see Section 2.10) and when shunted by inevitable stray capacitance limit bandwidth in applications requiring wide bandwidth.

Most general purpose operational amplifiers are designed to supply a load at their output terminal which should be no less than 2 kΩ, and it should be remembered that it is the parallel combination of an external load and feedback resistor which constitutes the effective load connected to the output terminal of an amplifier. In inverting circuit configurations input resistors which are connected to the inverting input terminal of the amplifier should be selected so as not significantly to load the input signal source which is connected to them.

Figure 4.1 *Basic voltage scaling applications (see Section 9.6 for offset balancing techniques)*
(a) Inverting amplifier (b) Follower with gain (c) Unity gain follower

4.2 Voltage scalers and impedance conversion

Ideal forms of the basic voltage scaling and impedance conversion applications for an operational amplifier have already been dealt with in Chapter 1. The circuits are for convenience shown again in Figure 4.1. The great attraction of all operational amplifier applications lies in the ability to set a precise operation with a minimum number of precise components, and in Figures 4.1(a) and (b) closed loop gain is determined by simply selecting two resistor values. The accuracy with which it is possible to set a precise gain is almost entirely dependent upon the tolerance in the resistor values used.

134

The inverting circuit in Figure 4.1(a) can be given any gain from zero upwards; the lower limit of the gain for the follower circuit Figure 4.1(b) is unity. In both configurations the upper limit to the gain value which it is practicable to use is determined by the requirement for maintaining an adequate loop gain so as to minimise gain error (see Section 2.3.1). It should also be remembered that closed loop bandwidth decreases with increase in closed loop gain. If high closed loop gains are required it is often better to connect two amplifier circuits in cascade rather than to use a single amplifier circuit.

Both the inverter and the follower configuration feature low output impedance (a characteristic of negative voltage feedback, see Section 2.2.2); the main performance difference between them, apart from signal inversion, lies in their input impedance. In the case of the inverter the signal source driving the circuit is loaded by the resistor R_1; the follower on the other hand presents a very high input impedance which ensures negligible loading in most applications. The unity gain follower in Figure 4.1(c) is often used to great advantage as a buffer stage to prevent interaction between a signal source and load, e.g. for unloading potentiometers, or buffering voltage references and potentiostats.

The follower achieves a high input impedance without the use of large value resistors. This is an advantage in applications requiring wide bandwidth since large value resistors used in any circuit inevitably interact with stray circuit capacitance to cause bandwidth limitations. In Section 2.3.3 the effective input impedance of the follower configuration was shown to be equal to the differential input impedance of the operational amplifier multiplied by the loop gain in the circuit $(Z_{in}\beta A_{OL})$. Practical operational amplifiers have their common mode input impedance Z_{cm} between their non-inverting input terminal and earth; this shunts $Z_{in}\beta A_{OL}$ and is normally less than $Z_{in}\beta A_{OL}$. The effective input impedance of the follower configuration is thus Z_{cm}. The high input impedance of the follower configuration makes it a better choice than the inverting configuration for use in many applications, but practical points which do not favour its use are that it is subject to common mode errors (see Section 2.11) and that the input voltage applied to the follower must not be allowed to exceed the maximum common mode voltage for the amplifier. However, these points do not usually impose too serious a restriction.

The main limitation associated with the inverting configuration is that its input impedance is effectively equal to the value of the input resistor R_1. If the application calls for a high input impedance to minimise signal source loading, this demands a large value for the resistor R_1 and an even larger value for R_2 assuming that closed loop gains greater than unity are required. Large resistor values inevitably give increased offset errors due to amplifier bias current, and stray capacity in parallel with a large feedback resistor limits bandwidth. For example, assume that it is required to use the inverting configuration for a circuit with closed loop gain 100 and input resistance 1 MΩ. In Figure 4.1(a) R_1 = 1 MΩ and R_2 = 100 MΩ is required. Stray capacitance C_s in parallel with R_f would limit the closed loop bandwidth to a frequency $f = 1/(2\pi C_s R_2)$. With C_s say 2 pF the closed loop bandwidth would be limited to 800 Hz, a severe restriction!

Figure 4.2 Resistive T network avoids the use of high value feedback resistor in an inverter circuit

Stable very high value resistors are not freely available and if the inverting configuration must be used the necessity for the use of a very large feedback resistor can be overcome by the use of a T resistance network as shown in Figure 4.2, but this is at the expense of a reduction in loop gain and an increase in noise gain $1/\beta$.

4.2.1 VARIABLE SCALE FACTOR (VARIABLE GAIN CONTROL)

Potentiometers instead of fixed value resistors may be used in order to allow a variable scale factor (variable gain control). The arrangement shown in Figure 4.3(a) allows a variation of scaling factor from zero to a very high value; however, the scaling factor does not vary linearly with

Figure 4.3 Variable scale factor (a) Nonlinear gain control (b) Linear gain control

respect to potentiometer rotation and the circuit has the disadvantage that the input impedance falls as the gain is increased. The circuit of Figure 4.3(b) gives a narrower range of scale factor variation from zero to R_2/R_1 but gain variation is linear with respect to potentiometer setting and input impedance remains constant at R_1. It should be remembered that changing the closed loop gain inevitably changes the closed loop bandwidth; also changes in the values of gain setting resistors produce a change in the offset error due to amplifier bias current. Offset errors due to bias current can be minimised by using a low bias current f.e.t. input amplifier type.

Figure 4.4 Programmable gain operational amplifier

4.2.2 SWITCHED SCALING FACTOR

Circuit arrangements can be made to switch in a series of different
values of scale setting resistors, thus allowing the switching of the scale
factor between pre-set values. Switching can be performed by a manual
control of a mechanical switch, by an electro-mechanical switch or by
means of some form of solid state switch. The circuit given in Figure 4.4
illustrates the use of a monolithic quad JFET switch in a programmable
gain operational amplifier circuit. The external pin connections only of
the switch are given; readers interested in its internal operation are
referred to the manufacturer's (National) data sheet. The gain select
inputs operate on normal TTL logic levels. The operational amplifier
shown in the circuit is a high input impedance f.e.t. input amplifier
type.

4.2.3 VOLTAGE CONTROLLED SCALING FACTOR

Voltage control of an operational amplifier scaling factor requires a
voltage controlled resistive element. Junction gate f.e.t.'s when operated

Figure 4.5   Voltage controlled scaling factor

below pinch off behave as linear resistances with channel resistance r_{ds} determined by the value of the gate source voltage; for small values of drain source voltage they exhibit a bilateral characteristic.

Linear voltage control of gain can be obtained by using a feedback arrangement between the drain and gate of the f.e.t., and the voltage swing actually across the f.e.t. can be kept small by including it in a T network as shown in the circuit of Figure 4.5. The effective resistance of the resistive T when connected to the operational amplifier summing point is

$$R_e = R_2 + R_3 + \frac{R_2 R_3}{r_{ds}}$$

When r_{ds} is the drain source resistance of the f.e.t. is determined by the relationship

$$r_{ds} = \frac{r_o}{1 - \dfrac{V_c}{2V_p}}$$

r_o is the drain source resistance for $V_{ds} = 0$; $I_{ds} = 0$
V_p is the pinch off voltage.

Substitution gives

$$R_e = R_2 + R_3 + \frac{R_2 R_3}{r_o} \left[1 - \frac{V_c}{2V_p} \right]$$

a linear function of V_c.

The closed loop signal gain of the circuit, $-R_e/R_1$, also varies linearly with the value of V_c. The range of gain variation obtainable depends upon the r_o of the f.e.t. used in the circuit. A practical circuit with the component values shown in Figure 4.5 gives the gain control shown by the graph.

4.3 Voltage summation

The voltage summing property of an ideal operational amplifier has been treated in Section 1.3.2; points governing the behaviour of a practical summing circuit are now discussed. A summing configuration is shown in Figure 4.6, in which the input signals are effectively isolated from one another by the virtual earth at the inverting input terminal of the amplifier. In order to minimise offset error due to amplifier bias current input, resistor values should be no greater than those necessary to prevent signal source loading. If large values of input resistor are necessary it is advisable to employ a low bias current f.e.t. input operational amplifier in order to minimise offset error. In the ideal circuit

See Section 9.6 for offset balancing techniques

$$e_o = -R_f\left[\frac{e_1}{R_1} + \frac{e_2}{R_2} + \frac{e_3}{R_3}\right]$$

$R_c = R_3 /\!/ R_2 /\!/ R_1 /\!/ R_f$

Optional for bias current compensation

Figure 4.6 Voltage summation

there is no limit to the number of input voltages which can be summed but in the practical circuit the number of inputs is limited by the requirement for maintaining an adequate loop gain. In assessing loop gain, closed loop signal bandwidth and drift error, account must be taken of all paths to the inverting input terminal of the amplifier (see Sections 2.3, 2.5, 2.10). Note that the closed loop gain $1/\beta$ for the circuit is

$$\frac{1}{\beta} = 1 + \frac{R_f}{R_1 \parallel R_2 \parallel R_3} \tag{4.1}$$

4.4 Differential input amplifier configurations (voltage subtractors)

A differential input operational amplifier allows input signals to be applied simultaneously to both input terminals so as to give a subtractor type of operation. Figure 4.7 shows the type of circuit configuration which is employed. An ideal analysis of this circuit was given in Section 1.3.5; some of its practical limitations are now discussed.

140

A prime requirement of a differential input amplifier circuit is that it should have a high c.m.r.r. According to the ideal performance equation of the circuit in Figure 4.7 the output is zero if the two input

Figure 4.7 Differential input configuration

signals e_1 and e_2 are equal. The ideal circuit has an infinite c.m.r.r.—not so the practical circuit. In a practical circuit any mismatch in the resistor ratio values connected to the amplifier input terminals causes a common mode signal ($e_1 = e_2 = e_{cm}$) to inject a differential signal to the amplifier. This differential signal is amplified to give rise to a non-zero output signal; c.m.r.r. is thus degraded.

In assessing the common mode characteristics of operational amplifier differential input circuit configurations care must be taken in distinguishing between the characteristics of the circuit and those of the operational amplifier used in it. The c.m.r.r. of the circuit is defined as

$$\text{c.m.r.r.} = \frac{\text{Differential gain of circuit}}{\text{Common mode gain of circuit}} \qquad (4.2)$$

and in the circuit of Figure 4.7 c.m.r.r. depends both upon resistor matching and upon the c.m.r.r. of the operational amplifier. The c.m.r.r. of the circuit due to resistor mismatch using resistor values with tolerance x is in the worst case

$$\text{c.m.r.r. (due to resistor tolerance)} \cong \frac{1 + \dfrac{R_2}{R_1}}{4x} \qquad (4.3)$$

(See Appendix A3)

141

For example: A one-amplifier differential input circuit with differential gain 10 ($R_2/R_1 = 10$), using resistors of 1% tolerance ($x = 0.01$) in the worst case has

$$c.m.r.r. \text{ (due to resistor tolerance)} \cong \frac{11}{0.04} = 2.75$$

$$\text{or} \cong 49 \text{ dB}$$

The overall c.m.r.r. of the circuit due to both resistor mismatch and the finite c.m.r.r. of the operational amplifier is

$$\text{Total c.m.r.r.} = \frac{c.m.r.r._{(R)} \times c.m.r.r._{(A)}}{c.m.r.r._{(R)} \pm c.m.r.r._{(A)}} \tag{4.4}$$

(See Appendix A3)

The common mode errors due to the two effects may be of the same or opposite sign so that the total c.m.r.r. may be greater than or less than the c.m.r.r. of the amplifier used in the circuit. It is of course possible to trim one of the external resistors in Figure 4.7 so that the common mode gain due to resistor tolerance is equal in magnitude but opposite in sign to the common mode gain of the circuit, due to the non-infinite c.m.r.r. of the operational amplifier. In theory an infinite c.m.r.r. can be attained in this way. In practice resistor trimming can give a 10 to 100 times increase in c.m.r.r. for the circuit over the c.m.r.r. of the amplifier used in it. A high c.m.r.r. achieved in this way is unfortunately not maintained: resistor values change with temperature, and also the c.m.r.r. of an operational amplifier does not remain stable.

In many applications a requirement of differential input amplifiers is that they have a high differential and common mode input imped-ance. The input impedance of the circuit in Figure 4.7 is determined by the resistor values used in the circuit; its differential input resistance is $2R_1$ and it has an effective common mode input resistance at each input point of $R_1 + R_2$. If large resistor values are used in the circuit to step up input resistance values stray capacitance inevitably has a more pronounced effect, causing a degradation in c.m.r.r. at the higher frequencies. Also high resistance values give an increased offset error because of amplifier bias current.

Despite the limitations of the simple subtractor circuit it is often used, because of its simplicity, in non-critical differential measurement applications. Improved performance can be obtained with circuit configurations which use more than one operational amplifier, or ready built differential input amplifiers specifically designed for differential measurements can be used. Such amplifiers are available both in modular and integrated circuit form and are referred to as instrumentation amplifiers, measurement amplifiers or data amplifiers[1].

$$e_o = e_2 \left[1 + \frac{R_2}{R_1} \right] \quad {}^{e_1} \left[1 + \frac{R_1}{R_2} \right] \frac{R_2}{R_1}$$

$$= (e_2 - e_1) \left[1 + \frac{R_2}{R_1} \right]$$

Figure 4.8 High input impedance differential amplifier configuration

The circuit shown in Figure 4.8 is a differential input amplifier configuration suitable for use in differential measurement applications. It uses two coupled followers to attain a high input impedance without the use of high value resistors, and provides the possibility of gain setting with a single resistor. The ideal performance equation for the circuit can be derived by treating each amplifier and its associated feedback circuits separately. Amplifier A_2 has two input signals applied to it: e_2 which appears at the output multiplied by $1/\beta$, and the output amplifier A_1 which appears at the output of amplifier A_2 multiplied by the signal gain $-R_2/R_1$ (see Section 2.3.1).

A practical circuit based upon Figure 4.8 has a c.m.r.r. which depends upon both resistor tolerances and upon the c.m.r.r. of the operational amplifiers used in it. The input common mode range for the circuit is

143

equal to that of the operational amplifiers. With the gain setting resistor R_3 in circuit the output voltage in the ideal case is determined by the equation

$$e_o = (e_2 - e_1)\left[1 + \frac{R_2}{R_1} + 2\,\frac{R_2}{R_3}\right] \qquad (4.5)$$

In deriving this equation it should be remembered that with R_3 in circuit $1/\beta$ for amplifier A_2 is now determined by R_2 and the parallel combination of R_1 and R_3.

Another high input impedance, cross coupled follower, differential input circuit configuration which is often used is shown in Figure 4.9.

Figure 4.9 *High input impedance differential amplifier configuration*

The differential input stage formed by the two followers produces a differential output voltage in response to a differential input signal. Assuming that the amplifiers in the input stage take no current at their input terminals the same current must flow through the three resistors. If we make the further usual assumption of negligible voltage between amplifier input terminals then this current

$$I = \frac{e_{o_1} - e_1}{R_2} = \frac{e_1 - e_2}{R_1} = \frac{e_2 - e_{o_2}}{R_2}$$

Thus
$$e_{o_1} = \left(1 + \frac{R_2}{R_1}\right)e_1 - \frac{R_2}{R_1}e_2$$

and
$$e_{o_2} = \left(1 + \frac{R_2}{R_1}\right) e_2 - \frac{R_2}{R_1} e_1$$

The differential output of the input stage is

$$e_{o_1} - e_{o_2} = (e_1 - e_2)\left[1 + 2\frac{R_2}{R_1}\right] \qquad (4.6)$$

Note that if $e_1 = e_2 = e_{cm}$ then $e_{o_1} = e_{o_2} = e_{cm}$. The input stage passes common mode input signals at unity gain; separately connected follower circuits would pass both common mode and differential signals at the same gain.

An isolated load such as a meter can be driven directly by the differential output from the input stage with a theoretically infinite c.m.r.r. unaffected by resistor tolerance and the possibility of gain setting by means of a single resistor value (R_1). In practice c.m.r.r. is not infinite because of differences in the internal common mode errors of the two amplifiers. Dual operational amplifiers can be used in this type of circuit with the possibility of drift error cancellation if the temperature drift coefficient on the two amplifiers match and track. Monolithic dual operational amplifiers have the advantage of maintaining both amplifiers at the same temperature. However, despite the monolithic construction, the matching of amplifier parameters is unfortunately completely random. Improved performance can be obtained by using dual amplifier devices (e.g. Precision Monolithics, mono OP-10) in which two separately matched operational amplifier chips are assembled into a single dual-in-line package.

In applications in which an earth referred load has to be driven the differential output produced by the cross coupled followers can be converted into a single ended output using the simple one-amplifier subtractor circuit of Figure 4.7. The overall c.m.r.r. obtained is greater than that of the subtractor circuit by a factor equal to the differential gain of the cross coupled follower input stage. The resistor values in the subtractor should be well matched but they do not need to be of large value. The input common mode range of both the circuits of Figure 4.8 and 4.9 is limited to that of the operational amplifiers used in the circuits. A differential input circuit configuration using two inverting amplifiers (see Appendix A1, Figure A2) can be given a larger input common mode range but with the disadvantage of the lower input resistance of the inverter configuration. In applications requiring a high

Figure 4.10 Current to voltage converter—stability analysis

impedance differential measurement in the presence of large or potentially dangerous common mode signals consideration should be given to the use of a so-called isolation amplifier[2,3,4].

4.5 Current scaling

The operational amplifier circuits considered thus far are suitable for scaling input signal voltages. In many measurement systems there is a

146

need to scale the output from signal sources, in which it is the current produced by the source which is of importance—for example many light-sensitive transducers provide an output current proportional to light intensity. The circuits considered in this section are designed to process an input signal current rather than an input signal voltage.

4.5.1 CURRENT TO VOLTAGE CONVERSION

In Section 1.3.1 it was shown that an ideal operational amplifier can provide an ideal zero voltage drop current to voltage conversion of a current to earth. Some of the considerations which must be taken into account in implementing a practical current to voltage converter are now discussed. These considerations relate to the possibility of closed loop instability and to the estimation and reduction of the drift errors which determine conversion accuracy. In practice the stability problem does not usually present too serious a difficulty; an externally connected capacitor C_f (see Section 2.7.3) connected in parallel with the scaling resistor R_f normally assures closed loop stability. The importance of offset is dependent upon the size of the current to be measured and the processing accuracy required.

In Figure 4.10 an operational amplifier current to voltage converter is supplied by a current source. The stability problem, if it is a problem, arises because of source capacitance. Source capacitance causes a phase lag in the feedback signal at the higher frequencies which can lead to insufficient phase margin. Closed loop stability is most conveniently examined in terms of the appropriate Bode plots; the Bode plot for $1/\beta$ is superimposed upon the open loop Bode plot in order to examine the frequency dependence of the magnitude and phase of the loop gain (see Section 2.5).

The value of $1/\beta$ for the circuit of Figure 4.10 without the capacitor C_f in the circuit is

$$\frac{1}{\beta} = \left[1 + \frac{R_f}{R_s} \right] \left[1 + j\omega C_s R_s \parallel R_f \right] \tag{4.7}$$

This breaks up at the angular frequency $\omega_c = 1(C_s R_s \parallel R_f)$. If this frequency occurs before the frequency at which $1/\beta$ and A_{OL} intersect the two plots will have a rate of closure of 40 dB/decade and there may be insufficient phase margin. Note that at frequency ω_1 the phase shift in β is $\theta = \tan^{-1}(\omega_1/\omega_c)$ and the phase margin in the circuit is $90 - \theta$ (see Section 2.6.1).

If C_f is connected in circuit it introduces a phase lead into the feedback loop which offsets the lag due to C_s. With C_f in circuit the value of $1/\beta$ becomes

$$\frac{1}{\beta} = \left[1 + \frac{R_f}{R_s}\right] \frac{1 + j\omega(C_s + C_f)R_s \parallel R_f}{1 + j\omega C_f R_f} \tag{4.8}$$

The $(1/\beta)$ log f plot breaks back at the angular frequency $1/(C_f R_f)$ and if this frequency is suitably chosen the $1/\beta$ and A_{OL} plots close at 20 dB/decade thus ensuring an adequate phase margin. The closed loop signal bandwidth is fixed by the value used for C_f at the frequency $f = 1/(2\pi C_f R_f)$.

In many practical applications of the current to voltage converter when it is supplied by a true current source R_s will be greater than the value of the scale setting resistor R_f making the value of $1/\beta$ approximately unity at low frequencies. If the current to be measured is from a source of lower impedance than R_f the noise gain $1/\beta$ will be greater, and the loop gain smaller with a consequent decrease in accuracy and an increase in drift error due to amplifier input offset voltage temperature dependence.

Offset and drift error may be estimated by applying the general method outlined in Section 2.10.4. An expression for the total equivalent input offset voltage is

$$E_{os} = \pm V_{io} + I_B^- R_f \parallel R_s$$

This appears at the output multiplied by $1/\beta$.

$$\text{Output offset voltage} = \left[1 + \frac{R_f}{R_s}\right] E_{os}$$

In order to assess accuracy this may be referred to the input (by dividing by R_f) as an equivalent input error current.

$$I_{os} = \left[1 + \frac{R_f}{R_s}\right] \frac{E_{os}}{R_f} = \pm \frac{V_{io}}{R_s \parallel R_f} + I_B^-$$

Amplifier bias current is normally the main error component. The large resistor values which are commonly used in the current to voltage converter make the error due to amplifier input offset voltage negligible.

Initial offset can be zeroed by means of a high value resistor used to feed a small adjustable current to the inverting input terminal of the amplifier.

The temperature drift of the amplifier bias current is then the limiting factor determining accuracy. A low bias current operational amplifier should be chosen for accurate measurements of small currents, e.g. an f.e.t. input type or a varactor bridge amplifier for extremely small currents. Measurement of currents in the pico amp range requires particular attention to the avoidance of stray leakage currents if the performance capabilities of low bias current amplifiers are to be realised (see Section 9.4).

The high resistor values which are necessary in order to set the scaling factor in small current measurements is sometimes a problem, because stable high value resistors are not freely available. Sensitivity can be increased without using very high value resistors by using a resistive T network as shown in Figure 4.11, but note that this is at the expense of a decrease in loop gain and an increase in offset and noise gain.

$$e_n = -I_{in} R_f \left[1 + \frac{R_2}{R_1} \right]$$

Figure 4.11   Resistive T network gives increased sensitivity without high value feedback resistor

Current input amplifier configurations are sometimes used to overcome the problem of the finite resistance of moving coil galvanometers when used for current measurement, or to increase the sensitivity of such meters. Possible circuit configurations are shown in Figure 4.12. Battery operation of the operational amplifier allows non-earth referred measurements to be made as in Figure 4.12(c).

4.5.2 CURRENT SUMMATION

The basic current to voltage converter circuit of Figure 4.11 can be used to sum currents to earth from separate signals sources; all that is required

*Figure 4.12 Current measurement circuits (a) Simple current measurement
(b) Current measurement with increased sensitivity
(c) Current measurement not referred to earth (battery supplies)*

is to add the extra input current paths to the inverting input terminal of
the amplifier. The principle is illustrated by the circuit shown in Figure
4.13; in order to ensure adequate phase margin the value required for
the feedback capacity C_f is now governed by the total capacitance to
ground at the inverting input terminal ($C_{s1} + C_{s2}$ etc.).

4.5.3 CURRENT DIFFERENCE TO VOLTAGE CONVERSION

In many current measurement applications it is desirable that the
measurement should not introduce a voltage drop into the measurement

$$e_0 = -[I_2 + I_1] R_f$$

$$e_0 = [I_2 - I_1] R_f$$

Figure 4.14 Current difference to voltage conversion

circuit. This condition is satisfied if the current is supplied to the inverting input terminal of an operational amplifier and its non-inverting
input terminal is earthed. A current difference measurement requires
the use of two operational amplifiers in order to satisfy the zero voltage
drop criterion. The circuit shown in Figure 4.14 combines the summing
property of one operational amplifier with a current inversion performed by a second amplifier. Amplifier A_2, with equal value resistors
connected between its output and its two input terminals, must cause
equal currents to flow towards its two input terminals in order to
maintain them at the same potential. The inverted current is supplied

to the summing amplifier via a very high effective output impedance obtained as a result of the positive feedback applied to amplifier A_2.

In cases where a voltage intrusion into the measurement circuit is allowable a single operational amplifier can be used to perform a current difference conversion. The circuit shown in Figure 4.15 gives an output

Figure 4.15 Single operational amplifier for current difference to voltage conversion

voltage which is proportional to the difference in the two input currents, I_1 and I_2, but note that in this circuit a voltage drop $I_2 R$ is introduced into the measurement circuit; this voltage represents a common mode input to the amplifier. Subtraction of equal input currents requires accurate matching of resistor values.

The circuit can be used to convert the small difference between two photomultiplier currents into a proportional voltage. Used in this way the voltage $I_2 R$ represents only a small proportion of the voltage applied to the photomultiplier tubes, so the voltage across a photomultiplier tube will change by only a very small percentage even for an appreciable difference current.

4.6 Voltage to current conversion

Some loads require a current drive rather than a voltage drive and in such cases an operational amplifier circuit configuration is required which will give a linear voltage to current conversion. Voltage control-led current sources are very useful in a variety of measurement applications, such as resistance measurement and transistor testing, and they are used to supply three-electrode electrochemical cells working under controlled current conditions and to drive coil loads for the production of magnetic fields. There are several ways in which an operational amplifier may be used to produce a voltage to current conversion. The circuit configuration adopted is in many cases determined by the operating requirements of the load; for example, is the load to be earthed or can it float, is a unidirectional or bidirectional current drive required?

4.6.1 VOLTAGE TO CURRENT CONVERTERS—FLOATING LOAD

The simplest current to voltage converter circuits are those for which the load is allowed to float; basic inverting and non-inverting voltage to current converters are illustrated in Figure 4.16. In each case the ideal

Figure 4.16 Simple voltage to current converter (load floating)

performance equation $I = e_{in}/R_1$ follows directly from the usual ideal amplifier assumptions. In the inverting configuration the input signal source must supply a current equal to the load current; in the non-inverting circuit negligible current is drawn from the signal source but common mode limitations and errors must be considered.

  In all voltage to current conversions the amplifier used in the circuit must be capable of providing the desired maximum load current, and the amplifier output voltage required for maximum load current must not exceed the amplifier output voltage rating. Remember that amplifier output limits can always be increased by some form of booster circuit (see Section 9.9).

  Inductive loads (coil driving) require special attention both in the matter of not exceeding an amplifier's maximum output limits and in achieving closed loop stability. An inductive load introduces an extra phase lag in the feedback loop, which can lead to an inadequate phase margin even when the amplifier used in the circuit is frequency compensated for unity gain operation. Closed loop stability can often be achieved by connecting a resistor in series with the inductive load and a lead capacitor directly between amplifier output and phase inverting input. Bandwidth is inevitably limited by these added components.

4.6.2 VOLTAGE TO CURRENT CONVERTERS—EARTHED LOAD

Comparatively simple circuits are possible for current drive of an earthed load if either the controlling input signal voltage or the power supply to the amplifier can be floated. The input signal must float in the circuit of Figure 4.17; negative feedback forces the differential input terminals

Figure 4.17   Voltage to current converter (floating signal source)

of the amplifier to be at the same potential and in doing so produces a voltage across the resistor R which is equal to e_{in}, within a margin of error determined by the ratio of e_{in} to the amplifier input error voltage. The current through R, except for the small amplifier bias current, all passes through the load, and there is negligible loading of the input voltage signal. Note that the voltage which appears across the load represents a common mode input voltage to the amplifier and common mode limitations and errors must therefore be considered.

Figure 4.18 illustrates the circuit configuration in which both input signal voltage and load are earthed but in which the power supply common is not connected to signal earth, and the amplifier must be used with floating supplies (e.g. battery power). The circuit can be used with inverting only amplifiers (e.g. chopper stabilised amplifiers for very low drift). It has a very high input impedance and imposes negligible loading on the input signal source. The input terminals of the amplifier remain at the potential of the power supply common so that common mode errors do not arise.

Figure 4.18 Voltage to current converter (floating power supplies)

4.6.3 VOLTAGE TO CURRENT CONVERTER—EARTHED LOAD AND POWER SUPPLIES

The circuit shown in Figure 4.19 can be used to supply a bidirectional current to an earthed load, the current being controlled directly by a

Figure 4.19 Voltage to current converter (earthed load and power supplies)

single ended input voltage. The ideal performance equation for the circuit is readily derived by making use of the usual ideal amplifier assumptions. Thus the signal at the inverting input terminal is

$$e_- = e_{in} \frac{R_2}{R_1 + R_2} + e_o \frac{R_1}{R_1 + R_2}$$

and the signal at the non-inverting input is

$$e_+ = \left[e_o - I_o R_5 \right] \frac{R_4}{R_3 + R_4}$$

It is assumed that $[R_4 + R_3] \gg R_5 // R_L$. The amplifier forces $e_- = e_+$. Resistor values are chosen so that $R_2/R_1 = R_3/R_4$.

Making these substitutions the performance equation simplifies to

$$I_o = - \frac{e_{in}}{R_5} \frac{R_2}{R_1}$$

The offset error for the circuit when referred to the signal input is

$$V_{in} \text{ (offset)} = \left[1 + \frac{R_1}{R_2} \right] E_{os}$$

where $E_{os} = \pm V_{io} + I_b^- R_s^- - I_b^+ R_s^+$ (see Section 2.10.4).

The load current is supplied by a very high effective output impedance. The value of this output impedance and the stability of load current against fluctuations in load impedance is very much dependent upon accurate matching of resistor ratios in the circuit. It is of course possible experimentally to trim, say, the value of resistor R_4 (by the use of a small potentiometer in series with it) in order to get near constant output current with variations in load.

4.6.4 UNIDIRECTIONAL CURRENT SOURCES AND SINKS

The voltage to current sources considered thus far have provided an output current which can flow in either direction. In applications requiring only a single polarity of output current, current sources can be formed using simply a transistor, a resistor and an operational amplifier. Circuits are shown in Figure 4.20; feedback around the operational amplifier forces the current through a current sensing resistor R_1 to take on a value such that $I_1 R_1 = e_{in}$. The current I_1 is the emitter

current of the transistor less the very small bias current of the amplifier. The collector current of the transistor which is almost equal to
its emitter current forms the stable output current to the load.

$$I_o = \alpha\left[\frac{e_{in} \pm V_{io}}{R_1} + I_b\right]$$

Departures from linearity in voltage to current conversion may be
expected to occur at low current levels because the current gain α of a
bipolar transistor falls at low values of collector current. Output currents
greater than the current output capability of the amplifier are possible
since the amplifier need only supply the base current to the output
transistor. A current limit is set by transistor saturation caused by the
voltage which appears across the load.

Figure 4.20 Current source and sinks

The linearity dependence on transistor current gain exhibited by the
current sources of Figure 4.20 can be overcome by using an f.e.t. in
place of the bipolar transistor, but output current is then limited to
the I_{DSS} of the f.e.t. The output current limit can be overcome by combining an f.e.t. and a bipolar transistor as shown in Figure 4.21. In this
circuit virtually all the current through the sensing resistor R_1 flows as
the output current, the only error contributions being the very small
gate leakage current and the amplifier bias current. If a pnp transistor
and an n channel f.e.t. are used in Figure 4.21 with a negative input voltage the circuit functions as a current source to a load returned to a
negative supply.

157

$V_S + ve$

$I_0 = \dfrac{e_{in}}{R_1}$

e_{in}

R_2

$R_2 I_{DSS} \gg 0.5V$

R_1

Figure 4.21 Precise current sink

4.7 A.C. amplifiers

Operational amplifiers are basically high gain d.c. amplifiers but they
are equally suitable for applications not requiring a d.c. response. When
used for a.c. amplification d.c. blocking capacitors are placed in the
signal path and the amplifier offset and drift specifications are not as
important. It is often possible to operate the amplifier with a single
power supply, using a split zener, or a resistive network biasing technique and thus reducing the requirement for separate positive and
negative power supplies.

4.7.1 PHASE-INVERTING A.C. AMPLIFIER

Figure 4.22 illustrates the basic inverting amplifier with a capacitor C_1
connected in series with the input resistor. Bias current to the inverting
input terminal of the amplifier is supplied through the feedback resistor
R_2. The gain of the amplifier is R_2/R_1 with the low frequency 3 dB
fall in gain occurring at a frequency $1/(2\pi C_1 R_1)$. The upper frequency
limit of this circuit will depend on the compensated open loop frequency
response of the particular amplifier type in use.

Figure 4.22 Phase inverting a.c. amplifier

Figure 4.23 Non-inverting a.c. amplifier

4.7.2 NON-INVERTING A.C. AMPLIFIER

The circuit illustrated in Figure 4.23 is basically the follower with gain
with the addition of d.c. blocking capacitors and the d.c. bias path R_3.
The closed loop gain of the circuit is $1 + (R_2/R_1)$. The closed loop low
frequency response will show two breaks at $f_1 = 1/(2\pi C_1 R_1)$ and
$f_2 = 1/(2\pi C_2 R_3)$. The input impedance of the circuit is determined by
the bias resistor R_3.

4.7.3 HIGH INPUT IMPEDANCE A.C. AMPLIFIER

The non-inverting amplifier, being a voltage follower, is intrinsically
capable of a high input impedance. Input impedance in the simple

159

Figure 4.24 High input impedance a.c. amplifier

follower of Figure 4.23 is reduced by the requirement for the d.c.
biasing path R_3. In the circuit illustrated in Figure 4.24 a high effective input impedance is obtained because of positive feedback applied
via R_2, C_1 and R_1 to the lower end of R_3. The technique of raising the
apparent value of an impedance by driving its low potential end with a
voltage in phase with, and almost as large as, the voltage at its high
potential end is known as 'bootstrapping'. The effective value of R_3 is
increased in this way by a factor equal to the loop gain.

REFERENCES

1. CLAYTON, G. B., *Linear Integrated Circuit Applications*, Macmillan (1975).
   Linear I.C. Applications Handbook, Tab Books (1977)
2. 'Medical Isolation Amplifier', Analog Dialogue, **5**, No. 2 (1971)
3. 'Getting Best Results with Isolation Amplifiers', Analog Dialogue, 5, No. 3
   (1971)
4. OTT, W. E., 'Instrumentation Amplifiers', Burr Brown Research Corporation
   Application Note A.N. 75

Exercises

4.1   In the circuit shown in Figure 4.6 input signals e_1, e_2 and e_3 are
      applied through input resistors 100 kΩ, 50 kΩ, and 10 kΩ
      respectively; the feedback resistor has a value 100 kΩ. Write down
      the ideal expression for the output signal. If the operational ampli-
      fier is assumed ideal except for a finite open loop gain of 80 dB
      what is the percentage error involved in the output sum?

4.2 An internally frequency compensated operational amplifier has an open loop gain 100 dB, unity gain frequency 4×10^6 Hz, input offset voltage 2 mV and bias current 100 pA. It is used in the circuit of Figure 4.2 with $R_1 = 1$ MΩ, $R_f = 1$ MΩ, $R_3 = 1$ kΩ, $R_2 = 100$ kΩ. Find:
(a) the signal gain,
(b) $1/\beta$,
(c) the closed loop bandwidth,
(d) the output offset.
(Consult Sections 2.5 and 2.10.4.)

4.3 Resistors $R_1 = 10$ kΩ, $R_2 = 1$ MΩ, with tolerance 1% are used in the circuit of Figure 4.7. The operational amplifier has $A_{OL} = 100$ dB, c.m.r.r. $= 80$ dB, unity gain frequency $f_1 = 10^6$ Hz, input offset voltage $V_{io} = 2$ mV, input difference current $I_{io} = 50$ nA. Find:
(a) the worst case c.m.r.r. of the circuit (use equations 4.3 and 4.4);
(b) the closed loop bandwidth (consult Section 2.5);
(c) the output offset (consult Section 2.10.4).

4.4 Resistors $R_1 = 1$ kΩ, $R_2 = 100$ kΩ, of 2% tolerance are used in the circuit of Figure 4.8. What is the worst case c.m.r.r. of the circuit due to resistor mismatch? If the operational amplifiers have an open loop gain bandwidth product of 4×10^6 Hz what is the closed loop signal bandwidth?
(See Section 2.5.)

4.5 An internally frequency compensated operational amplifier with unity gain frequency 10^6 Hz is used as a current to voltage converter and is supplied by a current source of very high internal resistance and capacitance $C_s = 5$ pF. A feedback resistor of value 1 MΩ is used; initially no feedback capacitor is connected but the circuit is found to be very lightly damped. Explain this fact and estimate the phase margin in the circuit. The problem of the lightly damped response is overcome by connecting a capacitor of value 10 pF in parallel with the feedback resistor. Explain the action of this capacitor and estimate the phase margin and signal bandwidth with the capacitor connected. Illustrate your answer with appropriate Bode plots.
(Consult Sections 4.5.1 and 2.6.)

4.6 A current to voltage converter has a feedback resistor of value 1 MΩ. Initial offset in the circuit is balanced by means of an adjustable current bias supplied through a resistor of value 10 MΩ. Assuming a temperature change of $10°C$ estimate the smallest current which can be converted with error no greater than 1%

(a) using a bipolar transistor input operational amplifier with $\Delta I_B/\Delta T = 1$ nA/$°C$ and $\Delta V_{io}/\Delta T = 10\ \mu V/°C$;

(b) using an f.e.t. input operational amplifier with $I_B = 50$ pA, doubling for a $10°C$ rise in temperature and $\Delta V_{io}/\Delta T = 40\ \mu V/°C$. Assume the input signal source has a resistance $R_s = 10$ MΩ. Repeat the question assuming a source resistance $R_s = 100$ kΩ. (Consult Section 4.5.1.)

CHAPTER FIVE

NONLINEAR CIRCUITS

In the basic amplifier circuits discussed in the previous chapter relationships between input and output signals are linear and independent of frequency, (at least over the range of frequencies of interest). Linear frequency independent relationships arise because of the use of linear resistors for input and feedback components. Nonlinear components give nonlinear relationships, reactive components give frequency dependent relationships.

Nonlinear amplifiers find many applications in computation and signal processing. In such applications the functional relationship required between input and output can be arbitrary, or alternatively some definite mathematical function may be required, for example, $e_o = e_i^2$, $c_o = \sin e_i$, $e_o = \log e_i$. In particular a logarithmic conversion has many uses. Computing operations such as multiplication, division, and the taking of powers or roots may all be performed using logarithmic amplifiers. Instrumentation systems, capable of accurately measuring both large and small signals on a single measurement channel, are sometimes required; in such cases a logarithmic amplifier provides a convenient method of compressing the wide range data. Conversely, an antilog relationship allows an expansion of narrow range data.

5.1 Amplifiers with defined nonlinearity

Defined nonlinear amplification requires the use of a circuit element with a voltage–current characteristic exhibiting the desired nonlinearity.

Figure 5.1 Operational amplifier with nonlinear input path

The element is connected as either the input or feedback path in an operational feedback circuit. In Figure 5.1 a nonlinear element is shown replacing the normal input resistor used in the inverting amplifier circuit. The usual summing point restraints applied to the circuit give

$$I_i = f(e_i) \text{ and } I_i = I_f = -\frac{e_o}{R}$$

Thus
$$e_o = -Rf(e_i)$$

Figure 5.2   Operational amplifier with nonlinear feedback path

In Figure 5.2 the positions of resistor and nonlinear element are interchanged. The amplifier output voltage drives the nonlinear element and the feedback current is thus related to the output voltage in the defined nonlinear manner.

$$I_f = -f(e_o)$$

and
$$I_i = I_f = \frac{e_i}{R}$$

Thus
$$e_o = -f^{-1}\left(\frac{e_i}{R}\right)$$

The circuit develops the required inverse function f^{-1}.

The chief difficulty associated with defined nonlinear amplification lies in the practical realisation of a circuit element exhibiting the desired nonlinear characteristic. Such elements are required to adhere accurately to the characteristic over the widest possible range of current and should ideally be insensitive to temperature changes.

The techniques used to achieve nonlinear amplification generally fall into two categories. In one method the desired nonlinear response is synthesised by a network, the network itself consisting of a combination of essentially linear elements. The other method makes use of the inherent nonlinearity of a particular device.

5.2 Synthesised nonlinear response

Any nonlinear function can be approximated by a series of straight line segments tangential to the desired function. Thus the functional relationship $e_o = -f(e_i)$, where f represents a desired nonlinearity, can be approximated in this way by using an operational amplifier to sum a series of currents, each current being linearly related to the input voltage e_i.

The process is illustrated by the graph in Figure 5.3 in which the

$$I_n = k_n(e_i - e_n)$$

$$k_n = \frac{1}{R_{an}}$$

$$e_n = \frac{R_{an}}{R_{bn}} E_{ref}$$

$$e_o = -R \lessgtr I_n$$

$$e_o = -RI_1$$

$$e_o = -R(I_1 + I_2)$$

$$e_o = -R(I_1 + I_2 + I_3)$$

Figure 5.3 Synthesised nonlinear response

165

currents are, $I_1 = k_1 (e_i - e_1)$, for $e_i > e_1$, $I_2 = k_2 (e_i - e_2)$, for $e_i > e_2$, $I_3 = k_3 (e_i - e_3)$, for $e_i > e_3$, etc. The break points, e_1, e_2, e_3 . . ., are each set by a diode, a resistive divider, and a reference voltage supply. The input voltage is connected to these several networks as shown by the circuit in Figure 5.3. The reference voltage polarity and the diode orientations illustrated are appropriate to a positive input signal.

Diode $D1$ becomes forward biased when the input voltage exceeds the first break point $e_1 = (R_{a_1}/R_{b_1}) E_{ref}$. Feedback from the amplifier output, through the resistor R, holds the inverting input terminal of the amplifier at earth potential. Neglecting the diode voltage drop, the current through diode $D1$ for values of the input voltage greater than the first break voltage e_1 is thus,

$$I_1 = \frac{1}{R_{a_1}} (e_i - e_1)$$

Similar reasoning gives the values of the currents through D_2, D_3, . . . D_n, as

$$I_n = \frac{1}{R_{a_n}} (e_i - e_n)$$

The values for the break voltages are given by

$$e_n = \frac{R_{a_n}}{R_{b_n}} E_{ref}$$

The slopes of the straight line segments used to approximate the desired function are

$$S_n = R \left(\frac{1}{R_{a_1}} + \frac{1}{R_{a_2}} + \ldots + \frac{1}{R_{a_n}} \right)$$

Negative input signals may be handled by using additional input networks with diode and reference voltage polarities reversed. The use of an additional amplifier to invert the polarity of the input signal, followed by input networks with appropriate diode and reference voltage polarities, allows non-monotonic functions to be generated.

Our simple treatment has neglected diode voltage drops; practical diodes exhibit a non-zero forward voltage drop with the added complication of a temperature dependence. By using additional amplifiers and

$$e_n = \frac{R_c}{R_{b_n}} E_{ref}$$

$$S_n = R\left[\frac{1}{R_{a_1}} + \frac{1}{R_{a_2}} + \cdots + \frac{1}{R_{a_n}}\right]$$

Figure 5.4 Nonlinear amplifier, break points stabilised

at the expense of added complication diode effects can be reduced to negligible proportions giving breakpoint voltages which change insignificantly with temperature.

A method of reducing diode effects is illustrated by the circuit shown in Figure 5.4. The operational amplifier diode combinations used in the input networks act essentially as precision rectifiers (see Section 8.7). Break point voltages are determined by E_{ref}, resistors R_c and resistors $R_{b_1}, R_{b_2}, \ldots R_{b_n}$, and

$$e_n = \frac{R_c}{R_{b_n}} E_{\text{ref}}$$

The slopes of the line segments are determined by resistors $R_{a_1}, R_{a_2}, \ldots R_{a_n}$, and

$$S_n = R\left(\frac{1}{R_{a_1}} + \frac{1}{R_{a_2}} + \ldots + \frac{1}{R_{a_n}}\right)$$

Figure 5.5 *Nonlinear amplifier with temperature compensated break points*

The circuit shown in Figure 5.5 illustrates another method of producing temperature stable break points which may be applicable in some cases. The external transistors should all be of the same type; a transistor type which exhibits a high current gain should be used.

The gain of the circuit for small output signals is R_2/R_1; transistors *Tr*2 and *Tr*3 although they are conducting, do not feed back any appreciable amount of current to the amplifier summing point. When the output voltage rises to a certain level (set by R_3, R_4 and V_S), transistor *Tr*2 saturates; this effectively connects R_3 in parallel with

168

R_2 and makes the gain of the circuit become $(R_2//R_3)/R_1$. The gain is similarly further reduced by the connection of R_5 in parallel with R_3 and R_2 when the output voltage rises to a level which causes saturation of transistor T_3; this level is determined by R_5, R_6 and V_S.

Temperature compensation is achieved in the circuit by the inclusion of transistors $Tr1$ and $Tr4$. Transistor $Tr1$ is used to temperature compensate the base–emitter voltages of $Tr2$ and $Tr3$ so as to keep the voltage across the feedback resistors R_3 and R_5 equal to the output voltage across the feedback resistor R_2. Any change in the base–emitter voltages of $Tr3$ and $Tr2$ alters the values of the output voltage needed to cause saturation of the transistors; transistor $Tr4$ is used to temperature compensate for this effect.

It is worth noting that temperature compensated networks for the synthesis of specific functions, and variable functions, are commercially available for those who require the convenience of ready built modules.

5.3 Logarithmic conversion with an inherently logarithmic device

Nonlinear effects in many devices, including thermionic diodes, thermistors, solid state diodes and transistors, have been used in order to obtain logarithmic amplification[1]. The discussion presented here is restricted to a consideration of the use of solid state diodes and transistors.

The logging performance obtained with an operational amplifier log element combination is influenced both by the characteristics of the amplifier and by the characteristics of the logging element. An understanding of accuracy limitations requires some knowledge of the action of the log device itself.

Shockley's first order theory for a single pn junction gives the relationship

$$I = I_o \left(\exp \frac{qV}{kT} - 1 \right) \tag{5.1}$$

I is the current through the junction (A)
I_o is the theoretical reverse saturation current (A)
V is the voltage across the junction
q is the magnitude of the electronic charge (1.6×10^{-19} C)
k is Boltzmann's constant (1.38×10^{-23} J/K)
T is the temperature in K.

Substituting values of constants gives $kT/q \approx 26$ mV at $27°$C; thus for values of V greater than say 100 mV the exponential term in equation 5.1 predominates and we may write

$$I \approx I_o \exp \frac{qV}{kT} \quad (V > 100 \text{ mV})$$

taking logarithms
$$\log_e \frac{I}{I_o} = \frac{qV}{kT}$$

or
$$V = 2.3 \frac{kT}{q} \log_{10} \frac{I}{I_o} \tag{5.2}$$

According to equation 5.2 a plot of $\log_{10} I$ against V gives a straight line of slope $2.3 \, kT/q$ V/decade of current change. (Note the factor $2.3 \, kT/q \approx 60$ mV at $27°$C.)

A diode, which is assumed to obey equation 5.2, is shown connected as the feedback element in the circuit illustrated in Figure 5.6. Referring to this circuit and assuming ideal amplifier performance

$$V = - \frac{R_1}{R_1 + R_2} \, e_o = 2.3 \frac{kT}{q} \log_{10} \frac{I}{I_o}$$

or
$$e_o = - \frac{R_1 + R_2}{R_1} \, 2.3 \frac{kT}{q} \log_{10} \frac{I}{I_o} \tag{5.3}$$

The input current in the circuit shown is $I = e_i/R$. In the derivation of equation 5.3 we neglect the loading effect of the current I on the resistive divider R_1, R_2. This divider is used to set a convenient scaling factor; 60 mV/decade of current change is a somewhat inconvenient factor and 1 V/decade of current change is usually preferred.

The circuit given in Figure 5.6 is attractively simple but is unfortunately rather limited in performance. Even assuming the availability of diodes which accurately obey equation 5.1 there remains the problem of temperature dependence which limits the usefulness of the simple circuit to a controlled temperature environment. The scaling factor $E_o = 2.3 \, kT/q$ is linearly dependent on temperature; it has a positive temperature coefficient of $0.3\%/°$C. This temperature dependence can be compensated by replacing the resistor R_1 in Figure 5.6 with a temperature sensitive resistor having a temperature coefficient which closely matches that of E_o.

170

Figure 5.6   Log amplifier with a diode as log element

A more marked and distinctly nonlinear temperature dependence is exhibited by the quantity I_o; it approximately doubles for every $10°C$ change in temperature. Temperature compensation can be obtained by using matched diodes in two separate logging amplifiers. The term I_o is eliminated by following this arrangement with a third amplifier connected as a subtractor. Further details of the method will be given in connection with the temperature compensation of transistor logarithmic elements.

A more serious problem associated with the use of diodes as log elements is that most diodes do not accurately obey equation 5.1. The derivation of this equation is based upon a single diffusion mechanism of current flow. It would appear however that there are several mechanisms operative and that diode current is more accurately represented as the sum of several components[1]. The current components each have the form

$$I_j = I_{oj} \left(\exp \frac{qV}{m_j kT} - i \right) \tag{5.4}$$

where m_j can take values between 1 and 4.

The log characteristics of a range of general purpose silicon diodes were measured in the author's laboratory[2]. The $V/\log_{10} I$ plots obtained were in general smooth curves which fitted asymptotically to two straight lines. A typical example is shown in Figure 5.7; the two straight lines in this case have slopes corresponding to values of m equal to 1.78 and 1.55.

Another factor which introduces departures from diode logging accuracy is the resistance associated with the bulk semiconducting

171

Figure 5.7 Typical V/log I plot for a general purpose diode

material. At the higher currents the voltage drop across this resistance becomes significant, causing a reduction in that part of the applied diode voltage which actually appears across the junction.

The above factors make general purpose diodes unsuitable for accurate logarithmic conversion except over a very much restricted logging range (3 current decades at the most). Temperature compensation requires the selection of matched diodes (matched m components), and this presents an added difficulty. So-called 'log diodes' are available which are said to exhibit a 7 decade current logging range[3]; they are expensive. Transistors, which we will now consider, appear to be the most convenient logarithmic elements for accurate logarithmic conversion.

Figure 5.8 Simple transistor model and sign conventions

172

A bipolar transistor consists essentially of two interacting pn junctions; the circuit symbol and a simple model for an npn transistor are illustrated in Figure 5.8. According to several sources[1,4] the collector current of a transistor can be accurately represented by the equation

$$I_C = \alpha_F I_{ES} \left(\exp \frac{-qV_E}{kT} - 1 \right) - I_{CS} \left(\exp \frac{-qV_C}{kT} - 1 \right)$$

$$- \Sigma I_{CS_j} \left(\exp \frac{-qV_C}{m_j kT} - 1 \right) \tag{5.5}$$

α_F is the current transfer ratio between emitter and collector; it is very nearly unity

I_{CS} is the collector reverse saturation current with the emitter shorted to the base

I_{ES} is the emitter reverse saturation current with the collector shorted to the base

m_j is a constant which takes on values between 1 and 4, dependent on the transistor type.

The sign convention adopted is shown in Figure 5.8. Equation 5.5 is appropriate for an npn transistor, the senses of the Vs and the Is being reversed for a pnp device.

The first term in equation 5.5 represents that part of the emitter current, comprised of minority carriers in the base, which diffuses to the collector. The second and third terms are analogous to the diode current equations (equations 5.1 and 5.4); they give the collector current for the emitter shorted to the base.

The adoption of a circuit configuration which makes $V_C = 0$ causes all but the first term in equation 5.5 to become zero and the collector current is then given by the equation

$$I_C = \alpha_F I_{ES} \left(\exp \frac{-qV_E}{kT} - 1 \right) \tag{5.6}$$

This is analogous to the 'ideal' diode relationship of equation 5.1. Note that the $m \neq 1$ components of collector current become zero. The emitter $m \neq 1$ current components behave largely as majority carriers in the base and as such do not diffuse to the collector. I_{ES} is typically of the order 10^{-13} A and α_F is very nearly unity so that for values of collector current $I_C \gg I_{ES}$ the exponential term in equation 5.6 predominates; assuming this predominance and taking logs give the relationship

$$- V_E = 2.3 \times \frac{kT}{q} \log_{10} \frac{I_C}{I_o} \qquad (5.7)$$

where we write $I_o = \alpha_F I_{ES}$.

Note that α_F should not be confused with the commonly used grounded base current gain $\alpha = I_c/I_E$. The emitter current of a transistor includes the collector current and $m_j \neq 1$ components (see equation 5.9); α is always less than α_F, it is a function of emitter voltage and it decreases substantially at low values of collector current. However, α_F remains essentially constant over the range of collector currents for which equation 5.7 is valid.

Figure 5.9 Log amplifier, transdiode configuration

The $V_C = 0$ condition may be obtained by connecting the collector of the transistor to the summing point of an operational amplifier. This connection is made in the circuit shown in Figure 5.9; the base of the transistor is connected to earth. The circuit illustrates the so-called transdiode (Patterson diode) logging configuration; the amplifier output terminal is connected to the emitter and the amplifier output provides the driving voltage ($e_o = V_E$). An alternative arrangement is illustrated in Figure 5.10; in this circuit the collector and base are connected together and the transistor acts as a diode.

174

The transdiode configuration of Figure 5.9 is capable of the widest range of log conversion of input current. Accurate log conversion requires that α_F remain constant over a wide range of current values. Silicon diffused transistors of the planar type have this characteristic; a logging capability over a current range of up to 10 decades has been reported for such transistors[1]. The upper end of the useful current range is determined by semiconductor bulk resistance effects and is usually between 1 and 10 mA.

The earthed base used in the transdiode configuration allows only single polarity input signals; reverse polarity requires the use of a complementary transistor type. In the transdiode configuration the transistor has an active gain which is frequency dependent; since the transistor is connected inside the feedback loop this introduces closed loop stability problems.

$$e_o = -2 \cdot 3 \frac{kT}{q} \log_{10} \left[\frac{I_i}{I_o(1 + \frac{1}{h_{FE}})} \right]$$

$$I_i = \frac{e_i}{R}$$

Figure 5.10 Log amplifier, diode connected transistor

The diode configuration, Figure 5.10, although not capable of as wide a logging range as the transdiode circuit, is in many respects more versatile than the transdiode circuit. Since it is a two terminal device its polarity can be reversed to allow a reversed input polarity. Several diodes may, if required, be connected in series for greater output voltages and it allows the possibility of using both input terminals of a differential input amplifier to create simple log of current ratio circuits. Connecting the base of the transistor directly to the collector shorts out the gain of the device thus simplifying the problem of achieving closed loop stability.

175

In the diode configuration the feedback current, I_f, is not exactly equal to I_C. But,

$$I_f = I_C + I_B$$

$$= I_C \left(1 + \frac{1}{h_{FE}} \right)$$

where $I_B = I_C/h_{FE}$ is the base current drawn by the transistor and h_{FE} is the common emitter d.c. current gain. Equation 5.7 becomes

$$-e_o = -V_E = 2.3 \frac{kT}{q} \log_{10} \left[\frac{I_i}{\alpha_F I_{ES}} \frac{1}{1 + \dfrac{1}{h_{FE}}} \right] \tag{5.8}$$

Transistors with a large value of h_{FE} should be used in order to reduce the error term. The fall in h_{FE} which occurs at low current levels sets the lower level of the input current at which the configuration departs significantly from logging accuracy. The logging range obtainable[5] is typically within the range 10^{-3} A to 10^{-9} A.

The curves illustrated in Figure 5.11[2] show the logging characteristics

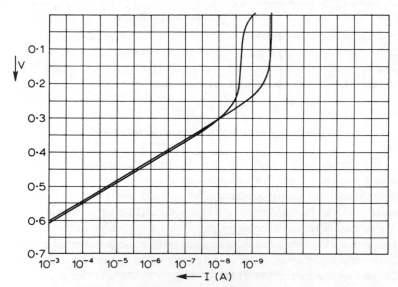

Figure 5.11 *Logging characteristics of diode connected transistors*
(Type 2N 3707)

of diode connected type 2N 3707 transistors. The curves shown are for
the two transistors which exhibited the smallest and greatest logging
range out of a batch of several of the same type of transistors tested.

A third transistor logging configuration which is sometimes used is
illustrated in Figure 5.12. The most useful feature of this connection is
the reduced loading on the operational amplifier output; the current
drawn from the output of the amplifier by the feedback loop is the
small base current of the transistor. Disadvantages of the circuit are lack
of reversibility, the necessity for a separate bias voltage supply for the
collector and an accurate logging range which is less than that given by
the other configurations[5].

The current fed back to the amplifier summing point is the emitter
current of the transistor (npn) which is given by the equation

$$I_E = \alpha_R I_{CS} \left(\exp \frac{-qV_C}{kT} - 1 \right) - I_{ES} \left(\exp \frac{-qV_C}{kT} - 1 \right)$$

$$- \Sigma I_{ESj} \left(\exp \frac{-qV_E}{m_j kT} - 1 \right) \qquad (5.9)$$

α_R is the reverse current gain of the transistor ($\alpha_R \approx 0.2$). The collector
is usually taken to a reverse bias of order 1 V so that $V_C \neq 0$ and the
first term in the equation although small does contribute an error. A
more significant error is contributed by the $m_j \neq 1$ components of
current represented by the third term of the equation. The useful logging range with this configuration[5] is typically within the range 10^{-5} A
to 10^{-8} A.

Figure 5.12 Log amplifier, transistor connection

5.4 Log amplifiers; practical design considerations

A successful operational amplifier log converter design requires proper attention to certain practical considerations. A general treatment of practical considerations involved in operational applications is given in Chapter 9. The reader wishing to implement a practical log amplifier design is advised to consult this more general treatment in addition to the material to be presented here.

The following are some of the more important points requiring attention in a practical logarithmic converter. The designer must ensure closed loop stability. The effect of the method used to achieve closed loop stability on the output slewing rate must be examined. Amplifier offsets must be balanced out if the full capability of the amplifier is to be exploited; in a practical circuit logging range is usually determined by amplifier offsets rather than by the dynamic range of the transistor logging device itself. The relative importance of voltage and current offset is determined by the magnitude of the source resistance (see Section 2.10); an amplifier type should be chosen to suit the particular application. The transistor logging device must be protected against possible damage caused by the inadvertent application of an excessive reverse polarity signal. A means of compensating for the pronounced temperature dependence of the logging transistor must be employed if the circuit is to be used in any but a strictly controlled temperature environment.

5.4.1 CLOSED LOOP STABILITY

It will be remembered (see Section 2.6) that in operational feedback circuits, stable (non-oscillatory) closed loop operation requires that the loop gain βA_{VOL} should be less than unity at frequencies for which the phase shift round the loop reaches $180°$. The condition implies that on a Bode plot the intersection of $1/\beta$ and A_{VOL} should occur with a rate of closure of less than 40 dB/decade.

In our treatment of feedback circuits thus far we have assumed the feedback fraction β to be determined by purely resistive components making $1/\beta$ real at all frequencies and never less than unity. Under such conditions an amplifier open loop response characterised by a 20 dB/decade roll off down to unity gain ensures closed loop stability for all values of input and feedback resistors. In practical circuits such an open loop response does not, unfortunately, always ensure closed loop stability. Capacitance between amplifier summing point and earth, due to strays, or amplifier input capacity, or both, causes a phase lag in the feedback fraction β at the higher frequencies, (with a corresponding phase lead in $1/\beta$). Capacitive loading at the amplifier output can cause an additional phase lag. Both effects can lead to instability.

The problem of stability in logarithmic amplifiers is further complicated by the nonlinear nature of the feedback. The feedback is greater and therefore $1/\beta$ is smaller at the higher input currents. In examining stability criteria it is convenient to assume an operational amplifier with a finite open loop gain characterised by a 20 dB/decade roll off down to unity gain. The effects of other departures of the amplifier from the ideal amplifier of Section 1.2 are initially neglected.

Figure 5.13 Bode plot for transdiode configuration

Since the feedback fraction β is dependent on the operating current, we examine stability in terms of a small signal feedback ratio. The small signal feedback ratio $\Delta e_f/\Delta e_o$ is assumed to be defined about some d.c. operating current I_C. Referring to the circuit shown in Figure 5.13 the

179

current fed back to the amplifier summing point I_f is equal to the collector current of the transistor I_C. Assuming a predominance of the exponential term in equation 5.6 we may write this equation as

$$I_f = \alpha_F I_{ES} \exp \frac{-qe_o}{kT} \tag{5.10}$$

Differentiating equation 5.10 with respect to e_o gives the small signal feedback resistance r_E.

Thus
$$\frac{\partial I_f}{\partial e_o} = \frac{-qI_f}{kT} = -\frac{1}{r_E}$$

and
$$r_E = \frac{kT}{qI_f} \approx \frac{1}{40I_f} \tag{5.11}$$

Note that for an operating current of 10^{-3} A, $r_E = 25\ \Omega$, but when the operating current is say 10^{-9} A, $r_E = 25$ MΩ.

A change in the output voltage Δe_o results in a change in the feedback current $\Delta I_f = -\Delta e_o/r_E$. This in turn causes a change $-\Delta I_f Z_1$ in the voltage fed back to the amplifier summing point. We may thus write the value of the small signal feedback ratio as

$$\beta = \frac{Z_1}{r_E} \tag{5.12}$$

Z_1 is the impedance between amplifier summing point and earth. In Figure 5.13 $Z_1 = R_1/(1 + j\omega C_1 R_1)$ where C_1 is the total capacitance between amplifier summing point and earth; C_1 is taken to include the capacitance between the collector and base of the transistor. The shunting effect of the collector output resistance is neglected.

Substituting for Z_1 gives

$$\frac{1}{\beta} = \frac{r_E}{R_1} (1 + j\omega C_1 R_1) \tag{5.13}$$

We note that at the higher operating currents it is possible for $1/\beta$ to be considerably less than unity ($r_E < R_1$). This feature is peculiar to the transdiode configuration; in other operational feedback circuits the lower limit of $1/\beta$ is unity. We can understand this effect if we remember that in the transdiode configuration the transistor acts as a common base amplifier for feedback signals and, as such, it can provide a voltage gain which is greater than unity.

Values of $1/\beta$ for different operating currents are shown in Figure 5.13. For the purpose of the discussion component values are chosen to simplify the arithmetic. The operational amplifier is assumed to have a unity gain bandwidth product of $10^7/(2\pi)$ Hz. We see immediately that the circuit fails to satisfy the closed loop stability criterion for operating currents greater than 10^{-6} A. One solution to the stability problem is to connect a capacitor C_2 between the operational amplifier output terminal and the summing point. This capacitor breaks with r_E at an angular frequency $\omega_2 = 1/(C_2 r_E)$ to cause attenuation in $1/\beta$. But remember that the value of r_E depends on the level of the operating current. The magnitude of C_2 required to ensure closed loop stability at the higher operating currents places a severe restriction on the bandwidth and output slewing rate at the lower levels of operating current. For example, to make $\omega_2 = 10^6$ rad/s at an operating current of 1 mA requires a value of C_2 equal to 0.04 μF. This value of C_2 makes $\omega_2 = 1$ rad/s at an operating current of 10^{-9} A, a time constant of 1 s.

Another practical difficulty arises because of the finite open loop output impedance of the operational amplifier. This inevitably causes a reduction in open loop gain when the amplifier is used to supply a

Figure 5.14 Bode plots for stable closed loop operation

low value load resistor. At an operating current of 1 mA, r_E = 25 Ω, and such a small value is likely to have a marked effect on the gain bandwidth product of the amplifier.

A remedy is to connect a resistor R_E in series with the emitter of the transistor. In addition to reducing the loading on the operational amplifier output R_E allows the use of smaller values for C_2. This in turn gives the system a wider bandwidth and an increased slewing rate at the lower levels of input current. The arrangement is illustrated in Figure 5.14. Closed loop stability is again conveniently examined in terms of a small signal value of the feedback fraction β. Referring to Figure 5.14,

$$- \Delta I_f = \frac{\Delta e_o}{R_E + r_E} + (\Delta e_o - \Delta e_f) j\omega C_2$$

e_f is the feedback voltage developed between the amplifier summing point and earth.

$$\Delta e_f = - \Delta I_f \frac{R_1}{1 + j\omega C_1 R_1}$$

The small signal feedback ratio

$$\beta = \frac{\Delta e_f}{\Delta e_o}$$

Manipulation of the above equations gives

$$\frac{1}{\beta} = \frac{R_2}{R_1} \frac{1 + j\omega(C_1 + C_2) R_1}{1 + j\omega C_2 R_2} \tag{5.14}$$

where $R_2 = R_E + r_E$.

The larger the value used for R_E the smaller is the value of C_2 required to ensure closed loop stability at the higher operating currents. The break out frequency for $1/\beta$, $\omega_2 = 1/(C_2 R_2)$ (at the higher operating currents) should be made to occur at least an octave before the intercept of $1/\beta$ with A_{VOL}. The maximum value which may be used for R_E is limited by the maximum output voltage swing of the amplifier, bearing in mind that the maximum output voltage across the logging transistor is approximately 0.6 V. Thus R_E should be chosen so that

$$V_{o\,max} - 0.6 > (I_{L\,max} + I_{C\,max}) R_E \tag{5.15}$$

Values of $1/\beta$ obtained from equation 5.14 are shown by the Bode plots in Figure 5.14.

If the transcendiode configuration is used for current logging, supplied by a current source, $R_1 \rightarrow \infty$ and equation 5.14 becomes

$$\frac{1}{\beta} \approx \frac{R_2 j\omega(C_1 + C_2)}{1 + j\omega C_2 R_2} \tag{5.16}$$

Values of $1/\beta$ given by equation 5.16 are shown by the Bode plots in Figure 5.15.

Figure 5.15 Bode plots, current logging

In both Figures 5.14 and 5.15, at high frequencies, the small signal value of $1/\beta$ tends to the value $1 + C_1/C_2$. In both cases the break out frequency for $1/\beta$ at the higher operating currents is

$$\omega_2 \approx \frac{1}{C_2 R_E} \quad (R_E \gg r_E)$$

A suitable choice for $C_2 R_E$ ensures closed loop stability at the higher operating currents. The use of the maximum value of R_E allowed by equation 5.15 permits the smallest value of C_2 and hence gives the fastest slewing rate at the low current levels. The low current value of ω_2 is $\omega_2' \approx 1/(C_2 r_E)$.

183

In the diode logging configuration the gain of the transistor is shorted out which means that $1/\beta$ cannot be less than unity. The circuit used with an amplifier having a 20 dB/decade roll off may, dependent on the value of $1/(C_1 R_1)$ relative to the unity gain bandwidth product of the amplifier, be closed loop stable without the addition of a stabilising network. If the simple circuit is not closed loop stable it may be stabilised in the same way as the transdiode circuit. Operating currents which make r_E less than the rated load of the amplifier, in any case, indicate the use of a resistor R_E connected in series with the emitter.

Bode plots for a diode logging configuration are illustrated in Figure 5.16. In the circuit shown the small signal value of $1/\beta$ is given by the relationship

$$\frac{1}{\beta} = 1 + \frac{Z_2}{Z_1}$$

where $Z_2 = R_2/(1 + j\omega C_2 R_2)$, $Z_1 = R_1/(1 + j\omega C_1 R_1)$ and $R_2 = R_E + r_E$.
Substituting for Z_2 and Z_1 gives

$$\frac{1}{\beta} = \frac{1 + \dfrac{R_2}{R_1} + j\omega(C_2 + C_1) R_2}{1 + j\omega C_2 R_2} \tag{5.17}$$

At high currents, $(r_E \ll R_E)$, the break frequency in $1/\beta$ is

$$\omega_1 \cong \frac{1 + \dfrac{R_E}{R_1}}{(C_2 + C_1) R_E}$$

and the break out frequency is

$$\omega_2 \cong \frac{1}{C_2 R_E}$$

At low current

$$\omega_1' \approx \frac{1}{(C_1 + C_2) R_1} \quad \text{and} \quad \omega_2' \approx \frac{1}{C_2 r_E}$$

Stability considerations in the transistor logging configuration of Figure 5.12 are similar to those encountered in the diode configuration although the problem of the loading of the amplifier output does not arise. We will not consider this connection in any detail.

184

Figure 5.16 Bode plots, diode configuration

5.4.2 OFFSET ERRORS

The lower limit to the logging range in an operational amplifier log converter is, in many cases, determined by amplifier offsets rather than by the dynamic range of the transistor log element. In the circuit for the transdiode configuration illustrated in Figure 5.17 we represent amplifier offsets in terms of equivalent input generators; the amplifier is assumed to have infinite open loop gain.

Referring to Figure 5.17 we see that the voltage at the summing point of the amplifier

$$e_{sp} = V_{io} - I_b^+ R_1$$

A finite voltage at the summing point makes $V_C \neq 0$ giving the possibility of a logging error through the I_{cs} terms of equation 5.5. Clearly any appreciable forward collector bias (V_C negative in the case of an npn transistor) must be avoided. Dependent on the magnitude of the bias current I_b^+ it may be advisable to omit the bias current compensating resistor and return the non-inverting input terminal of the amplifier directly to earth. Reverse collector bias (V_C positive in the case of an npn transistor) can only contribute a small error since $I_{cs} < 10^{-12}$ A.

If then we assume that the I_{CS} terms in equation 5.5 can be neglected we may use equation 5.7. Thus

$$V_E = e_o = -E_o \log_{10} \frac{I_C}{I_o}$$

Where we write

$$E_o = 2.3 \frac{kT}{q} \text{ and } I_o = \alpha_F I_{ES}$$

Figure 5.17 Offset errors, transdiode configuration

In Figure 5.17,

$$I_C = \frac{e_i - e_{sp}}{R_1} - I_b^-$$

$$= \frac{e_i - V_{io}}{R_1} - I_{io}$$

Therefore, $$e_o = -E_o \log_{10} \frac{\dfrac{e_i - V_{io}}{R_1} - I_{io}}{I_o} \tag{5.18}$$

We may use equation 5.18 to estimate the offset error for the transdiode logging configuration. If the bias current compensating resistor is omitted from the circuit we replace I_{io} in the equation by I_b.

186

Figure 5.18 Offset errors, diode configuration

A similar analysis may be carried out for the diode logging configuration illustrated in Figure 5.18; in this circuit,

$$V_E = e_o - e_{sp}$$
$$= e_o - V_{io}$$

A bias current compensating resistor does not balance out bias currents in this configuration (because of its effect on e_{sp}) and we return the non-inverting input terminal of the amplifier directly to earth.

Now
$$I_c = \frac{e_i - V_{io}}{R_1} - I_b^-$$

Making use of equation 5.8 and neglecting the $1/h_{FE}$ term gives

$$e_o = V_{io} - E_o \log_{10} \frac{\dfrac{e_i - V_{io}}{R_1} - I_b^-}{I_o} \tag{5.19}$$

5.4.3 BALANCING OFFSETS

In a practical log amplifier initial amplifier offsets may be balanced out and errors are then due to offset drifts. Separate biasing of voltage and current offsets are advisable for minimum errors and maximum logging range. Voltage biasing may be performed internally, or by a voltage bias applied to the non-inverting input terminal of the amplifier; the method recommended for the particular amplifier should be used. A current bias is applied to the inverting input terminal of the amplifier; Figure 5.19 illustrates a typical biasing arrangement. When balancing offsets, the logging element should be replaced by a high value resistor in the feedback path, a resistor of the same order of magnitude as the highest value of r_E to be encountered should be used. Input offset voltage is balanced first with the inverting input terminal of the amplifier shorted to earth. This short is removed and the bias current is then adjusted to

Figure 5.19 Offset balancing

zero the amplifier output. In practice it is advisable to inject a bias current slightly larger than I_b^- in order that the collector current for zero input shall not be zero (see slewing rate considerations in previous section). A collector rest current equal to say 1% of the smallest input current to be measured should be suitable. Component values used in the biasing networks should be chosen to allow balancing of the maximum specified values of V_{io} and I_b^-. Once the adjustements have been made R_f is of course replaced by the logging element.

5.4.4 CIRCUIT PROTECTION

A small input signal of the wrong polarity applied to a logging circuit can cause a large reverse emitter bias, with possible destruction of the

Diode circuit   Transdiode circuit

Figure 5.20 Protection against inverse polarity

logging transistor. It is advisable to provide logging elements with protection against excessive inverse voltage. Examples of such protective circuits are illustrated in Figure 5.20.

5.4.5 TEMPERATURE COMPENSATION

Transistor logging elements have an inherent temperature dependence which makes a single transistor log converter inaccurate unless the temperature is kept very constant. The main effect is due to the variation with temperature of the term $I_o = \alpha_F I_{ES}$, this approximately doubles for every $10^\circ C$ change in temperature. A less significant effect is due to the linear temperature dependence of the multiplying factor

$$E_o = 2.3 \frac{kT}{q}$$

which causes the slope of the logging characteristic to change with temperature by 0.3% per degree C in the vicinity of $27^\circ C$.

The use of matched transistors enables cancellation of the I_o terms. Consider two transistors with saturation currents I_{o1} and I_{o2}. We may write

$$V_{E_1} = -E_o \log_{10} \frac{I_{C_1}}{I_{o_1}}$$

and

$$V_{E_2} = -E_o \log_{10} \frac{I_{C_2}}{I_{o_2}}$$

189

This gives
$$V_{E_2} - V_{E_1} = E_o \log_{10} \frac{I_{C_1}}{I_{C_2}} \cdot \frac{I_{o_2}}{I_{o_1}}$$

Matched transistors make $I_{o_1} = I_{o_2}$ and

$$V_{E_2} - V_{E_1} = E_o \log_{10} \frac{I_{C_1}}{I_{C_2}} \qquad (5.20)$$

Thus a circuit arrangement which performs the subtraction operation (in effect taking the logarithm of a current ratio) replaces the uncontrollable I_o term with a fixed adjustable reference current I_{C_2}. Even if the transistors are not perfectly matched it is generally found that, for transistors of the same type, the ratio I_{o_1}/I_{o_2} remains fairly constant with change in temperature resulting in a fixed offset that can easily be balanced out.

The linear temperature dependence of the scaling factor E_o can be compensated by using an operational amplifier with a temperature sensitive feedback divider (see Figure 5.6 and equation 5.3). This system, by introducing gain, can at the same time be used to give a more convenient scaling factor. $E_o = 60$ mV at $27°C$; a scaling factor of 1 V per decade of current change is normally preferred.

It is worth noting that pairs of transistors matched for logarithmic operation, together with a resistive divider suitably calibrated and temperature compensated, are available in modular form (e.g., Analog Devices, Models 751 N and 751 P).

5.5 Some practical log and antilog circuit configurations

In this section circuits suitable as a basis for implementing practical log and antilog applications are discussed. Because of the temperature dependence of transistor logging elements circuits which use a single logging transistor are unsuitable for use except in non-critical applications and then only when ambient temperature variations are small. Most practical circuits employ matched logging transistors and achieve temperature compensation using the method outlined in the previous section. Commonly available silicon planar transistors exhibit a logging characteristic, but those designed for use at low values of collector current should be chosen if a wide logging range is required. Care should be taken to ensure that logging transistors are maintained at the same temperature; the use of dual transistors ensures matching and thermal tracking. The temperature sensitive resistor used to compensate for the temperature dependence of the scaling factor needs to be maintained at the same temperature as the logging transistor.

5.5.1 TEMPERATURE COMPENSATED LOG OF VOLTAGE AND CURRENT

The circuit for a temperature compensated log converter is shown in Figure 5.21. Using moderate performance general purpose operational amplifiers and assuming care in design and construction the circuit may be expected to provide logging of positive input voltages (use pnp transistors for negative input signals) in the range 10 mV to 10 V, with an accuracy over the whole range of the order 3% referred to the input for temperature changes of $\pm 10^{\circ}$C. Errors are greatest at the lower end of the logging range. The use of more specialised low drift operational amplifiers improves accuracy and extends the lower limit of the useful logging range.

The circuit uses two operational amplifiers and two logging transistors. The output of amplifier A_1, attenuated by the resistive divider R_3, R_4, provides the emitter base differential voltage between transistors T_1 and T_2. Neglecting the base current loading imposed by transistor T_2 the following relationship holds:

$$V_{E_1} - V_{E_2} = e_o \frac{R_3}{R_3 + R_4} \tag{5.21}$$

Negative feedback around amplifier A_2 forces V_{E_2} to take on that value which causes the collector current $I_{C_2} = I_{ref}$ to flow in transistor T_2. The collector current $I_{C_1} = I_{in}$ flows through transistor T_1 because of V_{E_1} imposed by negative feedback around amplifier A_1.

Substituting V_E values from equation 5.7 into equation 5.21 and rearranging gives the following circuit performance equation

$$e_o = - \frac{R_3 + R_4}{R_3} 2.3 \frac{kT}{q} \log_{10} \frac{I_{in}}{I_{ref}} \frac{I_{o_2}}{I_{o_1}} \tag{5.22}$$

where $I_{ref} = V_s/R_2$ and $I_{in} = e_{in}/R_1$ in a voltage logger. In a current logging application I_{in} is supplied directly to the inverting input terminal of amplifier A_1. There are different considerations regarding the offset balancing of voltage and current logging circuits, which will be discussed shortly.

191

Figure 5.21   Temperature compensated log of voltage converter

Output zero crossing

Readers who are unfamiliar with log amplifiers should carefully consider the significance of the circuit performance equation 5.22. Zero input signal does not give a zero output (log 0 is not defined). The output signal given by the circuit of Figure 5.21 is proportional to the log of the current ratio $I_{in}/I_{ref} \times I_{o2}/I_{o1}$. $\log_{10} 1 = 0$ and if $I_{o1} = I_{o2}$ zero output occurs when $I_{in} = I_{ref}$.

The input voltage for zero crossing of the output is under the control of the designer by his choice of I_{ref}. In Figure 5.21 $I_{ref} = V_s/R_2$ and zero crossing of the output may be adjusted by choice of R_2. Wide range logging applications require a very low drift operational amplifier for A_1; an amplifier of more modest drift performance can be used for A_2 by making I_{ref} greater than the smallest value of I_{in}. If $I_{in} \ll I_{ref}$ the base drive to T_2 forward biases its collector base junction, causing an effective change in the reference current in the performance equation 5.22. The extent to which this contributes an appreciable error is dependent upon the I_{cs} terms in equation 5.5 and upon the value of I_{ref}.

For example: Assume $I_{in} = 10^{-9}$ A, $I_{ref} = 10^{-4}$ A, $I_{cs} = 10^{-12}$ A and $R_3/(R_3 + R_4) = 1/16$. There are five decades difference in I_{in} and I_{ref}, the output voltage is + 5 V and a bias 5/16 V is applied to the collector base junction of T_2. If we take only the $m = 1$ term of the collector current equation (equation 5.5)

$$I_{cs} \left[\exp \frac{-qV_c}{kT} - 1 \right] \cong 10^{-12} \exp \frac{5}{16 \times 0.025}$$

$$\cong 10^{-12} \, 2.6 \times 10^5$$

Expressed as a percentage of I_{ref} this represents 0.26% error—perhaps not very significant but if the operating conditions anticipated are even more extreme the collector base voltage of T_2 can be held near zero by returning the non-inverting input terminal of amplifier A_2 to the base of transistor T_2 instead of to earth. A current source is then required to supply the reference current I_{ref} instead of the resistor R_2.

Scaling factor

The scaling factor for the circuit

$$K = 2.303 \frac{kT}{q} \frac{R_3 + R_4}{R_3} \tag{5.23}$$

193

may be set at any convenient value by choice of resistor values R_3 and R_4. A scaling factor of unity is often convenient; it corresponds to a 1 V output change for each decade change in the input current. Substitution of constants gives

$$K = 59 \times 10^{-3} \frac{R_3 + R_4}{R_3} \text{ at } 25°C$$

If $R_3 = 1 \text{ k}\Omega$ a value of $R_4 = 15.9 \text{ k}\Omega$ is required to make the scaling factor unity. Trimming the value of R_4 provides a convenient method for adjusting the scaling factor. The scaling factor varies by 0.33% per degree C. If such a variation is not tolerable the scale factor may be compensated by using a temperature sensitive resistor for R_3 with resistance varying directly with temperature. A suitable resistor is Tel Labs Type Q81 or alternatively a 430 Ω sensistor in series with a 570 Ω resistor may be used. The effectiveness of temperature compensation is dependent upon maintaining the logging transistors and the temperature compensating resistor at the same temperature.

Amplifier offsets, offset balancing
In log converters operational amplifier offset errors normally govern the lower limit of the smallest input signal for accurate conversion. Amplifier input offset voltage and its temperature drift normally set this limit in voltage logging applications, but in current logging circuits supplied by high impedance current sources it is amplifier bias current and its temperature drift which is the limiting performance parameter.

Voltage logging
Amplifier input offset voltage and bias current give rise to an equivalent input offset voltage:

$$E_{os} = \pm V_{io} + I_b^- R_s^- - I_b^+ R_s^+ \text{ (see Section 2.10.4)}$$

In a voltage logging circuit bias current can be compensated by making $R_s^- = R_s^+$ (the function of resistor R_5 in Figure 5.21), where R_s^- is effectively equal to R_1 plus the resistance of the signal source. E_{os} acts as a signal in series with the input signal source and the output voltage given by the circuit may be expressed as

$$e_o = - K \log_{10} \frac{e_{in} \pm E_{os}}{R_1 I_{ref}} \tag{5.24}$$

Consideration of the implications of equation 5.24 may be used to arrive at a convenient adjustment procedure for practically trimming offsets. If the temperature drift coefficients of moderate performance operational amplifiers are examined it would seem that at a very conservative estimate E_{os} could quite easily be trimmed to within 100 μV of zero (a considerably closer trim for low drift operational amplifiers). In equation 5.24 substitute $e_{in} = 0$, $E_{os} = 10^{-4}$, $R_1 = 10^4$, $I_{ref} = 10^{-4}$, $K = 1$. This gives

$$e_o = -1 \log_{10} 10^{-4} = +4 V$$

The initial offset of amplifier A_1 can thus be adjusted to within 100 μV of zero by setting e_{in} zero and adjusting the trim potentiometer P for an output of +4 V. Note that 100 μV offset represents an error of 1% when compared with the smallest input signal of 10 mV. After performing the initial offset adjustment subsequent temperature drift of E_{os} can further degrade accuracy.

Current logging

In current logging applications input current is supplied directly to the inverting input terminal of amplifier A_1. It is impracticable to use a bias current compensating resistor at the non-inverting input terminal of the amplifier because of the likely very high output impedance of the signal current source. The equivalent input offset voltage is

$$E_{os} = \pm V_{io} + I_b^- R_s$$

where R_s is the output resistance of the signal source.

The input error is more usefully expressed as an equivalent input offset current by dividing the equation by R_s. Thus

$$I_{os} = \frac{E_{os}}{R_s} = \pm \frac{V_{io}}{R_s} + I_b^-$$

A very large value for R_s makes the effect of amplifier input offset voltage V_{io} negligible, and accuracy at the lower levels of the input current range depends upon the amplifier bias current. In current logging applications in which the source impedance is not very high it may be necessary to balance out the initial value of V_{io}. The amplifier offset balance potentiometer can be used or an adjustable voltage bias may be applied to the non-inverting input terminal of the amplifier. In making the adjustment the inverting input should be connected to earth through a 10 kΩ resistor and the procedure for offset trimming outlined in the previous section for voltage logging should be carried out.

Figure 5.22 Testing the dynamic response of a log converter

The use of low offset drift f.e.t. input operational amplifiers in the circuit of Figure 5.21 makes it suitable for both voltage and current logging (e.g., amplifiers of the LF 155/255/355 series). Logging of very small input currents requires the use of more specialised low bias current amplifiers, e.g. Analog Devices AD 42J.

Closed loop stability and dynamic response
The dynamic response of a log amplifier is directly determined by the components used to achieve closed loop stability (C_1, C_2, R_e). Log amplifiers have a nonlinear feedback path; the small signal feedback fraction varies with the level of the input signal and this makes the

transient response and small signal bandwidth dependent upon the
signal level. The effect can be investigated in the circuit of Figure 5.21
by superimposing an input signal variation on top of a d.c. bias using
the test arrangement shown in Figure 5.22. The results shown in this
figure were obtained by adjusting the amplitude of the input squarewave and the value of the d.c. bias so as to make the output step
between various levels of its full range. Note that the response time
for a 1 V output step depends upon the input signal level, and that the
response time for increasing input signals is less than that for decreasing signals. The time taken for the output to slew through its full output range is dominated by the time taken to slew through the range
corresponding to the smallest decade of the input signal. Measured
response times for the circuit obtained by observing the output steps
with expanded time scale are as follows:

e_{in}	e_o	Input increasing	Input decreasing
1 V to 10 V	0 to − 1 V	lightly damped	
100 mV to 1 V	0 to + 1 V	22 s	92 s
10 mV to 100 mV	+1 V to + 2 V	230 s	1 ms
1 mV to 10 mV	+2 V to + 3 V	2.2 ms	9.6 ms

The test arrangement of Figure 5.22 can be used to measure the
small signal 3 dB bandwidth of the circuit for sinusoidal signals. Log
amplifiers accept only single polarity input signals; the sinusoidal
signal must be superimposed upon a steady d.c. bias. The small signal
bandwidth like the transient response time depends upon input
signal level. The output signal is of course non-sinusoidal.

A theoretical analysis of the factors governing the closed loop
stability of the log circuit of Figure 5.21 is somewhat more complex
than for the single transistor logging configuration previously
analysed (Section 5.4.1). Considerations governing the stability of
the amplifier A_2 are similar to those previously discussed but amplifier
A_1 is different—its feedback loop encloses both transistors T_1 and T_2.

Adequate phase margin for amplifier A_2 is ensured by choosing C_2
and R_E so that the frequency

$$f_{C_2} = \frac{1}{2\pi C_2 [R_E + r_{E_2}]}$$

occurs at least an octave before the loop gain becomes unity. The value

Figure 5.23 *Log of voltage or current ratio*

of r_{E_2} depends upon the value of the fixed reference current I_{ref} (see Section 5.4.1). The closed loop small signal 3 dB bandwidth of amplifier A_2 is equal to the frequency f_{C_2}.

Now consider the feedback loop round amplifier A_1. At frequencies much less than f_{C_2} amplifier A_2 holds V_{E_2} constant at a value determined by the reference current; changes in the output signal of amplifier A_1, attenuated by the divider $R_3 R_4$, are in effect applied directly to the emitter of T_1. The effective feedback resistance round amplifier A_1 is thus:

$$ r_{eff} = \frac{R_4 + R_3}{R_3} \, r_{E_1} $$

But at frequencies approaching f_{C_2} attenuation of amplifier A_2 gain causes an effective increase in the impedance of this path and introduces a phase lag into it. Critical damping of amplifier A_1 response for input signals at the upper end of the range requires a choice of the lead capacitor C_1 so that the break frequency $f_{C_1} = 1/(2\pi C_1 r_{eff})$ occurs well below f_{C_2}. In practice C_1 is often made less than that required for critical damping at the upper end of the input range. The lightly damped response at the higher input signal levels is accepted in order that the response time at the lower levels of the input signal should not be excessive. Response times can be decreased by using amplifiers with higher unity gain bandwidth, allowing the use of smaller frequency compensating capacitors, but a faster response can be obtained by using the log ratio circuit given in the next Section.

5.5.2 LOG OF VOLTAGE AND CURRENT RATIOS

The circuit discussed in the previous Section (Figure 5.21) gives an output which is proportional to the current ratio

$$ \frac{I_1}{I_{ref}} \frac{I_{o2}}{I_{o1}} $$

If an offset balance potentiometer is applied to amplifier A_2 and $I_{ref} = I_2$ is supplied by a second signal source the circuit performs a log ratio operation, but there are certain disadvantages to the arrangement. The input signal I_2 has a limited dynamic range because the collector base voltage of T_2 is not maintained at zero; also the zero crossing of the output which should occur when $I_1 = I_2$ depends upon the transistor I_o matching.

199

Log of ratio applications in which both inputs must be capable of the same wide dynamic range require the use of two separate logging amplifiers followed by a subtractor amplifier. Such a circuit is shown in Figure 5.23; the dynamic range obtainable with it depends upon the drift performance of the amplifiers used in the circuit. In a log of voltage ratio application it is the input offset voltage drift which is normally the most significant factor, but in a log of current ratio application it is the amplifier bias current which is the significant error source. Using the low offset voltage drift amplifiers of the LF 155/255/355 series the circuit is equally suitable for log of voltage or current ratio.

The performance equation for the circuit of Figure 5.23 is readily derived. Amplifier A_1 produces an output voltage

$$V_{o_1} = -2.3 \frac{kT}{q} \log_{10} \frac{I_1}{I_{o_1}}$$

and amplifier A_2 produces an output voltage

$$V_{o_2} = -2.3 \frac{kT}{q} \log_{10} \frac{I_2}{I_{o_2}}$$

The subtractor connected amplifier A_3 develops an output voltage e_o such that

$$e_o \frac{R_3}{R_3 + R_4} = V_{o_1} - V_{o_2}$$

Substitution gives

$$e_o = \frac{R_3 + R_4}{R_3} \frac{kT}{q} \log_{10} \frac{I_2}{I_1} \frac{I_{o_1}}{I_{o_2}}$$

In a log of voltage ratio application $I_1 = V_1/R_1$ and $I_2 = V_2/R_2$, but in a log of current ratio I_1 and I_2 are supplied directly to the inverting input terminals of A_1 and A_2 respectively.

Closed loop stability and dynamic response
Factors governing the closed loop stability of A_1 and A_2 have been discussed in Section 5.4.1. The time constants C_1, R_E of the compensating networks are chosen to ensure an adequate phase margin at the upper end of the input signal range. Values used determine the output response time for changes in input signals. At the higher input current

200

levels response time is determined primarily by $C R_E$ but at the lower current levels when the transistor $r_E \gg R_E$ response times become dependent upon input signal level. Note that the presence of R_E allows the use of smaller values for the lead capacitor C_1 and gives the circuit a faster response at the lower input signal levels than the previous circuit in Figure 5.21.

SETTING UP PROCEDURE

Adjust Subtractor. Lift the input signals to the subtractor and apply a
  low frequency sinewave input to both, adjust potentiometer P_1 for
  zero output from amplifier A_3.
Adjust for I_o mismatch and E_{os} of A_3.
Reconnect the inputs to A_3, make $V_1 = V_2 = 10$ V, adjust potentiometer P_2 for zero output of A_3.
E_{os} balance for $A_1 A_2$.
Earth V_2, make $V_1 = 10$ V, trim A_2 offset balance P_3 to make the
  output voltage from A_3 5 V. This sets the initial offset of amplifier
  A_1 to within 100 μV of zero. Repeat the procedure setting
  $V_2 = 10$ V and earthing V_1. Trim A_1 offset P_4 to get -5 V out.
This completes the initial adjustment procedure.

5.5.3 SINGLE AMPLIFIER LOG OF CURRENT RATIO CIRCUITS

In current logging applications where input currents are supplied by high impedance current sources a single operational amplifier circuit configuration can be used. Examples of such circuits are shown in Figure 5.24. In Figure 5.24(a) diode connection of the logging transistors restricts the lower level of the input signal currents to about 10^{-9} A; the upper level for I_2 is determined by the loading effect of this current on the potential divider. The lower level for I_2 is extended in Figure 5.24(b) by transdiode connection of T_2 but the extension is obtained at the expense of a restriction in the lower range for I_1 because of a loading which the base current of T_2 imposes upon I_1. Note that both circuits introduce a voltage drop into the measurement circuit.

5.5.4 ACCURACY OF LOG CONVERTERS

The accuracy of linear amplifier systems is often characterised by specifying output error as a percentage of full scale output; clearly percentage errors increase for input signals which give less than full

Figure 5.24 Single amplifier log of current ratio

scale output. In the case of log converters equal ratios of input signals give rise to equal incremental changes in output voltage and it is not usual to specify errors as a percentage of output—remember that zero output does not correspond to zero input. Errors in log amplifiers are normally expressed with respect to the input signal, and except at the extremes of the logging range the error with respect to input remains almost constant throughout the logging range. This property is particularly useful in processing wide range signals: consider a current logging amplifier designed to operate over say the five decades from

10^{-9} A to 10^{-4} A. The percentage measurement error involved in measuring the 10^{-9} A would be little different from that involved in measuring the current of 10^{-4} A. Imagine trying to measure 10^{-9} A with a linear current amplifier which gave a full scale output for an input signal of 10^{-4} A!

It is not difficult to relate respect to input percentage errors to the equivalent output voltage errors. If the respect to input error is x% low in a log converter the equivalent output voltage error is

$$- K \log_{10} \left[1 - \frac{x}{100} \right]$$

and if the respect to input error is x% high the output voltage error is

$$- K \log_{10} \left[1 + \frac{x}{100} \right]$$

The output is in error by differing amounts dependent upon whether the respect to input percentage error is low or high; the differences are not very significant. For example, if $K = 1$ and $x = 1$% substitution of values gives the output error as 4.32 mV low if x is 1% high, and 4.36 mV high if x is 1% low.

There are a variety of possible error contributions in a practical log converter: there are the tolerances and possible changes in the circuit parameters governing scaling factors and zero crossing, and there is amplifier offset drift. Even with these error sources removed by trimming and calibration there remains an error due to deviations from the ideal logging relationship (equation 5.7) exhibited by the transistors used as the logging element. This latter error is called the log conformity error; it is greatest at the extremes of the current logging range. Log conformity errors at the upper end of the logging range are caused by the bulk resistance effects of the transistor; they are dependent upon the current carrying capacity of the particular transistor type used and they normally set the upper limit of the range for accurate logging. In a practical log converter, log conformity error may typically be expected to contribute an error of around 0.5% with respect to input over the middle four decades of the logging range of the transistor.

Assuming that proper attention has been paid to scale factor stability and reference current stability and that thermal gradients between the logging and temperature compensating transistors are avoided, the lower level of the logging range in a practical converter is

Figure 5.25 Temperature compensated antilog converter

normally determined by amplifier offset drift. If a very low bias current operational amplifier is used for current logging the lower level for accurate operation is around 100 pA because of log conformity errors.

5.5.5 TEMPERATURE COMPENSATED ANTILOG CONVERTER

Circuitry of the type used in the temperature compensated log converter of Figure 5.21 can be rearranged to give a circuit which will perform the antilog conversion. Such a circuit is shown in Figure 5.25.

204

The input signal to the circuit, attenuated by the resistive divider $R_3 R_4$, provides the emitter base differential voltage between transistors T_1 and T_2 and

$$e_{in} \frac{R_3}{R_3 + R_4} = V_{E_2} - V_{E_1}$$

Negative feedback around amplifier A_1 forces V_{E_1} to take on that value which will cause the current $I_1 = I_{ref}$ to flow as a collector current in transistor T_1. If I_1 is held constant V_{E_1} is constant and V_{E_2} varies directly with the input signal. The voltage V_{E_2} determines the collector current which flows in transistor T_2. Negative feedback around amplifier A_2 forces this current to flow through resistor R_2 and amplifier A_2 develops an output voltage

$$e_o = I_2 R_2$$

Substitution of V_E values from equation 5.7 gives

$$e_{in} \frac{R_3}{R_3 + R_4} = \frac{kT}{q} \log_e \frac{I_i \, I_{o2}}{I_2 \, I_{o1}} \tag{5.25}$$

where $I_1 = I_{ref} = V_s/R_1$ and $I_2 = e_o/R_2$.

Antilogging and rearranging equation 5.25 give

$$e_o = R_2 \, I_{ref} \frac{I_{o2}}{I_{o1}} \exp\left(-\frac{q}{kT} \frac{R_3}{R_3 + R_4} e_{in}\right) \tag{5.26}$$

The equation can be interpreted in terms of any base other than the exponential by the use of the mathematical identity

$$a^x = b^{x \, \log_b a}$$

Expressed in terms of the normal base 10 the circuit performance equation becomes

$$e_o = R_2 \, I_{ref} \frac{I_{o2}}{I_{o1}} 10^{-e_{in}/K} \tag{5.27}$$

where $K = 2.3 \, \dfrac{kT}{q} \dfrac{(R_3 + R_4)}{R_3}$

Balance A_2 offset. Set e_1 sufficiently positive to cut off transistor T_2 completely (say e_1 = + 5V) and adjust the offset trim on A_2 for zero output.

Trim multiplying constant. The multiplying constant $M = R_2 I_{ref}$ I_{o_2}/I_{o_1} may be set at any convenient value which allows output signals within the capability of the amplifier A_2. Set e_1 at zero and adjust I_{ref} (by adjusting the value of resistor R_1) to make the output voltage of A_2 exactly equal to the value of the desired multiplier factor M.

Trim value of base, apply an input signal of -1 V and trim R_4 to make the output of amplifier A_2 exactly bM V (10 M for base 10). Response curves for the antilog converter are shown in Figure 5.25 plotted in terms of log e_{out}/e_{in}; note that e_{in} can be positive or negative but the output is always single polarity. If negative output signals are desired pnp transistors should be used as the logging elements.

5.6 Log antilog circuits for computation

Operational amplifier log and antilog converters can be combined in order to generate a variety of both linear and nonlinear functions. The circuits are interconnected in such a way that they perform operations normally involved in logarithmic computations. Examples of such computations are

$$\text{antilog } [n \log x] = x^n \tag{5.28}$$

$$\text{antilog } [\log x + \log y - \log z] = \frac{xy}{z} \tag{5.29}$$

5.6.1 LOG ANTILOG MULTIPLIER DIVIDER

A circuit for a log antilog multiplier divider is given in Figure 5.26, which embodies the computations involved in equation 5.29. The performance equation for the circuit is readily derived; the sum of the emitter voltages of the logging transistors around the loop starting from the base of the transistor T_1 and ending on the base of the transistor T_3 must sum to zero, thus:

$$V_{E_1} + V_{E_2} - V_{E_4} - V_{E_3} = 0$$

The emitter voltages of the transistors are constrained to take on values governed by their collector currents and determined by the basic logging equation (equation 5.7). Substitution of values gives

$$\frac{kT}{q} \cdot \log_e \frac{I_1}{I_{o_1}} + \frac{kT}{q} \cdot \log_e \frac{I_2}{I_{o_2}} - \frac{kT}{q} \cdot \log_e \frac{I_4}{I_{o_4}} - \frac{kT}{q} \cdot \log_e \frac{I_3}{I_{o_3}} = 0$$

Figure 5.26 Log antilog multiplier/divider

207

Assuming that the transistors are all at the same temperature allows cancellation of the temperature dependent scaling factor and gives

$$I_4 = \frac{I_{o4}}{I_{o1}} \frac{I_{o3}}{I_{o2}} \frac{I_1 I_2}{I_3}$$

Amplifier A_4 forces the current I_4 to flow through the resistor R_4 and in doing so generates an output voltage

$$e_o = \left[R_4 \frac{I_{o4}}{I_{o1}} \frac{I_{o3}}{I_{o2}} \right] \frac{I_1 I_2}{I_3} \qquad (5.30)$$

Input current can be supplied directly to the inverting input terminal of amplifiers A_1, A_2 and A_3. Input voltages are applied through appropriate input resistors making

$$I_1 = \frac{V_1}{R_1}, \ I_2 = \frac{V_2}{R_2}, \ I_3 = \frac{V_3}{R_3}$$

Note that the combination of log and antilog circuits with all transistors at the same temperature provides temperature compensation of the scaling factor kT/q without the necessity for a temperature compensating resistor.

The scaling factor for the circuit with voltage inputs is

$$S = \frac{R_4 R_3}{R_1 R_2} \frac{I_{o4}}{I_{o1}} \frac{I_{o3}}{I_{o2}}$$

By making R_4 variable the effect of I_o mismatch can be balanced and the scaling factor can be set to unity. The adjustment is made by setting $V_1 = V_2 = V_3 = 10$ V and altering R_4 for 10 V output. Using the LF 356 amplifier types as suggested, the maximum input offsets without an offset balance are 10 mV in $V_1 V_2$ and V_3 and an output offset of 10 mV due to amplifier A_4. In multiplier operation with V_3 set at 10 V the maximum error if amplifier offsets are not balanced can be expected to be around 30 mV at the output (out of 10 V that is 0.3% of full scale). Errors are greater in divider operation at small values of V_3. If the greatest possible accuracy is required amplifier offsets should be balanced; the reader is left to work out an offset balance procedure for himself.

5.6.2 MULTI-FUNCTION CIRCUIT

A log ratio circuit combined with an antilog converter can be used to perform computations embodied in a performance equation of the form

$$e_o = e_3 \left[\frac{e_1}{e_2} \right]^n$$

n is an arbitrary power which can be set by choice of resistor values. The circuit principles underlying this type of computation are illustrated by the arrangements shown in Figure 5.27; it represents a combination of the temperature compensated log converter and the temperature compensated antilog converter previously discussed (Figures 5.21 and 5.25).

The performance equation for the circuit in Figure 5.27 can be derived by summing emitter voltages starting at the base of T_1 and ending at the base of T_4, thus:

$$n \left[V_{E_1} - V_{E_2} \right] = V_{E_4} - V_{E_3}$$

where

$$n = \frac{R_7}{R_6} \frac{R_5 + R_6 + (1 - x) R}{R_7 + R_8 + xR}$$

Substitution of V_E values from the basic logging equation (equation 5.7) and dividing out the scaling factor kT/q lead directly to the performance equation

$$e_o = \left[\frac{R_4}{R_3} \frac{I_{o4}}{I_{o3}} \left(\frac{R_2}{R_1} \frac{I_{o2}}{I_{o1}} \right)^n \right] e_3 \left[\frac{e_1}{e_2} \right]^n \qquad (5.31)$$

Mismatch in I_{o1} and I_{o2} can be balanced by setting $e_1 = e_2 = 10$ V say and trimming R_2 to make the output of A_1 zero. The scaling factor can then be set to unity and mismatch in I_{o4} and I_{o3} cancelled by trimming R_3 so that $e_o = 10$ V when $e_1 = e_2 = e_3 = 10$ V. A required power n can be set up by applying known input signals and adjusting the power setting control to give the calculated output.

For example: $n = 2$, set $e_3 = e_2 = 10$ V, $e_1 = 9$ V and adjust for $e_o = 8.1$ V. The performance equation is $e_o = 1/10 \, (e_1)^2$; $n = 3$, set $e_3 = e_2 = 10$ V, $e_1 = 9$ V and adjust for $e_o = 7.29$ V. The performance equation is $e_o = 1/100 \, (e_1)^3$.

209

$$e_0 = e_3 \left[\frac{e_1}{e_2} \right]^n$$

Figure 5.27 Multifunction log antilog circuit

If maximum processing accuracy is sought amplifier offsets should be balanced out using the techniques previously described.

Ready built circuit modules are available which are functionally equivalent to the circuit of Figure 5.27, e.g. Analog Devices model 433, Burr Brown models 4301, 4302. The multifunction operation is useful in a wide range of processing and computation applications, for example, linearising a function by raising a voltage or a voltage ratio to an arbitrary power, performing true r.m.s., vector sums, sine, cosine, or arc tangent conversion functions[6,7]. The reader with an immediate requirement for these operations is well advised to consider the purchase of a ready built module rather than undertaking a design from scratch. If one starts costing design time and effort the modules begin to seem distinctly more economical.

5.6.3 A SIMPLE LOG ANTILOG SQUARE ROOT CONVERTER

Figure 5.28 shows a circuit for the generation of square roots which is less complex than the multifunction circuit of the previous section. The performance equation for the circuit can be derived by summing emitter voltages starting at the base of T_1 and ending at the base of T_4, thus:

$$V_{E_1} + V_{E_2} - V_{E_3} - V_{E_4} = 0$$

Figure 5.28 Simple log antilog square root converter

The basic transistor log relationship

$$V_E = -\frac{kT}{q} \log_e \frac{I_c}{I_o}$$

is substituted and transistors are assumed to be at the same temperature allowing cancellation of the temperature dependent kT/q factor, giving

$$\frac{I_{c_3} I_{c_4}}{I_{c_1} I_{c_2}} \frac{I_{o_1} I_{o_2}}{I_{o_3} I_{o_4}} = 1$$

If transistor base currents are assumed negligible collector currents are fixed by the circuit at the following values:

$$I_{c_2} = \frac{e_{in}}{R_2}, I_{c_1} = \frac{V_s}{R_1} \text{ (a reference current)},$$

$$I_{c_3} = I_{c_4} = \frac{e_o}{R_3}$$

Substituting these values and rearranging give the circuit performance equation as

$$e_o = \sqrt{\left(\frac{V_s R_3^2}{R_1 R_2} \frac{I_{o_3} I_{o_4}}{I_{o_1} I_{o_2}} e_{in} \right)} \tag{5.32}$$

The scaling factor $(V_s R_3 / R_1 R_2) (I_{o_3} I_{o_4} / I_{o_1} I_{o_2})$ is conveniently set to 10 by adjustment of R_2. This balances mismatch in the I_o terms and the circuit gives 10 V out for 10 V in.

SETTING UP PROCEDURE
Set $e_{in} = +10$ V, adjust R_2 to make $e_o = 10$ V; this establishes the
  performance equation, $e_o = \sqrt{(10 e_{in})}$.
Adjustment of the offset voltage of A_1 to within approx. 10 μV can
  be obtained by setting $e_{in} = 0$ and adjusting P_1 for $e_{out} = 10$ mV.
  The effect of A_2 output offset errors (typically of the order of
  3 mV) can be neglected.

5.6.4 A LOG ANTILOG TRUE R.M.S. TO D.C. CONVERTER

The circuit configuration shown in Figure 5.29 gives a d.c. output signal proportional to the true r.m.s. value of an input signal, which may have a complex alternating waveform or an alternating wave superimposed

212

upon a d.c. level. It consists of a precise rectifier (see Section 8.10) which is used to provide unidirectional signals to a following log anti-log computing circuit.

As in the other log antilog circuits the performance equation can be derived by summing emitter voltages and using the basic transistor log relationship. Thus, starting at the base of T_1 and ending at the base of T_4 we have

$$V_{E_1} + V_{E_2} - V_{E_3} - V_{E_4} = 0$$

Substituting for V_E values and cancelling out the temperature dependent scaling factor after antilogging give

$$\frac{I_{c_1} I_{c_2}}{I_{c_3} I_{c_4}} \frac{I_{o_3} I_{o_4}}{I_{o_1} I_{o_2}} = 1 \qquad (5.33)$$

Neglecting transistor base currents

$$I_{c_1} = I_{c_2} = \frac{e_{in}}{R_1}$$

Amplifier A_3 performs a running averaging of the current I_{c_3} and provided the averaging time constant $C R_2$ is considerably longer than the period of any alternating input signal the output of A_3 is a steady voltage proportional to the average value of I_{c_3}. We write

$$e_o = \overline{I_{c_3}} R_2$$

where $\overline{I_{c_3}}$ represents the average value of the current I_{c_3}.

Amplifier A_4 forces the relationship

$$I_{c_4} = \frac{e_o}{R_3} = \overline{I_3} \frac{R_2}{R_3}$$

Substitution of current values in equation 5.33 gives

$$\overline{I_{c_3}}\, \overline{I_{c_3}} = \frac{R_3}{R_2 R_1^2} \frac{I_{o_3} I_{o_4}}{I_{o_1} I_{o_2}} e_{in}^2$$

or

$$[\overline{I_{c_3}}]^2 = \frac{R_3}{R_2 R_1^2} \frac{I_{o_3} I_{o_4}}{I_{o_1} I_{o_2}} \overline{e_{in}^2}$$

Thus:

$$\overline{I_{c_3}}\, R_2 = e_o = \sqrt{\left(\frac{R_3 R_2}{R_1^2} \frac{I_{o_3} I_{o_4}}{I_{o_1} I_{o_2}} \overline{e_{in}^2} \right)} \qquad (5.34)$$

Op Amps LF 356 / LF412   CA3086

(Sub)

Functional diagram of the CA 3086

Figure 5.29 Log antilog r.m.s. to d.c. converter

Mismatch in the I_o terms can be balanced and the scaling factor set to unity by adjustment of the resistor R_3.

PRACTICAL POINTS

In the circuit shown the four transistors are conveniently provided by an R.C.A. type CA3086 transistor array. One of the transistors in the array is unused and the substrate pin 13 is connected to the -15 V supply rail. Four separate transistors or two dual transistors could be used as an alternative to the transistor array provided that they are maintained at the same temperature. Operational amplifiers of the LF 156/256/356 series were used in the practical circuit (an f.e.t. input type). This amplifier type allows the use of an internal offset balancing potentiometer without degrading amplifier drift performance to any extent. If different amplifier types are used they may require the use of an alternative offset balancing technique.

SETTING UP PROCEDURE

Set $e_{in} = 0$ and adjust potentiometer P_1 to make e_o approximately 10 mV.

Make $e_{in} = +$ 10 V d.c.; adjust potentiometer P_2 to make the output read $+$ 10 V d.c.

A practical circuit constructed in breadboard form was found to give errors no greater than 1% of full scale (full scale input 10 V peak). With a sinusoidal input signal of peak value 10 V the output was 5% down at a frequency of 90 kHz. Accuracy at the lower level of input signals could no doubt be improved by separately balancing the offsets of all amplifiers.

5.7 A variable transconductance four quadrant multiplier

Log antilog multiplier circuits allow only single polarity signals. Multipliers designed for four quadrant operation make use of alternative methods to obtain the multiplier operation. A commonly used technique is the so-called variable transconductance princple. Four quadrant multipliers based upon this principle are now freely available in integrated circuit form and are finding increasing use in analogue signal processing applications[8]. Four quadrant variable transconductance multipliers do not provide the accuracy at low signal levels that log antilog multipliers give, but they allow operation with alternating signals and provide greater speed and bandwidth than log multipliers.

216

Figure 5.30 Variable transconductance four quadrant multiplier

T_1, T_2, T_3, T_4, T_5
Transistor array type CA3086

$$e_0 = \left[\frac{R_1 R_S}{R_x R_y R_S}\right] V_x V_y$$

$R_x = R'_x + P_1$
$R_y = R'_y + P_2$

As an alternative to buying a ready built transconductance multiplier it is possible to build a useful general purpose four quadrant multiplier out of an operational amplifier, a five transistor array and a few resistors[9]. A circuit is shown in Figure 5.30; it is based upon an offset linearised two quadrant multiplier cell[6]. Its action can be understood in terms of the basic model shown in Figures 5.31(a) and (b).

Current relationships which must hold for the circuit in 5.31(a) can be derived by summing emitter voltages. Starting at the emitter of T_1 and ending at the emitter of T_4 the emitter base voltages must sum to zero, thus:

$$V_{E_1} - V_{E_2} + V_{E_3} - V_{E_4} = 0$$

Neglecting the base currents drawn by the assumed matched transistors and assuming the collector base voltage of T_3 and T_2 are held near zero we make use of the basic log relationship between collector current and emitter voltage (equation 5.7) yielding after cancelling out the temperature dependent scaling factor and antilogging the relationship between the currents in the multiplier cell as

$$I_1 I_3 = I_2 I_4 \qquad (5.35)$$

In Figure 5.30 currents are supplied by input signal voltages V_x and V_y as shown by the simplified model in Figure 5.31(b). The tail reference current of the pair $T_2 T_3$ is made to vary with V_x.
Note that $I_1 = I_R + I_y$, $I_4 = I_R - I_y$ and $I_3 = 2[I_R + I_x] - I_2$. Substituting values in equation 5.34 and rearranging give

$$I_2 = I_R + I_x + I_y + \frac{I_x I_y}{I_R}$$

Also we have

$$I_2 = I_R + I_x + I_y + I_o$$

The output current which is converted to an output voltage by the operational amplifier is thus:

$$I_o = \frac{I_x I_y}{I_R}$$

Now $\qquad I_x = \dfrac{V_x}{R_x}, \; I_y = \dfrac{V_y}{R_y}, \; I_R = \dfrac{V_s}{R_1}$

Figure 5.31 Models of transconductance multiplier cell (a) Basic linearised transconductance cell (b) Model of offset transconductance cell

and the output voltage is determined by the relationship

$$I_o R_s = e_o = \left[\frac{R_1 R_s}{R_x R_y V_s} \right] V_x V_y$$

Component values are selected to give a scaling factor of 1/10. Input
signals $V_x V_y$ may be of either polarity; with $V_x = V_y = 10$ V the
multiplier gives its full scale output of 10 V. The bandwidth and output slew rate of the multiplier are determined by the dynamic response
characteristics of the operational amplifier which is used as the output
current to voltage converter.

SETTING UP PROCEDURE

x and y offsets. Earth V_Y making $V_y = 0$. Apply a 20 V peak to peak
100 Hz input to V_x and adjust P_1 for minimum a.c. output.

Make $V_x = 0$, apply a 20 V peak to peak 100 Hz input to V_y and adjust
P_2 for minimum a.c. output.

Output offset. Make $V_x = V_y = 0$, adjust P_3 for zero output.

Scale Factor. Make $V_x = V_y = 10$ V (d.c.), adjust P_4 for 10 V output.

219

REFERENCES

1. PATTERSON, W. L., Rev. Sci. Instrum., **34**, 1311 (1963)
2. ASHCROFT, K., Unpublished Project Report, Physics Dept., Liverpool Polytechnic
3. BARTOS, D. J., 'Using Logarithmic Current Elements', Product Note, Keithley Instruments Inc.
4. EBERS, J. J., and MOLL, J. L., I.R.E. Proceedings, **42**, 1761 (1964)
5. BOLASE, W., and DAVID, E., 'Design of temperature compensated log circuits employing transistors and operational amplifiers', Application Report, Analog Devices
6. Non Linear Circuits Handbook, Analog Devices Engineering Staff (1974)
7. TEEPLE, C. R. 'Analog Shaping', Burr Brown Research Corporation, Application Note AN.70
8. CLAYTON, G. B., Linear Integrated Circuit Applications, Macmillan (1975). Linear I.C. Applications Handbook, Tab Books (1977)
9. GILBERT, B., 'Analogue Multiplier', New Electronics, **10**, 38

Exercises

5.1 In Figure 5.3 the resistor R has a value 100 kΩ, reference voltage supplies of $\pm$ 10 V are available. Find the values of the resistors in the input network required to generate the function shown in Figure 5.31. Sketch the circuit. (Hint: use an additional amplifier to invert input signal polarity.)

5.2 The nonlinear amplifier shown in Figure 5.5 uses the following component values: R_1 = 10 kΩ, R_2 = 10 kΩ, R_3 = 10 kΩ, R_4 = 150 kΩ, R_5 = 5 kΩ, R_6 = 25 kΩ, V_S = + 15 V. Deduce the values of the breakpoint voltages and the slopes of the straight line segments in the amplifier response. Sketch the relationship between amplifier input and output voltage. Neglect the voltage drop across a saturated transistor.

5.3 Calculate the output voltage of the simple log amplifier shown in Figure 5.9 if the input current is 10^{-9} A and the ambient temperature is 27°C. Assume that $\alpha_F / I_{ES} = 10^{-12}$ A at 27°C and that the operational amplifier behaves ideally. What does the output voltage become if the temperature rises by 10°C?

5.4 In the circuit shown in Figure 5.14, C_1 = 100 pF, C_2 = 100 pF, R_1 = 50 kΩ, R_E = 10 kΩ. Sketch the Bode plots for $1/\beta$ for input currents of 10^{-4} A, 10^{-6} A and 10^{-8} A.

5.5 The basic transdiode log configuration, Figure 5.9, is used with an input resistor $R = 100$ kΩ. The operational amplifier has an input offset $V_{io} = 1$ mV and a bias current $I_b^- = 20$ nA. If no offset balancing is employed what is the smallest input signal voltage that can be logged with an error no greater than 2%? Assume that the temperature remains constant.

5.6 What values for resistors R_4 and R_2 are required in Figure 5.21 in order that the input voltage change from $+ 10$ mV to $+ 10$ V should cause the output voltage to change from 0 to $- 10$ V? Assume that the transistors are perfectly matched and that all other circuit parameters are as shown.

5.7 In the circuit of Figure 5.21 resistor R_2 is made 1.5 MΩ; other circuit parameters are as shown. There is a mismatch in the logging transistors such that $I_{o_1} = 0.7\,I_{o_2}$. Find the value of the input signal for which the output is zero. Neglect amplifier offsets.

Figure 5.32 Circuits for exercise 5.10

221

5.8 The circuit of Figure 5.21 is used as a log of current converter, the non-inverting input terminal of A_1 is connected to earth and an input current is supplied directly to this inverting input terminal. If A_1 is an f.e.t. input operational amplifier with bias current I_B = 50 pA and input offset voltage V_{io} = 2 mV, what is the smallest input current which can be converted with no more than 2% error if it is supplied by (a) a true current source; (b) a current source of internal resistance 1 MΩ?

5.9 In the circuit of Figure 5.25, I_{ref} = 0.1 mA, R_3 = 1 kΩ. What values are required for resistors R_4 and R_2 in order that the circuit should generate an output signal $e_o = 2^{-e\,in}$? Assume that the temperature is 300° K and that transistors are perfectly matched. (Boltzmann constant k = 1.38 $\times$ 10^{-23} J/°K, electronic charge q = 1.6 $\times$ 10^{-19} C).

5.10 Assuming ideal operational action and that transistors are matched and follow the log relationship (equation 5.7) find the relationships between output and input signals for the circuits shown in Figure 5.32. Discuss problems likely to be encountered in practical realisations of the circuits.

CHAPTER SIX

INTEGRATORS AND DIFFERENTIATORS

At the heart of most operational amplifier applications lies the ability of the amplifier to force precise mathematical relationships between input and output signals. This chapter is devoted to the theoretical and practical considerations involved in using operational amplifiers to perform the mathematical operations of integration and differentiation.

Much has been written about operational integrators used in analogue computing systems, but with the advent of low cost operational amplifiers operational integrators are now used in a wide variety of instrumentation systems. Operational integrators are used to perform timing functions, to measure charge, to generate linear ramps and triangular waves and in many other applications. In this chapter we consider integrator action and the factors which must be taken into account when connecting and using practical integrators. Many examples of integrator applications will be found in subsequent chapters of the book.

6.1 The basic integrator

An understanding of the action of a practical integrator circuit is helped by starting with a thorough grasp of the principles underlying the action of the ideal circuit, mentioned briefly in Section 1.3.6. Errors in practical circuits may then be understood in terms of departures from ideal behaviour. There are two main principles underlying the action of an ideal integrator: the first principle concerns the ideal amplifier summing point restraints. All current from signal sources arriving at the inverting input terminal of an ideal amplifier must exit through the feedback path. The output voltage of the amplifier takes on just that value needed to keep the inverting input terminal at the same potential as the non-inverting input terminal and prevents accumulation of charge at the inverting input terminal. This statement holds whatever kind of circuit element is placed in the feedback path.

The second principle concerns the relationship between the voltage across a capacitor and the charge upon its plates, namely, $V_c = q/C$.

For charge to exist on the plates of a capacitor charge must pass in some way on to its plates; charge flow represents a current and we may write

$$i = \frac{dq}{dt}$$

thus
$$V_c = \frac{\int i\, dt}{C}$$

The voltage across a capacitor is proportional to the integral with respect to time of the capacitor charging current.

The two principles applied to the basic integrator circuit of Figure 6.1 lead directly to the ideal performance equation. Thus, an input

Figure 6.1 The basic integrator

current $I_{in} = e_{in}/R$ arrives at the amplifier summing point; the output voltage causes this current to flow as a charging current into the feedback capacitor and at the same time takes on just that value required to maintain a virtual earth at the amplifier summing point.

$$e_o = -V_c = -\frac{\int I_{in}\, dt}{C}$$

or
$$e_o = -\frac{1}{CR}\int e_{in}\, dt \qquad (6.1)$$

The input impedance of the integrator circuit is equal to the resistance R; output impedance is low because of the negative feedback which is inherent in the circuit. CR is called the characteristic time of the integrator. It is sometimes useful to think of $1/CR$ as the integrator 'gain' in terms of V/s output for each volt of input signal. For example,

224

if C = 1 μF and R = 1 MΩ an input signal of + 1 V would cause a current of 1 μA to flow towards the amplifier summing point. To maintain this charging current into the capacitor C would require the output to decrease linearly with time at a rate of − 1 V/s. If during the integration process the input signal were switched to zero the input current would become zero and the output voltage of the ideal integrator would remain constant (hold) at any value it happened to have reached. If the input polarity were reversed (V_{in} = − 1 V) this would require the output voltage to increase linearly at a rate of 1 V/s to maintain the 1 μA current flow away from the amplifier summing point.

6.2 Integrator run, set and hold modes

In a practical integrator circuit it is necessary to provide some means of setting a desired initial value of the integrator output voltage at the start of the integration time. In some systems, it is also desirable to be

Figure 6.2 Integrator, run, set, hold modes

able to stop the integrator at any time and for the integrator output then to remain constant at the value it has reached at that time. The principles underlying the switching of an integrator between its various modes of operation are shown in Figure 6.2; manual switching, relay switching or some form of solid state switching can be used.

The switches put in the 'set' positon allow the initial value of the integrator output to be set at any desired value within the output capability of the amplifier.

$$e_{o(t=0)} = \frac{-R_2}{R_1} E_{ref}$$

The integrator output does not immediately take on this value when switched to the set mode; it approaches the value exponentially in accordance with the relationship

$$e_o = e_{o(t=0)} + (e_o' - e_{o(t=0)}) \exp \frac{-t}{R_2 C}$$

where e_o' is the value of e_o at the instant of switching to the set mode. Note the period of the set mode must be long enough for the exponential to decay.

When switched to the 'run' mode the circuit integrates the input voltage and

$$e_o = e_{o(t=0)} - \frac{1}{CR} \int_0^t e_{in} \, dt$$

If the integrator is switched to the 'hold' mode integration is stopped and ideally the output of the integrator then remains constant at any value it may have reached. In practice, in both the 'run' and 'hold' modes, drift causes an integrator error.

6.3 Integrator errors

The deviations from ideal behaviour that are exhibited by a practical integrator circuit are conveniently treated as errors. A firm understanding of the sources of error enables the designer to choose an amplifier and associated circuit to minimise errors.

6.3.1 OFFSET AND DRIFT ERRORS IN PRACTICAL INTEGRATORS

The greatest source of error in practical integrators is normally that due to offset and drift of the amplifier. Even with zero applied input signal amplifier input offset voltage and bias current cause a continuous charging of the feedback capacitor, and the output voltage of a

$$e_0 = -\frac{1}{CR}\int e_{in}\,dt \underbrace{\pm \frac{1}{CR}\int V_{io}\,dt + \frac{1}{C}\int I_B\,dt \pm V_{io}}_{\text{error due to } V_{io} \text{ and } I_B}$$

Figure 6.3 Equivalent circuit used for estimating error due to input offset voltage and bias current

practical free running integrator will change continuously until the amplifier output eventually drifts into either positive or negative saturation.

Integrator output voltage drift with time can be adjusted to zero under a particular set of conditions and at a particular time by cancelling the effects of the amplifier offsets with a suitable balance control (see Section 9.6). However, amplifier offsets are temperature dependent, supply voltage dependent, and they show a long term time dependence. This means that a zero output drift condition established with a balance control is not maintained and a free running integrator therefore always ends up in one of its saturated states.

Integrator error due to amplifier input offset voltage and bias current is readily deduced from the equivalent circuit shown in Figure 6.3. For the moment we assume that the open loop gain and open loop input impedance of the amplifier are infinite. We may write

$$I_f = I_{in} - I_b$$

where $I_{in} = (e_{in} \pm V_{os})/R$.

Now
$$V_c = \pm V_{io} - e_0 = \frac{\int I_f\,dt}{C}$$

Thus
$$e_0 = -\frac{1}{CR}\underbrace{\int e_{in}\,dt}_{\substack{\text{ideal perform-}\\\text{ance}\\\text{equation}}} \pm \underbrace{\frac{1}{CR}\int V_{io}\,dt + \frac{1}{C}\int I_b.dt \pm V_{io}}_{\text{error due to offset and bias current}} \quad (6.2)$$

The percentage error after a particular integration time may be written

$$\pm \frac{V_{io} + I_b R}{\overline{e_{in}}} \times 100\% \tag{6.3}$$

where $\overline{e_{in}}$ is the time average of the input signal over the integration period.

In some applications it is more useful to refer integrator offset errors to the output. Offsets cause the output of an integrator to have an error in the form of an output drift rate (a ramp) determined by the relationship

$$\frac{de_o}{dt}_{\text{(due to offset)}} = \pm \frac{V_{io}}{CR} + \frac{I_B}{C} \tag{6.4}$$

It should be noted that V_{io} and I_b are initial values plus accumulative drift. If initial values are balanced this leaves temperature drift values only.

The error component due to amplifier bias current can be reduced by connecting a resistor equal in value to that of the integrating resistor between the non-inverting input terminal of the amplifier and earth. With this resistor in circuit, values of the amplifier input difference current I_{io} should be substituted for I_b in the drift error equations.

An integrator when switched to the hold mode normally has R open circuit so that according to the error equations it is then bias current alone which accounts for drift in the hold mode. However, as will be shown later, finite open loop gain and finite amplifier input impedance give rise to an additional source of error in the hold mode.

Examination of equation 6.4 suggests that for a particular integrator characteristic time drift error is minimised by choice of a capacitor value large enough to make the bias current contribution to the drift negligibly small compared with the input offset value contribution. There are, however, other practical considerations which set limits to the capacitor value. A large value of C requires a correspondingly small value of R for a particular CR value. The input impedance of the integrator is set by the value of the resistor R; the minimum value which can be used is dependent upon signal loading error. Capacitor leakage represents an additional source of integrator drift. Large value high performance capacitors are expensive, and if the performance capabilities of an operational amplifier as an integrator are to be realised a capacitor whose dielectric leakage current is less than the amplifier bias current must be used. Long term low drift integrators favour the use of low current f.e.t. input amplifiers; they allow the bias current contribution to integrator drift to be made negligible without the use of excessively large capacitor values. When such amplifiers are used it is important to prevent leakage paths to the amplifier summing point degrading performance (see Section 9.4.3).

6.3.2 INTEGRATOR ERRORS DUE TO FINITE OPEN LOOP GAIN, FINITE INPUT IMPEDANCE AND FINITE BANDWIDTH

The ideal performance equation for an operational integrator (equation 6.1) was obtained from the assumption that the operational amplifier used in the circuit had infinite open loop gain and bandwidth. In all operational amplifier negative feedback circuits the extent to which a practical circuit performance departs from the ideal is governed by the loop gain βA_{OL} (see Section 2.2). The larger the loop gain the closer

Figure 6.4 Effect of finite open loop gain and input impedance

229

the practical circuit conforms to the ideal. In a practical integrator as we shall show, finite open loop gain causes integrator performance errors for very low frequency input signals, and finite bandwidth causes errors for high frequency input signals.

Integrator errors due to inadequate loop gain are discussed in terms of the circuit and Bode plots shown in Figure 6.4. The operational amplifier used in the circuit is assumed to have a finite differential input resistance R_d, finite open loop gain and a first order frequency response described by the relationship

$$A_{OL(jf)} = \frac{A_o}{1 + j\dfrac{f}{f_c}} \qquad \text{(see Section 2.4.1)}$$

The closed loop performance equation for the circuit expressed in the form: actual performance equation = ideal performance equation × gain error factor (see Section 2.3) is

$$\frac{e_{o(jf)}}{e_{I(jf)}} = -\frac{1}{j\,2\pi f\,CR}\left[\frac{1}{1 + \dfrac{1}{\beta A_{OL(jf)}}}\right] \qquad (6.5)$$

Provided that the magnitude of the loop gain is large the integrator performance closely approximates the ideal. The feedback fraction for the circuit is

$$\beta = \frac{R'}{R' + \dfrac{1}{j\,2\pi fC}}$$

where $R' = R\,R_d/(R + R_d)$ and $1/\beta = 1 + 1/(j\,2\pi fCR')$.

The intersection of the Bode plots for $1/\beta$ with the open loop response shows that the magnitude of the loop gain becomes unity at the frequencies $1/(2\pi A_o CR')$ and f_1. The integrator must thus be expected to perform near ideally at frequencies such that:

$$\frac{1}{2\pi A_o CR'} \ll f \ll f_1$$

Figure 6.5 Time lag in integrator step response due to finite open loop bandwidth

Errors at high frequencies due to finite open loop bandwidth
At frequencies approaching and exceeding the amplifier unity gain
frequency f_1

$$\beta \to 1, A_{OL(jf)} \to -j\,\frac{f_1}{f}$$

and equation 6.5 approximates to

$$\frac{e_{o(jf)}}{e_{1(jf)}} = -\frac{1}{j\,2\pi f CR}\left[\frac{1}{1+j\,\dfrac{f}{f_1}}\right] \qquad (6.6)$$

Equation 6.6 represents the equation for an ideal integrator cascaded
with a first order low pass function with break frequency equal to the
open loop unity gain frequency of the operational amplifier. It is the
attenuation and phase shift produced by this first order function which
represent the errors in the steady state sinusoidal response of the
integrator at frequencies approaching f_1. Also associated with this low
pass function the output of a practical integrator exhibits a time lag
in response to an input step signal as shown in Figure 6.5. The time
lag is inversely proportional to the open loop unity gain frequency
f_1 of the operational amplifier.

Errors at low frequencies due to finite open loop gain
The gain of an ideal integrator circuit continues to increase as the
signal frequency is decreased, but clearly in a practical integrator

circuit the gain cannot be greater than the open loop gain of the amplifier A_o ($A_o R_d/(R + R_d)$ if we allow for finite input resistance). At frequencies less than f_1/A_o, $A_{OL(Jf)} \rightarrow A_o$ and

$$\frac{1}{\beta A_{OL(Jf)}} = \frac{1}{A_o} + \frac{1}{j\,2\pi f A_o CR'} \cong \frac{1}{j\,2\pi f A_o CR'}$$

Equation 6.5 approximates to

$$\frac{e_{o(Jf)}}{e_{1(Jf)}} = -\frac{1}{j\,2\pi f\,CR}\left[\frac{\dfrac{1}{1 + \dfrac{1}{j\,2\pi f A_o CR'}}}{}\right]$$

$$= -\frac{A_o\,\dfrac{R'}{R}}{1 + j\,2\pi f A_o CR'} \tag{6.7}$$

for $f < f_c$.

An insight is gained into the operation of integrators at low frequencies by realising that equation 6.7 is equivalent to the response of an ideal operational amplifier with infinite gain and infinite input resistance but with a feedback impedance consisting of a capacitor C in parallel with a resistor $A_o R'$ as shown by the equivalent circuit in Figure 6.6.

Figure 6.6 Integrator low frequency equivalent circuit representing effect of finite open loop gain

Integrators are often supplied with a constant d.c. input voltage for the purpose of producing an output voltage which varies linearly with time (a linear ramp). In such applications low frequency errors, as set by finite open loop gain, cause departures from the ramp linearity. The output response of the low frequency equivalent circuit of Figure 6.6 to an input step voltage V_s is determined by the relationship

$$V_{o(t)} = A_o \frac{R'}{R} \cdot V_s \left[1 - \exp \left(- \frac{t}{A_o CR'} \right) \right] \qquad (6.8)$$

Expanding the exponential in equation 6.8 as a power series gives

$$V_{o(t)} = V_s \left[\frac{t}{CR} - \frac{t^2}{2A_o(CR')CR} + \ldots \right] \qquad (6.9)$$

The first term in equation 6.9 represents the ideal response (a linear ramp), the second and subsequent terms represent the error due to finite gain which causes a departure from linearity. The departure from linearity is governed principally by the second term; expressing this as a percentage of the ideal linear term gives

$$\text{Ramp nonlinearity error} = - \frac{t}{2 A_o CR'} \times 100\% \qquad (6.10)$$

Thus, when using an integrator to produce a linear ramp the period of the ramp T must be such that $T \ll 2A_o CR'$.

Finite open loop gain causes errors in hold mode
If during an integration process the input voltage to the integrator is switched to zero the output of the integrator should ideally remain constant (hold) at any value it may have reached, but finite open loop gain and finite input resistance in addition to amplifier offsets contribute errors in the hold mode.

$$V_{o(t)} = V_i \exp \left(\frac{-t}{CA_o R_d} \right) + I_B A_o R_d \left[1 - \exp \left(\frac{-t}{CA_o R_d} \right) \right] \pm V_{io}$$

Figure 6.7 Equivalent circuit used to find error in hold mode

233

Minimum error in the hold mode is obtained by open circuiting the input resistor R; this makes the effective leakage resistance A_oR', due to finite amplifier open loop gain and finite input resistance, equal to A_oR_d. This effective leakage resistance tends to discharge any fixed voltage stored across the integrating capacitor and the equivalent circuit shown in Figure 6.7 may be used to compute drift error in the hold mode. An expression for the output drift error in the hold mode can be obtained by assuming that when the integrator is switched to the hold mode the capacitor in the equivalent circuit of Figure 6.7 has an initial voltage V_1. The subsequent time variation of the amplifier output voltage is determined by the relationship

$$V_{o(t)} = V_i \exp\left(-\frac{t}{CA_oR_d}\right) + I_BA_oR_d\left[1 - \exp\left(-\frac{t}{CA_oR_d}\right)\right]$$
$$\pm V_{io} \qquad (6.11)$$

If we write $\exp[-t/(CA_oR_d)] = 1 - [t/(CA_oR_d)]$ (for $t \ll CA_oR_d$) the output drift value in the hold mode may be written as

$$\frac{dV_o}{dt_{(\text{hold mode})}} = \frac{V_i}{CA_oR_d} + \frac{I_B}{C} \qquad (6.12)$$

In a practical integrator the drift error in the hold mode is normally dominated by the effect of amplifier bias current and stray leakage currents, but equation 6.12 shows that even if all leakage paths were eliminated and amplifier bias current compensated an error would still remain because of the amplifier's finite open loop gain and finite input resistance.

6.3.3 SLEWING RATE ERRORS

In fast integrators which require an output voltage which changes rapidly with time, slew rate limitations, which set the maximum rate at which the amplifier output can change, can cause performance errors.

Slew rate limitations are inherent and arise from the basic mechanism of capacitor charging. Amplifier slew rates are normally specified for the amplifier used to drive a resistive load when it is the charging of the frequency compensating capacitor which is the slew rate determining factor (see Section 2.7.1). In integrator applications the feedback capacitor is charged by the output current of the amplifier; the output current limit of the amplifier may, in such cases, impose an output slew rate limit which is less than the published amplifier slew rate. Remember that the output current of the amplifier must supply any external load as well as the current taken by the feedback capacitor. For example, consider an amplifier with an output current limit of ± 5 mA used in an integrator circuit with a feedback capacitor of 0.01 μF. Even if all the output current were available to supply the feedback capacitor, the maximum rate of change of output voltage would be $I_o/C = (5 \times 10^{-3}/10^{-8}) \times 10^6 = 0.5$ V/μs.

6.4 Extensions to a basic integrator

There are a variety of external circuit modifications which can be made to the basic integrator circuit in order to change its response characteristics and extend its usefulness.

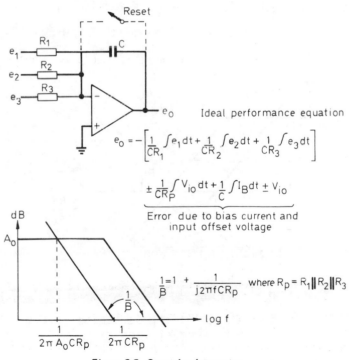

Figure 6.8 Summing integrator

235

6.4.1 SUMMING INTEGRATOR

The current summing property of the inverting input terminal of a differential input operational amplifier can be exploited to allow a single amplifier to perform both summation and integration at the same time. The principle is illustrated by the circuit shown in Figure 6.8. Note that by using different input resistor values the contributions to the output of the several inputs is weighted in inverse proportion to the resistor values. Considerations involved in determining performance errors are much the same as those outlined for the basic integrator, the only difference being that a resistor R_p equal to the parallel sum of all

Ideal performance equation

$$e_O = -\left[e_{in}\frac{R_2}{R_1} + \frac{1}{CR_1}\int e_{in}dt\right]$$

$$\underbrace{\frac{1}{CR_1}\int V_{io}dt + \frac{1}{C}\int I_B dt + \left[\pm V_{io} + I_B R_1\right]\left[1 + \frac{R_2}{R_1}\right]}_{\text{Error due to input offset voltage and bias current}}$$

$$\frac{1}{\beta} = \left[1 + \frac{R_2}{R_1}\right]\left[1 + \frac{1}{j\omega C(R_1 + R_2)}\right]$$

$$\frac{e_O}{e_{in}} = \frac{R_2}{R_1}\left[1 + \frac{1}{j\omega CR_2}\right]\left[\frac{1}{1 + \frac{1}{BA_{OL(j\omega)}}}\right]$$

Figure 6.9 Augmenting integrator

236

input resistors must be substituted in place of the single input resistor R in the error equations of the basic integrator. Note that low frequency errors due to finite gain occur at frequencies approaching and below the frequency at which the Bode plots for $1/\beta$ and the open loop gain intersect and for the summing integrator $1/\beta$ is determined by the parallel sum of all input resistors R_p:

$$\frac{1}{\beta} = 1 + \frac{1}{j\,2\pi f C R_p}$$

6.4.2 AUGMENTING INTEGRATOR

A resistor connected in series with the feedback capacitor of a basic integrator (Figure 6.9) makes the circuit produce a composite output consisting of a component proportional to the input signal added to a component proportional to the time integral of the input signal. The principle may also be adapted to the summing integrator of Figure 6.8 by connecting a resistor in series with the feedback capacitor in that circuit.

6.4.3 DIFFERENTIAL INTEGRATOR

The subtractor principle of Section 4.4 can be applied to give a circuit in which a single differential input operational amplifier produces an output signal proportional to the time integral of the difference between two input signals. A circuit for this purpose is shown in Figure 6.10; it may be used to integrate the output of a floating source

Ideal performance equation

$$e_0 = \frac{1}{CR} \int (e_2 - e_1)\, dt$$

$$\pm \frac{1}{CR} \int V_{io}\, dt + \frac{1}{C} \int I_{io}\, dt \pm V_{io}$$

Error due to input offset current and input offset voltage

Figure 6.10 Differential integrator

237

Figure 6.11 Differential integrator using two operational amplifiers

whilst rejecting common mode input signals. The ability of the circuit to reject common mode signals depends on the c.m.r.r. of the amplifier but in addition, it is also very much dependent upon an accurate matching of the time constants of the networks connected to the two input terminals.

Performance errors in a practical circuit are determined by amplifier input offset voltage, amplifier bias current and by finite amplifier open loop gain and bandwidth, in much the same way as for the basic integrator. The presence of two capacitors when a single amplifier is used to perform a differential integration operation increases the problems associated with providing a practical circuit with reset and hold modes, and it may be found more convenient to use the two-amplifier circuit shown in Figure 6.11 to perform the differential integrator operation. In this circuit one amplifier acts as a simple inverter and the other acts as a summing integrator. The c.m.r.r. of the circuit does not depend upon the c.m.r.r. of the amplifiers but it is still dependent upon accurate resistor matching.

6.4.4 CURRENT INTEGRATOR

It is sometimes desirable to form an output signal voltage proportional to the integral with respect to time of an input current rather than an input voltage (see also Section 8.1.2). The basic integrator is readily adapted to current integration by simply omitting the input resistor R.

A practical current integrator circuit is outlined in Figure 6.12. The circuit is suitable for integrating small currents to earth produced by very high impedance current sources; it produces a negligible voltage

Figure 6.12 Current integrator and Bode plots

intrusion into the measurement circuit. Assuming that offsets are nulled
and suitable precautions taken to avoid leakage the circuit may be
expected to provide accurate integration for very small input currents.
If external leakages are reduced to negligible proportions the accuracy
limitations are set by amplifier bias current drift. If the greatest possible
accuracy is required in order to integrate sub-picoamp input currents a
varactor bridge type amplifier should be used (e.g. Analog Devices
model 310). For larger input currents greater economy will be provided
by using an f.e.t. input operational amplifier.

6.4.5 INTEGRAL OF CURRENT SUM AND CURRENT DIFFERENCES

The current integrator circuit (Figure 6.12) is readily adapted to summation, all that is necessary being to supply the extra input currents
to the amplifier summing point.

239

A one-amplifier circuit can also be used to generate an output voltage proportional to the integral of a current difference; the principle is illustrated by the circuit shown in Figure 6.13. However,

Figure 6.13 Simple one-amplifier circuit for integral of current difference

with two capacitors, practical problems are involved in the provision of reset and capacitors must be accurately matched if c.m.r.r. is not to be degraded; also the circuit introduces a voltage drop, $1/C(\int I_2 dt)$, into the measurement circuit. Because of these difficulties it is usually more convenient to employ two amplifiers in order to perform the integration of a current difference, one amplifier acting as a current inverter and the other as a summing integrator. A circuit is illustrated in Figure 6.14. Offsets and drift of the amplifier types used in the practical circuit determine performance accuracy in much the same way as before.

6.5 Integrator reset

Unlike a normal amplifier circuit the output of an integrator does not return to zero when the input signal is made zero, and a practical integrator must therefore always be provided with some means of resetting its output voltage to zero (or some desired initial value). Switching of an integrator between its various modes of operation can be performed by mechanical switches or relays (see Section 6.2) but it is sometimes desirable to provide a solid state reset switch or to arrange that the integrator automatically resets when its output reaches some predetermined level.

Reset switches are connected in parallel with the integrator feedback

240

Figure 6.14 Two-amplifier circuit for integral of current difference

capacitor and in the run mode any leakage across the open switch adds to the error due to amplifier bias current. The extent to which switch leakage becomes a design consideration is dependent upon the magnitude of the input signal current to the integrator and the desired accuracy. If input currents are large compared to the leakage of a simple solid state switch then clearly a simple switch will suffice. If not, then a leakage reduction switch configuration must be sought.

Figure 6.15 Low leakage integrator reset

A low leakage reset switch can be implemented using two MOS f.e.t.'s as shown in Figure 6.15. When using p channel MOS switches the source substrate junction must not be allowed to become forward

241

biased; the substrate must therefore never be allowed to become negative with respect to the input signal. The leakage current of an MOS switch in the 'Off' state occurs mainly across the substrate to drain junction.

In Figure 6.15 a negative going reset pulse turns on T_1 and T_2 shorting the integrator capacitor and setting the output voltage to near zero (to a voltage $\cong - V_{in} \ 2 \ R_{ON}/R_1$, where R_{ON} is the low 'On' existence of the MOS switch, $R_{ON} \ll R_2$). When the switches turn off the leakage current of T_2 passes through resistor R_2; the small voltage across R_2 is blocked from the amplifier summing junction by T_1. T_1 has practically no voltage across its junctions because its substrate is earthed and leakage currents are hence negligibly small.

The reset switch of Figure 6.15 can be made to provide an automatic reset by the addition of a comparator (see Section 7.1) used to sense the integrator output and to provide the reset drive. In the circuit shown in Figure 6.16 both the integrator output voltage at which reset occurs and the level to which the integrator is reset can be independently adjusted. Reset occurs when the output voltage of the integrator reaches the comparator trip point. During reset the capacitor discharges until the integrator output voltage reaches the lower comparator trip point as determined by comparator hysteresis. Integrator errors are governed by amplifier offsets and bias currents, finite open loop gain and bandwidth as for the basic integrator. Finite comparator switching time and the bias current and offset voltage of the comparator amplifier introduce errors in the reset level and reset point but these can be compensated practically by choosing values of V_{ref} and V_z to give desired values of $V_{o(max)}$ and V_{reset}.

6.6 A.C. integrators

In integrator applications not requiring a response down to d.c. it is possible, by bounding the d.c. closed loop gain, to obviate the necessity for providing output reset. The circuit shown in Figure 6.17 has a steady state sinusoidal response governed by the relationship

$$\frac{e_o}{e_{in}} = - \frac{\dfrac{R_2}{1 + j \ 2\pi f C R_2}}{R_1} \left[\frac{1}{1 + \dfrac{1}{\beta A_{OL}}} \right] \qquad (6.13)$$

Reset occurs at $V_{O(max)} = \dfrac{V_{ref} R_4 + V_z R_3}{R_3 + R_4}$

Reset level $V_{reset} = \dfrac{V_{ref} R_4 - V_z R_3}{R_3 + R_4}$

Figure 6.16 Comparator provides automatic integrator reset

Figure 6.17 a.c. integrator

For $\beta A_{OL} \gg 1$ and for frequencies greater than $1/(2\pi R_2 C)$ the response approximates that of the ideal integrator

$$\frac{e_o}{e_{in}} \cong - \frac{1}{j\, 2\pi fCR_1}$$

At a frequency a decade away from $1/(2\pi R_2 C)$ the magnitude error is only 0.5%.

The presence of R_2 prevents integrator drift due to amplifier bias current and offset voltage from causing the amplifier to drift into saturation. Instead the output assumes a d.c. value of

$$V_{o(offset)} = \left[1 + \frac{R_2}{R_1} \right] \left[\pm V_{io} - I_B \frac{R_2 R_1}{R_1 + R_2} \right]$$

This output offset limits the dynamic range for a.c. output signals. As in all applications an offset balance can be used to cancel initial values of amplifier input offset voltage and bias current; output offset is then due to amplifier drift.

6.7 Differentiators

The differentiator operation, whilst not as widely employed as the integrator operation, is nevertheless sometimes useful in signal processing applications, for example in the measurement of ramp slopes and for showing up discontinuities in waveforms.

The factors which account for the more limited use of the differentiator as compared to the integrator are: (1) Differentiation unlike integration is a noise amplifying process. Noise problems are inherent in differentiators and are not just a defect of practical circuits. (2) Differences between the ideal differentiator circuit and the practical circuit are more marked than between the ideal and real integrator.

6.7.1 THE BASIC DIFFERENTIATOR

A simple differentiator circuit (Figure 6.18) is obtained by interchanging the position of the resistor and capacitor in the basic integrator circuit. The ideal performance equation for the simple differentiator is readily derived from the usual ideal amplifier assumptions. Since the

244

Figure 6.18 Basic differentiation and Bode plots

input signal is applied through a capacitor there is current flow to the amplifier summing point and a non-zero output voltage only when the input voltage changes.

The current to the amplifier summing point is

$$I_{in} = C \frac{d e_{in}}{dt}$$

and in the ideal case this must be equal to the current through the feedback resistor R, thus:

$$e_o = - I_f R$$

or

$$e_o = - CR \frac{d e_{in}}{dt} \tag{6.14}$$

245

The basic differentiator is unstable because of amplifier finite open loop bandwidth

The feedback fraction β for the simple differentiator circuit of Figure 6.18 is

$$\beta = \cfrac{\cfrac{1}{j\omega C}}{R + \cfrac{1}{j\omega C}} = \cfrac{1}{1 + j\omega CR}$$

and at angular frequencies greater than $1/CR$ the feedback signal lags behind the amplifier output signal by a phase angle approaching 90°. Because of the finite open loop bandwidth of practical operational amplifiers signals with frequency above the open loop bandwidth undergo an additional phase lag in their passage through the amplifier. The two phase lags can readily add up to 180° making the overall feedback positive rather than negative and resulting in circuit instability.

Bode plots of $1/\beta$ and A_{OL} for the simple differentiator are shown in Figure 6.18; the two intersect with a rate of closure of 40 dB/decade, indicating a near zero phase margin. The simple differentiator thus exhibits a marked gain peaking for signal frequencies approaching the frequency at which the magnitude of the loop gain is unity, and any transient disturbance in the circuit gives rise to output ringing. This very lightly damped response means that any additional phase shift in the feedback loop, say due to capacitive loading at the output, can cause the simple differentiative circuit to break out into sustained oscillations.

6.8 Practical considerations in differentiator design

6.8.1 BANDWIDTH LIMITS

Practical differentiators invariably employ some means of limiting the differentiator bandwidth in order to achieve closed loop stability. One method is to connect a resistor R_1 in series with the input capacitor as shown in Figure 6.19. With the value of R_1 suitably chosen, Bode plots for $1/\beta$ and A_{OL} intersect with a rate of closure of 20 dB/decade thus ensuring adequate stability phase margin. Resistor R_1 also serves to increase the effective input impedance of the differentiator circuit and

$$\frac{1}{B} = \frac{1 + j\omega C(R + R_1)}{1 + j\omega CR_1}$$

Figure 6.19   Input resistor R_1 increases phase margin in basic differentiation

Figure 6.20   Noise reduction in differentiation by restriction of high frequency gain

247

to reduce its high frequency gain. Note that the gain of a differentiator increases with increase in frequency. This means that in a practical differentiator circuit high frequency noise components are amplified even though the desired signal may not contain high frequency components; high frequency noise may thus obscure the desired signal at the output. Adding the resistor R_1 introduces a break in the differentiator 20 dB/decade rise at an angular frequency

$$\omega_b = \frac{1}{CR_1}$$

In order to obtain additional attenuation of high frequency noise components at frequencies above those of interest a capacitor C_1 may be connected in parallel with the differentiator feedback resistor as shown in Figure 6.20. With C_1 chosen so that $C_1 R = C R_1$ Bode plots are as shown in Figure 6.20. Note that the circuit acts as an integrator for frequencies greater than $1/(2\pi C_1 R)$. In both the circuits of Figure 6.19 and Figure 6.20 component values should be chosen to place the break point ω_b well enough above the maximum operating frequency to ensure the required accuracy.

6.8.2 OFFSET ERRORS IN A PRACTICAL DIFFERENTIATOR

Amplifier bias current and input offset voltage give rise to a d.c. offset error at the output of a practical differentiator circuit. The error is readily deduced from the equivalent circuit shown in Figure 6.21. For the purpose of offset error evaluation the open loop gain of the amplifier is assumed to be infinite. We may write

$$I_{in} = C \frac{de_{in}}{dt}$$

Figure 6.21   Equivalent circuit for evaluation of differentiator offset error

and
$$I_f = I_{in} - I_B = \pm \frac{V_{io} - e_o}{R}$$

Thus
$$e_o = - CR \frac{de_{in}}{dt} + I_B R \pm V_{io} \qquad (6.15)$$

ideal per-	error due to input
formance	offset voltage and
equation	bias current

The output offset error can be referred to the input of the differentiator where it represents an equivalent input error:

$$\frac{d V_{in}}{dt}_{(error)} = - \frac{I_B}{C} \pm \frac{V_{io}}{CR} \qquad (6.16)$$

Amplifier input offset voltage and bias current are initial values plus drift; if initial values are balanced this leaves drift values only. The error component due to amplifier bias current can be reduced by connecting a resistor, equal in magnitude to the feedback resistor, between the non-inverting input terminal of the amplifier and earth. With this resistor in circuit values of the amplifier input difference current I_{io} should be substituted for I_B in the error equations.

Equation 6.16 suggests that for a particular CR value the equivalent input error is reduced by making C as large as possible. But note that the current through the feedback resistor R is supplied by the amplifier; R must therefore not be so small as seriously to load the amplifier output. Also there are problems involved in large capacitor values (say C greater than 1 to 10 μF) as discussed in 6.3.1.

6.8.3 CHOICE OF DIFFERENTIATOR COMPONENT VALUES

The component values and the amplifier type to be used in a practical differentiator are dictated by the accuracy requirements assessed in terms of the expected frequency content and magnitude of the input signal. The starting point in a differentiator design is normally the choice of characteristic time CR. It is convenient to select a characteristic time such that the amplifier will give near full scale output for the maximum expected rate of change of the input signal.

Choose
$$CR = \frac{|V_o|_{max}}{\left(\dfrac{de_{in}}{dt} \right)_{max}} \qquad (6.17)$$

An output bounding circuit can be added to prevent unexpectedly fast input signals from driving the output into saturation limits. Select R say in the range 10–100 kΩ, that is not too low as to draw seriously on the amplifier output current, and then calculate the necessary value of C. If the calculation calls for a value of C greater than 1 to 10 μF it is usually

Figure 6.22 Summing differentiator

Figure 6.23 Differential differentiators

better to increase the value of R and use an amplifier type with smaller bias current if the equivalent input offset error (equation 6.16) is such as to reduce the accuracy below the design limit.

Ensure adequate closed loop phase margin and reduce high frequency noise by selecting a value for R_1 (Figure 6.19) or $C_1 R_1$ (Figure 6.20). Noise reduction dictates that the break in the 20 dB/decade increasing gain of the differentiator should be set no higher than that value required to ensure accurate differentiation of the highest input signal frequency of interest.

6.9 Modifications to the basic differentiator

Modifications comparable to those discussed in Section 6.4 can be made in order to vary the response characteristics of a differentiator circuit.

6.9.1 SUMMING DIFFERENTIATOR

The derivative of several input signals may be combined in a summing differentiator. A summing differentiator, shown in Figure 6.22, is formed by introducing additional capacitive input paths to the amplifier summing point. If required the individual derivatives may be weighted by simply proportioning the input capacitors accordingly. As in the case of the simple differentiator circuit, it is necessary to introduce a break in the differentiator response in order to achieve an adequate closed loop stability phase margin and to reduce high frequency noise. Remember that phase margin is determined by the rate of closure of $1/\beta$ and the open loop response; in assessing $1/\beta$ due allowance must be made for all paths to the amplifier summing point. Errors due to amplifier bias current and input offset voltage may be assessed in much the same way as for the basic differentiator circuit.

6.9.2 DIFFERENTIAL DIFFERENTIATOR

A circuit which produces an output proportional to the difference between the derivatives of two input signals is shown in Figure 6.23. It is formed by adding a circuit network to the non-inverting input terminal which is identical to that connected to the inverting input terminal in the basic differentiator circuit. Resistor R_1 is, as before, introduced for closed loop stability and to reduce noise. The circuit can be used to differentiate the output of a floating source whilst at the same time removing any common mode signal. Like all single amplifier differential circuits the circuit is subject to common mode errors and limitations and c.m.r.r. is dependent both upon component matching and the c.m.r.r. of the amplifier used in the circuit.

6.10   Analogue computation

A book on operational amplifiers would be incomplete without some mention of analogue computation. Early operational amplifiers were designed primarily for use in analogue computers; it is only the advent of low cost operational amplifiers that has allowed their extensive application in other fields. A detailed discussion of the principles of operation of analogue computers and their many possible applications to the solution of mathematical problems is beyond the intent of this text. We content ourselves with a brief statement of operating principles and include for the sake of interest some simple examples of the type of problem that can be solved. The reader interested in studying the subject further is referred to one of the many texts available.

Figure 6.24   Solution of simultaneous equations

In an analogue computer the numbers representing the variables in a problem are represented continuously by voltages. The mathematical operations involved in the solution of a problem are performed in the computer as analogous operations on these voltages. As a simple example we consider the analogue computer approach to the solution of simultaneous equations.

Consider the two equations

$$a_1x + b_1y = c_1$$

$$a_2x + b_2y = c_2$$

The mathematical operations involved in these equations are those of addition and multiplication by constant coefficients. These operations are readily performed on electrical signals. Addition can be performed by a summing amplifier. Multiplication by a coefficient greater than unity also requires the use of an amplifier; a simple resistive divider can be used to multiply by a coefficient less than unity.

The two equations may be rewritten as

$$\left.\begin{array}{l} x = \dfrac{c_1}{a_1} - \dfrac{b_1}{a_1}\, y \\[2ex] y = \dfrac{c_2}{b_2} - \dfrac{a_2}{b_2}\, x \end{array}\right\} \tag{6.18}$$

they may be solved using two summing amplifiers connected as shown in Figure 6.24. The output from amplifier 1 is a voltage analogous to x and the output from amplifier 2 is a voltage analogous to y. A fraction b_1/a_1 of y is taken using a variable potentiometer and this is summed (in amplifier 1) with a voltage representing $-c_1/a_1$ derived from a potentiometer and voltage source. Similar considerations apply to the summation performed by amplifier 2.

The amplifiers present a continuous solution for x and y. Scaling is set by the magnitude of the voltages used to represent the constants c_1 and c_2; the scaling factor chosen must ensure that the voltages representing the unknowns do not exceed the output capability of the amplifiers. Desired values of the coefficients are selected by adjustment of the potentiometers and the solutions are determined by measuring the voltages representing x and y. Time variation of c_1 and c_2 may be represented by appropriate time varying voltages in which case the time varying solutions x and y can be measured with an oscilloscope or chart recorder.

The use of additional amplifiers enables the principle outlined above to be extended to the solution of equations involving many more than two unknowns.

6.10.1 SOLUTION OF DIFFERENTIAL EQUATIONS

The operational integrators described earlier in this chapter can be used to solve many types of differential equation. As a simple example we consider the type of equation used to represent the law of radioactive decay:

$$\frac{dN}{dt} = - \lambda N \qquad (6.19)$$

where N represents the number of atoms of a radioactive element present in a system at any instant

λ is the decay constant of the element and represents the probability that any atom of the element will disintegrate in unit time.

Figure 6.25   Integrator used to solve simple differential equation

The equation may be solved by the circuit shown in Figure 6.25. Consider a voltage representing the value of dN/dt applied to an integrator; the output from the integrator will be a voltage representing $-N$. According to the equation which we wish to solve $dN/dt = -\lambda N$; if the output of the integrator is multiplied by λ and fed back

254

to the input the circuit will force the voltage representing N to vary in the manner described by the equation. The initial value of N is represented by a voltage set at the output of the integrator at the start of the integrating time; this voltage is set by choice of E_{ref}, R_2 and R_1. Time scaling is determined by choice of integrator time constant CR.

The circuit of Figure 6.25 illustrates the use of an analogue computer in the simulation of the response of a physical system. Any physical system whose behaviour can be described mathematically can be simulated with an analogue computer. The radioactive decay example may be developed further.

Consider a radioactive element A which decays with decay constant λ_a to form an element B which in turn decays with constant λ_b to form an element C, and so on. The growth and decay of the number of atoms of the different elements present in a system may be represented by the equations

$$\left. \begin{array}{l} \dfrac{dN_a}{dt} = - \lambda_a N_a \\[2mm] \dfrac{dN_b}{dt} = \lambda_a N_a - \lambda_b N_b \\[2mm] \dfrac{dN_c}{dt} = \lambda_b N_b - \lambda_c N_c \end{array} \right\} \qquad (6.20)$$

The behaviour of the system may be simulated by the arrangement of amplifiers shown in Figure 6.26. Typical solutions for the values of N_a and N_b as a function of time are shown by the oscillograph in Figure 6.27.

There are many examples of systems which may usefully be studied with an analogue computer; analogue computers have, for example, been used in the solution of rate equations in chemical kinetics. We consider one more example; the simulation of the behaviour of a mechanical system.

Consider the mechanical vibrations of a body of mass m on the end of a spring of force constant k in the presence of viscous damping described by the damping constant b. The differential equation for the displacement x of the body from its equilibrium position is

$$m \frac{d^2x}{dt^2} + b \frac{dx}{dt} + kx = 0 \qquad (6.21)$$

Figure 6.26 Analogue computer used to simulate radioactive decay

Figure 6.27 Solutions for N_a and N_b of equation 6.21

Rearranging,

$$\frac{d^2x}{dt^2} = -\frac{b}{m}\frac{dx}{dt} - \frac{k}{m}x \qquad (6.22)$$

The design of an analogue computer to solve this equation starts by assuming that a voltage representing d^2x/dt^2 is available. This is integrated to give $-dx/dt$ which is integrated again to give x. A potentiometer across the output of the first integrator is used to obtain the fraction b/m of $-dx/dt$; this is inverted and added to the fraction k/m of x from the output of the second integrator. According to equation 6.22 the output from the summing amplifier is equal to d^2x/dt^2; this signal is returned to the input of the first integrator where it was assumed to be originally.

Figure 6.28 Analogue computer used to simulate behaviour of a vibrating system

The circuit illustrated in Figure 6.28 shows the arrangement that may be used to perform the above operations. The circuit forces the voltages, which are used to represent the variables, to vary in the manner described by the equation to be solved. Voltages representing initial values of x and dx/dt are set at the output of the appropriate integrators. The computation is initiated by simultaneously switching both integrators into the run mode. Solutions for x, the displacement,

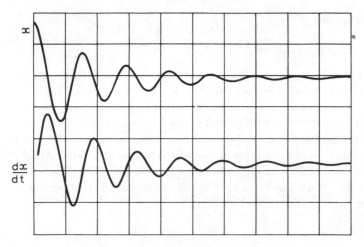

Figure 6.29   Computer solutions for initial conditions $\underset{t=0}{x} = x_1$ $|\dfrac{\mathrm{d}x}{\mathrm{d}t}|_{t=0} = 0$

and $\mathrm{d}x/\mathrm{d}t$, the velocity, may be displayed by an oscilloscope or chart recorder. The oscillograph illustrated in Figure 6.29 shows the type of solutions for the initial conditions

$$\underset{t=0}{x} = x_1 \text{ and } \dfrac{\mathrm{d}x}{\underset{t=0}{\mathrm{d}t}} = 0$$

6.11 A simple analogue computer

Where proposed usage does not justify the purchase of a commercial analogue computer it is possible, using available monolithic amplifiers, to assemble quickly an analogue system suitable for a variety of applications. The solutions illustrated in Figures 6.27 and 6.29 were obtained with such a system which will be described.

The basic building blocks of a simple analogue computer are summing amplifiers and integrators. Integrators have to be provided with a means of switching between modes and it is convenient to arrange for a repetitive display of solutions. The circuit for a summing amplifier is illustrated in Figure 6.30. Two unity gain inputs and a single gain of ten input are provided; any input not in use is connected to earth.

258

Figure 6.30 Summing amplifier for simple analogue computer

Figure 6.31 Integrator with reset for simple analogue computer

The circuit for the integrator used in the system is illustrated in Figure 6.31. A gated amplifier arrangement, rather than relay switching, is used to reset initial conditions. Transistor *Tr*1, connected as an emitter follower, constitutes the reset amplifier with a diode gate connecting it to the integrator summing point. When the diode gate is conducting, the base of *Tr*1 acts as a new summing point for the reset conditions so that the integrator network does not need to be disconnected. The initial value of the integrator output voltage is set by adjustment of the potentiometer *R*. The integrator is switched to the run mode when the diode gate is rendered non-conducting; integrator time constant is set by choice of integrating capacitor *C*. The integrator balance control is adjusted for zero drift with the integrator switched to the run mode and all inputs to the integrator connected to earth.

Figure 6.32   Circuit for switching integrators between run and reset modes

All integrators in use in the system need to be simultaneously switched into the run mode; the circuit used to accomplish this is illustrated in Figure 6.32. The circuit utilises a monolithic dual operational amplifier. Amplifier A_1 acts as a unity gain inverter; its input and output signals act as the switching signals to the diode gates. Two methods of operation are provided, 'manual' and 'repetitive'. In the manual mode all integrators are switched simultaneously with the switch S_1. In the repetitive mode of operation amplifier A_2, acting as a non-symmetrical astable multivibrator, switches the integrators repetitively between the reset and run modes. The 'run', or 'compute' time is adjustable by means of potentiometer R_1.

Exercises

6.1 A simple integrator circuit (Figure 6.1) uses $C = 0.1\ \mu F$, $R = 100\ k\Omega$, the input point of the circuit is connected to earth and the output drift rate is adjusted to zero by means of a suitable offset balance (Section 9.6). Find the output drift rate if the temperature changes by $10°C$ assuming that the operational amplifier has $\Delta V_{io}/\Delta T = 20\ \mu V/°C$ and $\Delta I_B/\Delta T = 0.5\ nA/°C$ (see Section 6.3.1).

6.2 Show how you would use a single operational amplifier to generate the relationship

$$e_o = - \int_0^t (e_1 + 2e_2 + 10e_3)\ dt$$

Find component values if the integrating capacitor has a value $1\ \mu F$. Assume ideal operational amplifier action.

　　If the operational amplifier has input offset voltage $V_{io} = 10\ \mu V$ and bias current $I_B = 2\ nA$ what is the integrator output drift rate? What is the drift rate if the integrator is switched to the 'hold' mode? (see Sections 6.3.1 and 6.4.1).

6.3 In the simple integrator circuit of Figure 6.1 $C = 0.1\ \mu F$ and $R = 100\ k\Omega$. If the open loop gain of the operational amplifier at zero frequency is 80 dB what is the lowest sinusoidal signal frequency for which the phase error will be no more than $5°$?

　　What is the amplitude error at this frequency?

　　What happens to the phase and amplitude errors if an additional input path in the form of a second $100\ k\Omega$ resistor is added to the circuit? (see Section 6.3.2).

6.4 A simple operational integrator with $C = 0.1\ \mu F$ and $R = 100\ k\Omega$ is to be used to produce a linear ramp of amplitude 10 V and duration 100 ms. What is the minimum value of the open loop gain of the amplifier to ensure that the departure from linearity of the ramp is less than 0.1%? (see equation 6.10).

6.5 An a.c. integrator (Figure 6.17) uses the component values $R_1 = 10\ k\Omega$, $R_2 = 1\ M\Omega$, $C = 0.1\ \mu F$. What is the lowest input signal frequency for which the magnitude of the output is in error by no more than 0.5%? (see Section 6.6).

6.6 An internally frequency compensated operational amplifier with open loop gain 100 dB and unity gain frequency 10^6 Hz is used as a simple differentiator (Figure 6.18) with $C = 0.1\ \mu F$ and $R = 100\ k\Omega$. The circuit is found to be very lightly damped; explain this fact and estimate the approximate value of the frequency at which the output 'rings' when the circuit is subjected to a transient disturbance. In order to overcome the stability problem a resistor $R_1 = 1\ k\Omega$ is connected in series with the input capacitor C. Explain the action of this resistor and estimate the highest frequency for which the magnitude of the output signal will be in error by no more than 0.5% with R_1 in circuit. Illustrate your answer with appropriate Bode plots (see Sections 6.7.1 and 6.8.1).

If the operational amplifier has a bias current $I_B = 0.1\ \mu A$ and input offset voltage $V_{io} = 4$ mV what will be the output offset in the circuit? (see Section 6.8.2).

CHAPTER SEVEN

SWITCHING AND POSITIVE FEEDBACK CIRCUITS

In the operational amplifier applications discussed thus far the amplifiers have been used in negative feedback circuits. Under normal conditions, when used in such a circuit, the amplifier output voltage takes on values which lie in between its positive and negative saturation limits. The amplifier has a high gain and the negative feedback forces the voltage between its differential input terminals to be very small at all times.

When an operational amplifier is used without feedback (open loop operation) the amplifier output will usually be in one of its saturated states; the application of a small differential input signal of appropriate polarity will cause the output to switch to its other saturated condition. In this chapter we will deal with applications which make use of this amplifier switching characteristic. We will also consider circuits in which operational amplifiers have positive feedback applied to them, either to increase their switching speed, or to give a circuit which produces controlled oscillation of definite frequency.

7.1 Comparators

A comparator is a device which is used to sense when a varying signal reaches some threshold value. Comparators find application in many electronic systems: for example, they may be used to sense when a linear ramp reaches some defined voltage level, or to indicate whether or not a pulse has an amplitude greater than a particular value. Provided that suitable output limiting is provided, comparator outputs may be used to drive digital logic circuits.

Not all operational amplifier types are suitable for comparator application; in addition to low offset and drift, rapid switching times are normally essential. Some amplifiers are designed specifically for comparator application, with fast response time and an output compatible with digital integrated circuit elements.

The simplest way of using a differential input amplifier to perform the comparator function is to apply the signal voltage directly to one of the input terminals of the amplifier and a reference voltage to the other. The principle is illustrated in Figure 7.1; the amplifier is used open loop; its output makes a transition between saturated states as

Figure 7.1 Differential amplifier used as comparator

the input signal passes through a value equal to E_{ref}. E_{ref} must not exceed the maximum common mode voltage for the particular amplifier. The polarity of the output transition may be changed by interchanging the position of signal and reference voltages.

Figure 7.2 Single ended amplifier used as comparator

The circuit in Figure 7.2 illustrates an alternative arrangement for comparator operation. Signal and reference voltages are applied to the same input terminal of the amplifier through appropriate resistors. The other input terminal is earthed and the circuit is thus not subject to common mode voltage limitations. The output transition occurs when

$$e_i = e_t = -E_{ref} \frac{R_1}{R_2}$$

the threshold voltage e_t can be set by choice of input resistors and the reference voltage may be any convenient voltage of opposite polarity to the input signal.

Figure 7.3 Regenerative comparator

In both the comparators shown, for the full output transition to take place the input voltage must swing past the threshold voltage by an amount

$$\frac{V_{o\,sat}^{+} - V_{o\,sat}^{-}}{{}^{'}A_{VOL}}$$

In the case of rapidly changing input signals the output transition time is dependent on amplifier characteristics. The switching time with slowly varying input signals depends on the rate of change of the input voltage. In such cases it is often advantageous to speed up the output transition by using some form of positive feedback.

A circuit which illustrates the principle of action of a regenerative comparator is shown in Figure 7.3. Positive feedback is applied via resistor R_2 connected between the amplifier output and the non-phase inverting input terminal. When e_i reaches the threshold voltage the positive feedback forces the amplifier to switch regeneratively between saturated states. The output transition time is made virtually independent of the rate of change of input voltage.

The circuit exhibits hysteresis, that is, the transition takes place for different values of e_i dependent on whether e_i is increasing or decreasing towards the threshold value. The transfer curve for the comparator is illustrated for a value of E_{ref} equal to zero. The threshold value for e_i at which the transition takes place has a value, neglecting offsets, equal to

$$V_{o\,sat} \frac{R_1}{R_1 + R_2}$$

265

$V_{o\ sat}$ can take on both its positive and negative saturation values and the amount of hysteresis is thus

$$V_H \approx (V_{o\ sat}^+ - V_{o\ sat}^-)\ \frac{R_1}{R_1 + R_2} \tag{7.1}$$

The amount of hysteresis is directly dependent on the magnitude of the positive feedback fraction,

$$\beta = \frac{R_1}{R_1 + R_2}$$

The comparator transfer curve illustrated is appropriate for an amplifier having a linear transfer curve between saturation limits. The transfer curve for a practical amplifier may be expected to show deviations from linearity as its saturation limits are approached (see Section 3.1). Amplifier gain as measured by the slope of the transfer curve decreases from its rated value through a range of values to zero as the output voltage approaches its saturation value. In a regenerative comparator switching in fact takes place when the loop gain (βA_{VOL}) becomes unity; thus with small values of β the switching points occur when the output voltage has a value measurably less than its saturation value. It is not advisable to use a regenerative comparator with a very small amount of hysteresis (small β) since this in practice invariably results in high frequency parasitic oscillations at switching.

In all comparator circuits outputs may be clamped to desired values rather than using saturation limiting; it is emphasised again that care must be taken to ensure that reference and input voltages do not exceed allowable limits for common mode and differential input signals.

Figure 7.4 Comparator with diode bounding

Figure 7.5 Comparison of sum of input signals

A modification of the simple comparator of Figure 7.2 is illustrated in Figure 7.4. Diodes are used to impose output bounds; the output voltage varies approximately logarithmically with the current into the amplifier summing point. The circuit incorporates a variable amount of hysteresis, which may be used, if required, to speed up the output transition for slowly varying input signals. In both the circuits of Figures 7.2 and 7.4 the state of the output depends essentially on the direction of the current flowing towards the amplifier summing point. The circuits may thus be used to compare the sum of several voltages against a reference merely by adding appropriate resistors to the amplifier summing point. The principle is illustrated by the circuit shown in Figure 7.5; the output transition in this circuit occurs when

$$\frac{e_1}{R_1} + \frac{e_2}{R_2} = -\frac{E_{ref}}{R_3}$$

The circuit includes a method of restricting the comparator output for compatibility with digital integrated circuits. Resistor R is included to limit the amplifier output current.

7.2 Multivibrators

Multivibrators are a group of circuits which have two states; they are used extensively in pulse systems. There are three types: astable multivibrators (free running), monostable multivibrators (one shot) and bistable multivibrators (flip-flop). In an astable multivibrator the two states of the circuit are momentarily stable and the circuit switches repetitively between these two states. The monostable multivibrator

267

has only one stable state; it can be made to change to its other state by the application of a suitable triggering pulse, but it then returns to its stable state after a time interval determined by circuit values. The bistable multivibrator has two stable states, in either of which it will remain indefinitely until appropriately triggered, when it will switch to the other state. Operational amplifiers arranged with appropriate positive feedback can be made to operate well as multivibrators of all three types.

7.2.1 ASTABLE MULTIVIBRATORS

The circuit illustrated in Figure 7.6 shows a differential input operational amplifier acting as a free running symmetrical multivibrator. The two states of the circuit between which it switches are those in which

Figure 7.6   Free running symmetrical multivibrator

the amplifier output is at positive and negative saturation. The amplifier output is thus a square wave; the period of the square wave is determined by the time constant CR and the feedback ratio established by the potential divider R_1, R_2.

The action of the circuit is conveniently described by reference to the waveforms illustrated in Figure 7.7. Starting at the time t' when the amplifier is in negative saturation, the voltage at terminal A is $\beta V_{o\,sat}^-$

where
$$\beta = \frac{R_1}{R_1 + R_2}$$

268

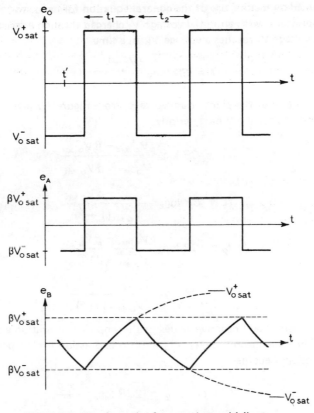

Figure 7.7 Waveforms for free running multivibrator

Terminal B is positive with respect to terminal A and its potential is decreasing as C charges down through R. When the potential difference between the two input terminals approaches zero the amplifier comes out of saturation; the positive feedback from the output to terminal A causes a regenerative switching which drives the amplifier to positive saturation. The voltage across a capacitor in series with a resistor cannot change instantaneously, and the potential at the terminal B therefore remains substantially constant during this rapid transition. Capacitor C now charges up through R and the potential at B rises exponentially; when it reaches $\beta V_o^+{}_{sat}$ the circuit switches back to the state in which the amplifier is in negative saturation.

An expression for calculating the period of the oscillations may be obtained by making use of the general equation for capacitor charging. A capacitor C with an initial voltage V_i charged through a resistor R by a voltage V_f reaches a voltage V_b in a time

$$t = CR \log_e \frac{V_f - V_i}{V_f - V_b} \tag{7.2}$$

Substitution of the appropriate voltages from Figure 7.7 gives the expressions for the timing periods,

$$t_1 = CR \log_e \frac{V_{o\ sat}^+ - \beta V_{o\ sat}^-}{V_{o\ sat}^+ - \beta V_{o\ sat}^+}$$

$$t_1 = CR \log_e \frac{V_{o\ sat}^+ - \beta V_{o\ sat}^-}{V_{o\ sat}^+ (1 - \beta)} \tag{7.3}$$

and

$$t_2 = CR \log_e \frac{V_{o\ sat}^- - \beta V_{o\ sat}^+}{V_{o\ sat}^- - \beta V_{o\ sat}^-}$$

$$t_2 = CR \log_e \frac{V_{o\ sat}^- - \beta V_{o\ sat}^+}{V_{o\ sat}^- (1 - \beta)} \tag{7.4}$$

If the positive and negative values of the amplifier saturation voltage have the same magnitude, $t_1 = t_2$ and the expression for the period of oscillations becomes

$$T = t_1 + t_2 = 2CR \log_e \frac{1 + \beta}{1 - \beta}$$

or

$$T = 2CR \log_e \left(1 + 2 \frac{R_1}{R_2}\right) \tag{7.5}$$

A free running multivibrator with a non-symmetrical waveform may be obtained by the use of a circuit of the type illustrated in Figure 7.8. In this circuit capacitor C charges up through diode D_1 and resistor R_3; diode D_2 is reverse biased during the timing period t_1 which is governed by the time constant CR_3. Capacitor C charges down through diode D_2 and resistor R_4; diode D_1 is reverse biased and the time constant CR_4 governs the period t_2. The waveforms are illustrated in Figure 7.9.

Astable multivibrators may be arranged so that their period is an exact multiple of the period of a synchronising signal. The synchronising signal is conveniently injected into the circuit at the non-phase inverting input terminal of the amplifier as indicated in Figure 7.8.

Figure 7.8 Non-symmetrical multivibrator

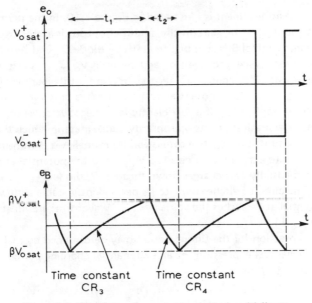

Figure 7.9 Waveforms for non-symmetrical multivibrator

7.2.2 MONOSTABLE MULTIVIBRATORS

The connection of a diode in parallel with the timing capacitor in an astable circuit may be used to prevent the phase inverting input terminal of the amplifier from going positive; this gives a monostable

271

Figure 7.10 Monostable multivibrator

circuit. The arrangement is illustrated in Figure 7.10. In the permanently stable state of this circuit the amplifier output is at positive saturation, terminal B is clamped to earth by diode D_1 and terminal A is positive with respect to earth by an amount $\beta V_{o\ sat}^+$. It is assumed that the resistor R_3 is much greater than R_1 so that its loading effect may be neglected. If the potential at the point A is brought down to earth by the application of a sufficiently large negative pulse the circuit switches regeneratively to its temporarily stable state in which the amplifier output is in negative saturation. Terminal A is then negative with respect to earth by an amount $\beta V_{o\ sat}^-$ and the potential at B falls exponentially as C charges down through R; diode D_1 is reverse biased. The circuit switches back to its permanently stable state when the potential at B reaches the value $\beta V_{o\ sat}^-$. Waveforms are illustrated in Figure 7.11.

The expression for the timing period may be obtained by making use of the general expression for an exponential charging period (equation 7.2), thus:

$$T = CR \log_e \frac{V_{o\ sat}^- - 0}{V_{o\ sat}^- - \beta V_{o\ sat}^-}$$

$$T = CR \log_e \frac{1}{1 - \beta}$$

Substitution for β gives

$$T = CR \log_e \left(1 + \frac{R_1}{R_2}\right) \tag{7.6}$$

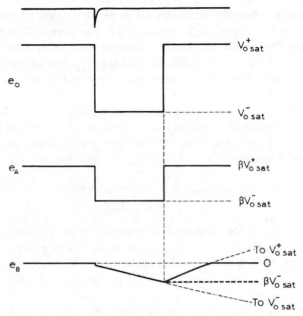

$V_o^+{}_{sat}$

e_o

$V_o^-{}_{sat}$

e_A

$\beta V_o^+{}_{sat}$

$\beta V_o^-{}_{sat}$

To $V_o^+{}_{sat}$

e_B

0

$\beta V_o^-{}_{sat}$

To $V_o^-{}_{sat}$

Figure 7.11   Waveforms for monostable multivibrator

Figure 7.12   Monostable multivibrator with period controlled by reference voltage

The circuit illustrated in Figure 7.12 shows an alternative arrangement for a monostable circuit. The timing period is controlled by the
magnitude of a negative reference voltage that is applied to the phase
inverting input terminal of the amplifier. The timing capacitor C connected between amplifier output and the non-phase inverting input
terminal A provides the necessary positive feedback path. In the
permanently stable state of the circuit the amplifier output is at
positive saturation, the input terminal A is at earth potential and input

273

terminal B is at the reference potential. A positive trigger of magnitude greater than E_{ref} applied to the terminal B brings the amplifier out of saturation. The circuit then switches regeneratively to its temporarily stable state in which the amplifier output is at negative saturation. The negative voltage step at the output is communicated to A by the capacitor C; the potential at A then rises exponentially as C charges up through R. The circuit switches back to its permanently stable state when the potential at A reaches E_{ref}. Waveforms are illustrated in Figure 7.13; the expression for the timing period may again be obtained by use of equation 7.2.

Thus, $$T = CR \log_e \frac{0 + (V_o^+{}_{sat} - V_o^-{}_{sat})}{0 - E_{ref}} \qquad (7.7)$$

The timing periods obtained with practical astable and monostable multivibrators, based on the circuits described, may be expected to show minor deviations from the values derived in the text. This is due primarily to the effect of amplifier offsets. The values of voltages and components used for practical circuits must ensure that amplifier limitations are not exceeded; the use of bounding circuits may involve some slight modification to the expressions derived for the timing periods.

7.2.3 BISTABLE MULTIVIBRATOR

The circuit shown in Figure 7.14 illustrates the use of an operational amplifier as a simple bistable multivibrator. The two stable conditions for this circuit are with the amplifier output at positive or negative saturation. It is held in one or other of these states by the positive feedback applied via resistors R_2 and R_1. A triggering pulse of suitable polarity applied to the phase inverting input terminal causes the circuit to regeneratively switch states.

7.3 Sine wave oscillators

An electronic oscillator is a device which continuously produces a repetitive time varying electrical signal. The characteristics of an oscillator performance which are normally of importance are the waveform, amplitude, and frequency of the signal it produces.

Figure 7.13 Waveforms for circuit of Figure 7.12

Figure 7.14 Bistable multivibrator

Operational amplifiers, suitably connected to external components, may be used as oscillators. The astable multivibrator, which was discussed in the previous section, is an example of such an application. Multivibrators produce non-sinusoidal oscillations; a different arrangement of the amplifier is necessary if sinusoidal oscillations are required.

The application of sufficient positive feedback will transform any amplifier into an oscillator; indeed, when using high gain, fast roll off amplifiers, it is all too easy to obtain inadvertent oscillations, because of undesired positive feedback, if proper attention is not paid to decoupling and frequency compensating techniques. When an amplifier circuit is designed specifically to produce oscillations a positive feedback loop is deliberately introduced into the circuit.

As an aid to understanding the action of so called 'feedback oscillators' consider the simple feedback loop illustrated in Figure 7.15.

275

The diagram shows an amplifier, gain A, with a fraction β of its output signal returned to its input via a feedback network. In the general case both A and β are frequency dependent and are represented mathematically as complex quantities. If the loop is broken at any point and a signal is injected there the signal reappears at the other end of the break multiplied by the loop gain βA. The condition that the circuit should produce continuous oscillations when the loop is closed is that the loop gain should be real, positive, and greater than unity. If this condition is satisfied any minute disturbance (for example noise) will start a signal the amplitude of which will grow continuously because of the amplification round the loop.

Figure 7.15 Representation of simple feedback loop

In order that the arrangement shown in Figure 7.15 should produce a sinusoidal oscillation of defined frequency the circuit components must be chosen so as to make the loop gain greater than unity only at the desired oscillation frequency. Values of loop gain greater than unity

Figure 7.16 Wien bridge oscillator

cause a continuous growth in signal amplitude and large amplitude signals eventually result in waveform distortion. For a stable amplitude of oscillations, with undistorted waveform, it is necessary to make the effective loop gain decrease with increase in signal amplitude. Oscillations then grow to some limiting stable amplitude at which the loop gain becomes exactly unity.

7.3.1 WIEN BRIDGE OSCILLATOR

The circuit shown in Figure 7.16 illustrates the use of an operational amplifier in a so-called Wien bridge oscillator. In this circuit feedback is applied between the output terminal and the non-phase inverting input terminal of the amplifier via the frequency dependent network Z_2, Z_1. The network produces zero phase shift at a frequency

$$f_o = \frac{1}{2\pi CR}$$

Oscillations thus take place at this frequency since the feedback is positive. The attenuation produced by the network

$$\left(\frac{Z_1}{Z_1 + Z_2}\right)$$

is one third at the frequency f_o. Negative feedback is applied to the amplifier via resistors R_2 and R_1 in order to reduce the loop gain to unity and so ensure a sinusoidal output waveform. If the amplifier had infinite open loop gain, oscillations would just be maintained for values of R_2 and R_1 such that

$$\frac{R_1}{R_1 + R_2} = \frac{1}{3}$$

In a practical circuit, in order to maintain a stable amplitude of oscillations, a nonlinear resistor is normally used for R_1; this makes the loop gain depend upon the amplitude of oscillations. Increase in amplitude of oscillations causes an increase in the current through R_1 which results in an increase in the value of R_1. An increase in the magnitude of R_1 means a greater amount of negative feedback and a consequent reduction in loop gain and signal amplitude.

The impedances Z_2, Z_1, R_2, R_1, in fact form the arms of a bridge network (a Wien bridge). It is the bridge unbalance voltage which constitutes the signal actually applied between the differential input terminals of the amplifier. Analysis of the bridge network shows that when $R_2 = 2R_1$ the bridge is balanced at a frequency

$$f_o = \frac{1}{2\pi CR}$$

In practice a small imbalance must always exist but the greater the open loop gain of the amplifier the closer is the bridge to balance and the greater is the frequency stability of the oscillator.

Figure 7.17 Wien bridge oscillator with f.e.t. amplitude stabilisation

The circuit illustrated in Figure 7.17 shows an alternative method of ensuring amplitude stability; a field effect transistor is used in place of the nonlinear resistor R_1. For small values of drain source voltage (below 'pinch off') a field effect transistor behaves very much like a linear resistor (R_{DS}) whose magnitude is determined by the voltage applied between the gate and source terminals of the f.e.t. In this circuit the oscillator output voltage is rectified by the diode D, filtered by $R_5 C_2$, and applied via potentiometer R_6 to the gate of the f.e.t. The arrangement ensures that R_{DS} takes on that value just necessary to maintain a particular amplitude of oscillation. The signal amplitude applied to the bridge must be small enough to ensure that the f.e.t. is working in the linear resistance region.

278

Figure 7.18 Quadrature oscillator

7.3.2 QUADRATURE OSCILLATOR

The circuit illustrated in Figure 7.18 may be used to generate two
sinusoidal signals in quadrature. The circuit uses two amplifiers: one
acts as a non-inverting integrator, the other as an inverting integrator
(see Sections 6.1.3 and 6.1.1). The two amplifiers are connected in
cascade to form a feedback loop. The feedback loop is represented by
the differential equations

$$RC \frac{de_s}{dt} = e_c, \quad RC \frac{de_c}{dt} = -e_s$$

The solution is represented by a sinusoidal oscillation of frequency

$$f = \frac{1}{2\pi CR}$$

In practice the resistor R_1 is made slightly larger than the other
resistors to ensure a sufficient positive feedback for oscillations. The
zener diodes, used to bound the output of the inverting integrator,
serve to stabilise the amplitude of oscillations.

7.4 Waveform generators

Signals with a wave shape other than sinusoidal are sometimes required.
Signal generators which provide a variety of waveforms are commonly

referred to as function generators; they are commercially available as complete test instruments. The variety of wave shapes provided generally determines the complexity and cost of the generator system. Facilities such as voltage control of frequency and the ability to provide a single wave or group of waves (triggered or gated operation) are sometimes made available.

In addition to their use by the electronics engineer function generators are applied in many other fields such as geophysics, biophysics and in education where they function as stimulators, testers, timers, etc. Function generator systems can be readily synthesised using operational amplifiers, an approach which is particularly useful when the need for a special purpose generator arises, or when a function generator is to be built in as an integral part of a piece of specialised equipment and a commercial function generator is inconvenient or prohibited by cost considerations.

Figure 7.19 Basic triangular squarewave generator

The basic wave shapes produced by most function generators are square and triangular. These waves can be shaped by nonlinear amplifiers or limiters to produce a variety of other wave forms, including a sinusoidal wave form.

There are two basic functions performed in a waveform generator system: a capacitor charging used to fix waveform periods and generate a triangular wave, and a comparator function used to sense capacitor voltage and switch between charge and discharge conditions. In the astable multivibrator circuit discussed in Section 7.2.1 both functions are performed by a single operational amplifier. The astable multivibrator gives a squarewave and a nonlinear triangular wave, but any loading of the triangular wave inevitably influences timing periods. In order to generate a linear triangular wave the capacitor must be charged with a constant current. The astable multivibrator can be modified for constant current charging (see Appendix A.1, Figure A.7) but it is more satisfactory to use two operational amplifiers in order to separate the linear capacitor charging function and the comparator function.

7.4.1 A BASIC TRIANGULAR SQUAREWAVE GENERATOR

A basic circuit for a triangular squarewave generator is given in Figure 7.19; it consists of an integrator and regenerative comparator connected in a positive feedback loop. Precise triangular waves are formed by integration of the squarewave which is fed back from the output of the comparator to the input of the integrator. With the comparator output at its positive saturation level the integrator output ramps down at the rate

$$- \frac{V_o^+}{CR} \text{ V/s}$$

until it reaches the lower trip point of the comparator:

$$- V_o^+ \frac{R_1}{R_2}$$

The comparator output then switches rapidly to its negative saturation level V_o^- and the integrator output then ramps up at the rate

$$- \frac{V_o^-}{CR} \text{ V/s}$$

When the integrator output reaches the upper trip point of the comparator:

$$- V_o^- \frac{R_1}{R_2}$$

the comparator again switches states and the process repeats. The waveform periods are determined by the relationships

$$T_1 = \frac{[V_o^+ - V_o^-] \dfrac{R_1}{R_2}}{\dfrac{V_o^+}{CR}} \; s \qquad (7.8)$$

$$T_2 = \frac{[V_o^+ - V_o^-] \dfrac{R_1}{R_2}}{- \dfrac{V_o^-}{CR}} \; s \qquad (7.9)$$

If the comparator positive and negative output limits have the same magnitude, $V_o^+ = - V_o^-$, $T_1 = T_2$ and the frequency of the oscillations is determined by the relationship

$$f = \frac{1}{T_1 + T_2} = \frac{R_2}{4 R_1 CR} \qquad (7.10)$$

Equation 7.10 has been derived from the assumption of ideal operational amplifier action. The performance limits of a practical circuit are determined by comparator slew rate and integrator bandwidth at the higher frequencies and by integrator drift at the lower frequencies. The slew rate of the amplifier used as the comparator makes the comparator transition times significant at the higher frequencies; finite integrator bandwidth (see Section 6.3.2) causes a time lag in the integrator output response to the sudden changes in its input signal level.

Bias current and input offset voltage of the amplifier A_1 used as an integrator give rise to an integrator output drift

$$\pm \frac{V_{io}}{CR} + \frac{I_B}{C} \qquad \text{(see Section 6.3.1)}$$

282

which, as the output of the comparator changes polarity, increases one
integration rate and decreases the other. The effect is only significant
at low frequencies. It causes a lack of symmetry in the generator
wave forms.

The effect of bias current and input offset voltage on the performance of A_2 is to introduce an equivalent error voltage at the non-
inverting input terminal of value

$$\pm \ V_{io} \ + \ I_B R_s \ (R_s = R_1 \| R_2)$$

This error voltage shifts both comparator trip points an equal amount
which in turn shifts the d.c. level of the triangular wave but leaves its
amplitude unchanged.

7.4.2 VARYING THE WAVEFORM CHARACTERISTICS OF THE BASIC GENERATOR

The waveform characteristics of the basic function generator system
of Figure 7.19 can be varied by strategically positioned potentiometers.
The circuit shown in Figure 7.20 gives one possible arrangement which
allows adjustment of frequency, waveform symmetry, triangular wave
d.c. offset and triangular wave amplitude. The circuit includes a zener
output limiting clamp on the comparator which sets the squarewave
amplitude at $\pm V_z$.

Adjustment of the timing resistor R controls the frequency and
does not alter other waveform characteristics. Potentiometer P_1
applies a voltage V_1 to the inverting input of the regenerative comparator amplifier A_2. This shifts both comparator trip points by an
amount V_1 / β, where β is the positive feedback fraction determined by
the setting of potentiometer P_2. The effect is to shift the d.c. level of
the triangular wave by an amount V_1 / β.

The setting of potentiometer P_2 determines the amount of hysteresis
in the regenerative comparator, which by its effect on the comparator
trip points controls the triangular wave amplitude. Change of triangular
wave amplitude is inevitably accompanied by a change in frequency;
decrease in triangular wave amplitude causes a proportional increase
in frequency.

Potentiometer P_3 applies a d.c. offset to the integrator resulting in
an increase in one timing period and a decrease in the other; in this
way it controls waveform symmetry but it also affects the frequency.

Figure 7.20 Waveform generator with control of waveform characteristics

Triangular wave amplitude control

P_2

A_2

Triangular wave d.c. level control

P_1

V_S+ V_S-

V_2

C

A_1

Symmetry control

P_3

V_S+ V_S-

R

Frequency control

Figure 7.21 Waveform generator with frequency unaffected by symmetry control

The circuit shown in Figure 7.21 is an alternative arrangement which allows a control waveform symmetry without at the same time altering the frequency. In the circuit resistor values R_1 and R_2, which control the comparator trip points, are chosen so as to give a triangular wave of amplitude approximately 10 V peak to peak allowing a single polarity triangular wave or ramp to be obtained by adjustment of the triangular wave offset control potentiometer P_1. The traces as given in Figure 7.21 show the control of wave form symmetry obtained by adjusting the symmetry control potentiometer P_3. Note that change of symmetry does not noticeably influence the frequency. There is however some interaction between the symmetry control potentiometer and the frequency control potentiometer at the extreme settings of the symmetry control, because of an unequal loading of the frequency control potentiometer on the run up and run down portion of

$$f = \frac{R_2}{40 \, R_1 CR} \, V_c$$

Figure 7.22 Four quadrant multiplier allows voltage control of frequency the triangular wave. If this interaction is not tolerable a follower can be used to buffer the output of the frequency control potentiometer. Temperature dependence of the forward voltage drops of diodes D_1 and D_2, which can be expected to cause frequency instability at the lower levels of frequency, are compensated by diodes D_3 and D_4.

7.4.3 WAVEFORM GENERATOR WITH VOLTAGE CONTROL OF FREQUENCY

It is often convenient to be able to control or modulate the frequency of a waveform generator with a control voltage. In order to achieve a voltage control of the frequency of the basic waveform generator of Figure 7.19 an arrangement must be sought whereby the magnitude of the current to the integrator can be varied in response to an externally applied control voltage. The arrangement must allow for sign inversion of the integrator current (brought about by the comparator transition) to allow the integrator output to ramp both up and down.

A four quadrant multiplier (see Section 5.7) connected between the comparator and the integrator input as shown in Figure 7.22 can be used to obtain a voltage control of the frequency of the basic waveform generator. The multiplier may be thought of as acting as a voltage controlled potentiometer. A four quadrant multiplier used in this way is quite practicable in these days of relatively inexpensive integrated circuit four quadrant multipliers (e.g. Analog Devices AD 533).

Assuming that the scaling factor of the multiplier is the normal 1/10, the square wave is multiplied by $V_c/10$ before being applied to the integrator. The equations for the waveform periods (equations 7.8 and 7.9) are in effect multiplied by $10/V_c$ and the expression for the frequency (equation 7.10) is multiplied by $V_c/10$. If the comparator positive and negative output limit are equal in magnitude the frequency of oscillations given by the circuit in Figure 7.22 is thus:

$$f = \frac{V_c}{10} \frac{R_2}{4 R_1 CR} \tag{7.11}$$

The accuracy of frequency control is mainly determined by the multiplier offset and nonlinearity. A symmetrical zener clamp is used at the output of the comparator so that the comparator output does not exceed ± 10 V full scale input for the multiplier. Other examples of voltage controlled waveform generators systems will be found in Appendix A.1.

Exercises

7.1 A regenerative comparator is required to give a negative output transition when an input signal rises through a value 3 V; the reverse transition is to take place when the input signal decreases through the value 2.7 V. The upper and lower limits of the comparator output are to be bound at $+5$ and -1 V respectively. Sketch a suitable circuit indicating appropriate component values.

7.2 A comparator is required to indicate when the average value of three input signals, weighted in the proportions, 1:2:3, is equal to 5 V. Sketch a suitable circuit.

7.3 The following component values are used in the circuit shown in Figure 7.6; $C = 0.1$ μF, $R = 50$ kΩ, $R_1 = 10$ kΩ, $R_2 = 50$ kΩ. The output voltage is bounded to the limits $+10$ V and -5 V. Calculate timing periods and sketch the circuit waveforms (analogous to Figure 7.7).

7.4 The circuit of Figure 7.8 is used to produce a non-symmetrical square wave of frequency 1 kHz and mark—space ratio 4:1. If $R_2 = 20$ kΩ, $R_1 = 5$ kΩ, $C = 0.01$ μF, and the output limits of the amplifier are fixed at ± 10 V what are the values of R_3 and R_4?

7.5 The following components are used in the circuit of Figure 7.12; $C = 0.01$ μF, $R = 10$ kΩ. The output limits of the amplifier are set at $+6$ V and -1 V. What range of values for the reference voltage are required for an output pulse of duration variable between 50 μs and 200 μs?

7.6 An operational integrator and a regenerative comparator are connected in a feedback loop as shown in Figure 7.19. The component values used are R = 10 kΩ, C = 0.1 μF, R_1 = 10 kΩ, R_2 = 20 kΩ, and the operational amplifiers have saturation output limits ± 12 V. Find the amplitude and time periods of the waveforms produced. Sketch the waveforms (assume ideal operational amplifier action). If the operational amplifier which is used as an integrator has input offset voltage V_{io} = 2 mV and bias current I_B = 0.6 μA calculate the waveform timing periods when resistor R is made 1 MΩ.

CHAPTER EIGHT

OPERATIONAL AMPLIFIERS IN INSTRUMENTATION APPLICATIONS

This chapter presents a miscellaneous collection of measurement and processing applications. Some of these are simply suggestions for the use of circuits given in previous chapters; others introduce new modes of operational amplifier use.

8.1 Transducer amplifiers

The first stage in any measurement system is often a transducer. This is a device for converting the quantity under investigation (the measurand) such as mechanical movement, temperature, pressure, force and the like into an electrical signal. Those readers requiring a simple introduction to a range of some of the more common transducers used in industry should refer to Appendix A4 of this book.

Operational amplifier based circuits are used extensively as transducer amplifiers to step up the power level of the electrical signal given by the transducer prior to subsequent signal manipulation. The choice of a suitable operational amplifier circuit configuration for use with a particular transducer requires an understanding of the electrical output characteristics of the transducer, in order that the amplifier input circuit may establish correct electrical operating conditions for the transducer. A knowledge of the expected frequency content of the transducer output signal is required in order that the amplifier may be designed to have a sufficiently wide bandwidth to allow a faithful reproduction of the signal waveform. Transducers which produce an output voltage proportional to the measurement variable require a high input impedance amplifier configuration (follower configuration) in order to minimise signal loading error, whereas transducers which give an output current which is proportional to the measurement variable (e.g. photodiodes) are conveniently processed using an operational amplifier configured as a current to voltage converter (see Section 4.5.1). Transducer measurements in which the transducer is remote from the amplifier are often bedevilled with the pick up of unwanted signals[1] and in such cases the use of a differential input amplifier circuit configuration can provide rejection of these unwanted signals (see Section 4.4).

8.1.1 BRIDGE AMPLIFIERS FOR RESISTIVE TRANSDUCERS

A wide range of transducers exist which consist essentially of a resistive element; resistive transducers are available which respond to temperature, light intensity and physical strain. When precise measurements are to be made using resistive transducers the transducers are normally included in the arms of a balanced bridge. Changes in the physical variable to which the transducer is sensitive cause an unbalance in the bridge, the extent of the unbalance being used to measure the change in the physical variable. Operational amplifiers are well suited for application in such balanced bridge circuits.

There are several ways in which an operational amplifier can be used to indicate the extent of a bridge unbalance; the most suitable configuration is dependent upon the nature of the particular application. Here are some of the points that have to be considered in choosing a particular circuit: earthed or floating bridge voltage supply; earthed or floating unknown resistor; output voltage linearly related to changes in the unknown resistor; sensitivity of the arrangement dependent on the bridge impedance level. This last point determines

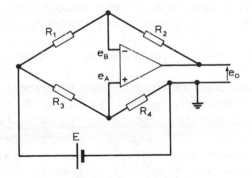

Figure 8.1   Bridge amplifier, earthed bridge supply

whether or not the arrangement is sensitive to changes in the ambient temperature affecting all the arms of the bridge.

The circuit illustrated in Figure 8.1 is basically an application of the subtractor amplifier (Section 4.3.1). Referring to Figure 8.1 and assuming that the amplifier behaves ideally, the following analysis holds:

$$\left.\begin{aligned} e_B &= e_o + \frac{(E - e_o)\,R_2}{R_1 + R_2} \\ e_A &= E\,\frac{R_4}{R_3 + R_4} \end{aligned}\right\} \tag{8.1}$$

But
$$e_A = e_B$$

Substitution and rearrangement give

$$e_o = \left[\frac{R_4 - \dfrac{R_2}{R_1}\,R_3}{R_4 + R_3}\right] E \tag{8.2}$$

There are two ways in which the circuit may be used dependent upon which arm of the bridge the unknown resistor is connected in. One method is to make $R_1 = R_o$ and $R_2 = R_x = R_o\,(1 + \alpha)$ the unknown, R_3 is made equal to R_4 both, say, some value R. Substituting these values in equation 8.2 gives

$$e_o = \frac{\alpha}{2} E \tag{8.3}$$

Used in this way the circuit gives an output voltage which is linearly dependent upon $(R_x - R_o)$, the difference between the unknown and the standard. Linearity is maintained for large deviations from bridge balance. Perhaps a disadvantage of the arrangement is that the unknown resistor is floating.

Another way of using the circuit is to place the unknown resistor in the position occupied by R_4. We make $R_3 = R_o$, $R_4 = R_x = R_o\,(1 + \alpha)$ and $R_1 = R_2 = R$. Substituting these values in equation 8.2 gives

$$e_o = \frac{\alpha}{2 + \alpha} E \tag{8.4}$$

With this arrangement the output is linear only for small deviations in the unknown ($\alpha \ll 2$). The connection is useful however when one end

292

of the unknown must be earthed. The amplifier output does not have to supply the current passing through the unknown resistor, thus large currents may be passed through the unknown if this is required by the application.

An advantage of both arrangements of the circuit is the earthed bridge supply and the independence of the output on bridge impedance levels. The circuit does not provide amplification however and the measurement of small resistance changes would probably require the use of an additional amplifier. Care must be taken to ensure that the maximum common mode voltage for the particular amplifier in use is not exceeded.

Figure 8.2 Bridge amplifier, single ended amplifier

A single ended input amplifier may be used in the circuit shown in Figure 8.2. Bridge unbalance causes a voltage

$$\frac{E\alpha}{4\left(1 + \dfrac{\alpha}{2}\right)}$$

to be developed across the bridge. In order to force the amplifier input voltage (e_ϵ) to zero the amplifier output voltage develops a voltage at A equal to the bridge unbalance voltage.

293

Thus,

$$e_o \frac{R_1}{R_1 + R_2} = \frac{E\alpha}{4\left(1 + \dfrac{\alpha}{2}\right)}$$

and

$$e_o = \frac{R_1 + R_2}{R_1} \frac{E\alpha}{4\left(1 + \dfrac{\alpha}{2}\right)} \qquad (8.5)$$

$$\left(\text{linear for } \frac{\alpha}{2} \ll 1\right)$$

The circuit provides amplification of bridge unbalance voltage and has the advantage of independence on bridge impedance levels. The necessity for a floating bridge supply may sometimes be a disadvantage.

Figure 8.3 Bridge amplifier, earthed or floating supply

Another bridge circuit is shown in Figure 8.3. In this circuit the feedback round the amplifier causes the opposing corners of the bridge to be at the same potential. The amplifier output voltage establishes the differential current into the bridge needed to maintain this condition. Analysis of the circuit follows.

Summing currents at B

$$\frac{E - e_B}{R_o} - \frac{e_B}{R_o} + \frac{e_o - e_B}{R} = 0$$

Summing currents at A

$$\frac{E - e_A}{R_o} - \frac{e_A}{R_o(1 + \alpha)} - \frac{e_A}{R} = 0$$

Equating $e_A = e_B$ and rearranging give

$$e_o = \frac{R}{R_o} E\alpha \frac{1}{(1 + \alpha)\left(1 + \frac{R_o}{R}\right) + 1} \qquad (8.6)$$

(linear for $\alpha \ll 1$)

The circuit may be used with an earthed bridge supply but it has the disadvantage of a sensitivity dependent on bridge impedance levels. The amplifier type used should be insensitive to the common mode voltage at the input.

8.1.2 CHARGE AMPLIFIERS

Some transducers have a very high output impedance which is essentially capacitive; examples are piezoelectric accelerometers, pressure transducers and capacitive microphones. Transducers of this kind operate by producing a charge which is proportional to the measurement variable; the charge can be converted into a voltage by using an operational amplifier connected as a current integrator. The current integrator or charge amplifier arrangement has the desirable feature that external capacitance in parallel with the transducer (including cable capacitance) is connected between the virtual earth at the amplifier summing point and earth. In the ideal case this stray capacitance is uncharged and changes in its value do not influence the output signal.

A theoretical charge amplifier circuit is shown in Figure 8.4(a). The output of the capacitive transducer is represented by an equivalent circuit consisting of a voltage source e_t in series with a capacitance C_t. The amplifier gives an output signal $e_o = -(C_t/C_f)e_t$. In a practical charge amplifier it is necessary to provide a d.c. path for amplifier bias current in the form of a resistor R_f connected between the amplifier output terminal and the inverting input, otherwise continuous charging of C_f causes the output to drift into saturation. The presence of the resistor R_f limits the lower bandwidth limit of the charge amplifier to the frequency

$$f_L = \frac{1}{2\pi C_f R_f}$$

Figure 8.4 Charge amplifier (a) Ideal charge amplifier (b) Practical charge amplifier and Bode plots

Bode plots for the practical circuit are illustrated in Figure 8.4(b). Note that for signal frequencies less than f_L the output is proportional to the differential of the input signal. A very large value for R_f is required if the amplifier output is to reproduce faithfully slow changes in the measurement variable, and this normally requires the use of a low bias current operational amplifier (an f.e.t. input amplifier or for the ultimate in low bias current a varactor bridge amplifier) in order that offset and drift error should not be excessive. The upper frequency closed loop bandwidth limit of the charge amplifier is determined by the frequency at which $1/\beta$ and the open loop gain plot intersect. If it is required to restrict the upper bandwidth limit this can be accomplished by connecting a resistor R_1 in series with the transducer as shown in Figure 8.4.

8.2 Resistance measurement

Operational amplifiers can be used to establish a constant current in an unknown resistor, the voltage across the resistor then being used as a direct and precise measurement of the unknown resistor value. A simple, yet nevertheless accurate, method of resistance measurement can be made with an operational amplifier connected in the basic inverting feedback configuration (see Sections 1.2 and 4.2). The principle is illustrated in Figure 8.5. The unknown resistor, which must not be

Figure 8.5 Operational amplifier allows in-circuit resistor measurement

directly earthed, is connected between the amplifier output and the inverting input. A standard resistor value connected at the input and supplied by a reference voltage forces a constant current through the unknown resistor and the value of the amplifier output voltage is directly proportional to the unknown resistor value. A particular feature of this technique is that it can be used for in-circuit testing of components. Since one end of the unknown resistor is connected to the virtual earth at the inverting input terminal of the amplifier any other path to earth at this end of the resistor will not affect the reading (except in so far as extra paths to earth at the inverting input decrease the loop gain and increase the noise gain $1/\beta$). Paths to earth at the amplifier output are simply driven by the amplifier output voltage without affecting the test current through the unknown resistor R_f.

8.3 Capacitance measurement

Many test circuits for capacitance measurement, particularly those for the measurement of small capacitance values, are bedevilled by the presence of stray capacitance to earth at the test point. This difficulty can be overcome by using a test circuit of the type shown in Figure 8.6. The capacitance whose value is to be measured is connected to the virtual earth at the summing point of an operational amplifier; stray capacitance to earth at this point has a negligibly small voltage across

$$C_x = \frac{V_o}{V_{in}} C_1$$

$$\frac{1}{B} = \frac{1 + j\omega(C_x + C_s)R_f}{1 + j\omega\,C_s\,R_f}$$

$$\frac{V_o}{V_{in}} = \frac{j\omega\,C_x R_f}{1 + j\omega\,C_s R_f}$$

Figure 8.6 Capacitance measurement

it and therefore does not affect the measurement. Feedback is via a known standard capacitor connected between the amplifier output and the inverting input terminal. The circuit is supplied with an alternating test signal and the amplifier produces an output signal of magnitude proportional to the ratio of the two capacitor values.

A resistor R_f is connected in parallel with the standard capacitor C_s in order to prevent output drift which would otherwise arise because of a continuous charging of C_s by amplifier bias current. Amplifier bias current through R_f causes a d.c. output offset but this is not significant provided that it does not limit the output swing. The frequency of the test signal which is used should lie between a decade higher than the break frequency $1/2\pi C_s R_f$ and a decade lower than the intersection of $1/\beta$ with the amplifier open loop response (see the Bode plots in Figure 8.6). The value of the standard capacitor C_s should be chosen so that it is of the same order of magnitude as the unknown capacitor C_x.

8.4 Hot wire anemometer with constant temperature operation[2]

Hot wire anemometers are often used to measure air speed, the underlying principle of operation resting upon the cooling of an electrically heated platinum element produced by the movement of air over the filament. An operational amplifier can be used to provide constant temperature operation of the filament, overcoming the danger of burnout possible with constant current operation and the relative insensitivity at low air speeds of constant voltage operation. In the arrangement illustrated in Figure 8.7 the heated filament is included as one arm of a balanced bridge. The bridge is supplied by the output voltage of the operational amplifier with a simple transistor emitter follower being used to boost the amplifier output current (a Darlington connected transistor pair can be used for greater currents). The amplifier output voltage changes in such a way as to force the input error voltage towards zero and in so doing it establishes the bridge balance condition

$$R_f = \frac{R_1 R_2}{R_3}$$

Platinum has a positive temperature coefficient of resistance: if air flow over the filament increases it looses more heat, but the amplifier

Figure 8.7 Operational amplifier forces constant filament temperature in hot wire anemometer

output voltage increases so as to increase the power dissipation in the filament and hold its resistance and hence its temperature constant.

The output voltage of the amplifier represents an amplified form of the filament voltage and gives a measure of the air speed over the filament. Output varies nonlinearly with air speed, sensitivity being greater at the lower speeds. The operating temperature of the filament is set by varying the value of resistor R_2; some experimentation is needed to find the setting which gives best sensitivity. It is possible for the circuit to remain inoperative when switched on. The emitter follower may not be brought into conduction because the operational amplifier output is in negative saturation; a positive offset applied to the non-inverting input of the operational amplifier ensures operation at switch-on.

Heated thermistors are sometimes used in air flow measurements; they too can be worked at constant temperature using an operational amplifier circuit arrangement similar to that of Figure 8.7. It is simply necessary to interchange input leads to the operational amplifier because thermistors have a negative temperature coefficient of resistance. Their resistance decreases with rise in temperature. Constant temperature operation makes for more rapid measurements of changes in flow since there is no thermal delay whilst a change in temperature of the sensing element takes place.

8.5 Chemical measurements

Operational amplifier circuits readily lend themselves to performing the individual functional operations necessary in many chemical instrumentation systems; operational amplifier based circuitry has been used in electrochemical instrumentation for a number of years. The majority of chemical instruments operate on d.c. and slowly varying signals well within the bandwidth range of even the slowest operational amplifiers, and today's low drift amplifiers are ideally suited for chemical instrumentation applications. Modern operational amplifiers allow an upgrading of the earlier circuit designs; also they are easier to use than the early operational amplifiers making it possible for the experimental chemist to construct his own measurement system even though he is comparatively inexperienced in electronic design.

Many of the applications presented in this and other chapters of the book will be found suitable for use in chemical instruments. The operational amplifier approach is applicable in all stages of the measurement system in providing suitable operating conditions for the transducer, and any subsequent modifying and conditioning of the signal which is required to provide a suitable input to a final read-out system. The designer must have a familiarity with the electrical characteristics of his chemical transducer and he needs to understand the significance of amplifier d.c. offset and drift specifications and their relationship to d.c. errors. Amplifier types which will enable the measurement system to achieve the required accuracy can then be chosen with confidence. Some transducers, e.g. glass electrodes used for pH measurements, have a very high internal impedance; they provide a signal potential which must not be allowed to supply any appreciable current. An operational amplifier connected in the high input impedance follower configuration provides an excellent method of isolating such transducers. Using a low bias current f.e.t. input operational amplifier (e.g. Analog Devices AD 523), current drain can be reduced to less than 1 pA. Signals from transducers which provide an output current, e.g. gas chromatographs, flame ionisation detectors, can be measured using an operational amplifier connected in the current to voltage converter configuration (see Section 4.5.1). Some transducers provide a floating output signal; this can be converted into a single ended signal reference to earth using either a single operational amplifier connected in the differential configuration or a fully developed differential amplifier configuration (see Section 4.4).

8.5.1 POTENTIOSTATS AND GALVANOSTATS

There are operational amplifier circuits which have been developed specifically for use in chemical instrumentation. Many electrochemical measurements involve the use of a three-electrode cell, and so-called potentiostats and galvanostats provide defined operating conditions during the measurement of potential/current variations at an electrode. Operational amplifiers can be configured so as to provide both potentiostatic and galvanostatic conditions.

In a potentiostat the three electrodes in the electrochemical cell are the so-called working electrode, which is to be maintained at a controlled potential, a reference electrode, and a counter electrode which is usually an inert metal such as platinum or mercury, and serves merely to allow current to flow through the cell. The requirements of the potentiostat are that:

1. A potential difference E (which may or may not be time invariant) is maintained between the working and reference electrode.
2. Only a negligibly small current flows through the reference electrode, as only under this 'potentiostatic' condition does a reference electrode behave as such.
3. The current flowing into the working electrode is measured.

Galvanostats use the same type of three-electrode cell but galvanostatic conditions require that a control current be made to flow between the working and counter electrodes and that the potential difference between the working and reference electrode be measured.

There are a variety of ways in which operational amplifiers can be used to impose potentiostatic and galvanostatic operation. Potentiostatic operation requires that the cell be driven by a low impedance voltage source, whereas galvanostatic operation requires a current source drive. Both systems require that the reference electrode be isolated by a high input impedance buffer. Since the ways in which operational amplifiers can be configured to provide either voltage drive or current drive are many and varied, so also are there many potentiostatic and galvanostatic arrangements.

A simple single amplifier potentiostat is shown in Figure 8.8. The working electrode is earthed and the control potential difference E is applied between the reference electrode and the inverting input terminal of the operational amplifier. The operational amplifier output is connect-

To recorder for
current measurement

Figure 8.8 Single-amplifier potentiostat arrangement

ed to the counter electrode, and in maintaining a virtual earth at the
non-inverting input of the amplifier feedback forces the potential
difference E to appear between working and reference electrodes. A
resistor connected in series with the amplifier output and the counter
electrode is used to sense and measure the cell current. Note that this
circuit has the very high input impedance characteristic of the follower
configuration. The large input impedance means that only a very small
current flows to the reference electrode. The use of a low bias current
f.e.t. input operational amplifier (e.g. Analog Devices AD 506L) can
reduce the reference electrode current to a few pA, or for very low
voltage drift of the reference electrode and low current, a chopper
stabilised amplifier can be used (e.g. Analog Devices type 233J).

In the circuit of Figure 8.8 a floating signal source is required to
provide the cell voltage. An alternative potentiostat which uses three
operational amplifiers is shown in Figure 8.9. The three-amplifier
configuration allows the summation of both a constant and a time
varying control voltage, and also allows for an offset of the output
current readout. Amplifier A_1, connected as a unity gain follower, is
used to isolate the reference electrode and prevent current being drawn
through it. Amplifier A_2 maintains a virtual earth at its summing point
because of negative feedback through the counter electrode. The virtual
earth condition requires that

$$\frac{E_{ref}}{R_1} + \frac{V_C}{R_2} + \frac{V_{C(t)}}{R_3} = 0$$

i.e.
$$E_{ref} = - V_C \frac{R_1}{R_2} + V_{C(t)} \frac{R_1}{R_3}$$

303

$$E_{ref} = -\left[V_C \frac{R_1}{R_2} + {}^.V_{C(t)} \frac{R_1}{R_3} \right]$$

$$V_o = -\left[I_{cell} - I_B \right] R_f$$

Figure 8.9 Three-amplifier potentiostat

A constant V_C and a time varying voltage $V_{C(t)}$ can be summed to give E_{ref}. Amplifier A_3 is connected as a current to voltage converter; the working electrode connected to its summing point is maintained at earth potential and a voltage proportional to the cell current is produced at the low output impedance of amplifier A_3.

A simple one-amplifier circuit for imposing control current galvanostatic conditions on an electrochemical cell is shown in Figure 8.10. The arrangement requires that the cell be allowed to float and requires

$$I_{cell} = \frac{V_{in}}{R}$$

Figure 8.10 Galvanostat arrangement—cell floating

a high input impedance measuring circuit in order to measure the potential difference between working and reference electrodes.

A galvanostat arrangement in which the working electrode is earthed is shown in Figure 8.11. In this circuit, amplifier A_1 connected as a current source configuration (see Section 4.6.2) must be operated with floating power supplies (battery operation for example). Amplifier A_2 connected as a unity gain follower isolates the reference electrode and reads out the voltage between working electrode and reference.

Figure 8.11 Galvanostat—earthed cell

In potentiostat or galvanostat applications requiring a slowly varying voltage or current, an operational amplifier connected as an integrator can be used to provide a slowly varying voltage ramp. Alternatively, one of the waveform generator circuits discussed in Section 7.4 can be used to provide the necessary control signal waveform.

8.6 Further examples of transducer interfacing

8.6.1 TEMPERATURE MEASUREMENT USING A THERMOCOUPLE

The typical circuit shown in Figure 8.13[3] may be used with a K-type (see Appendix A4) thermocouple for the measurement of temperatures in the range 0 to 100°C. If the component values are changed then this circuit may be used with other types of thermocouple at different temperatures.

Figure 8.12 Temperature measurement using a thermocouple and an operational amplifier

The cold junction is at ambient temperature and any variation in this is compensated by the change, about 2 mV/°C, in the forward voltage drop of diode D2. This voltage is reduced by a potential divider to the thermocouple output. The result, amplified by ICI and displayed on any 100 μA meter, is proportional to the temperature of the measurement junction of the thermocouple. The single voltage supply is intended to be from a standard 9 V battery and ICI should be any operational amplifier of the 741 type.

The meter zero is set by adjustment of R3; the span by R9 and the calibration procedure is as follows:

1. Place the measurement junction of the thermocouple in freezing water to give 0°C.
2. Adjust R3 to set a reading of zero on the meter.
3. Place the measurement junction of the thermocouple in boiling water to give 100°C.
4. Adjust R9 to give full scale deflection on the meter.
5. The circuit will then be calibrated to give a reading of 1 μA/°C.

306

8.6.2 LIGHT SENSITIVE SWITCHING

The circuit shown in Figure 8.13[4] is that of a very sensitive light-operated relay. The relay is energised when the light striking the light-dependent resistor, NORP12, exceeds a level determined by the setting of VR1. If R1 and NORP12 were interchanged, the relay would be energised in poor light conditions and would de-energise when the light intensity reached the pre-set level.

8.6.3 SENSING AN ANALOGUE LIGHT LEVEL

The circuit shown in Figure 8.14[5] uses a general purpose photodiode and an operational amplifier requiring only a low input bias current; typically an LM 308 or an FET input type. The input signal to the operational amplifier is obtained from the photodiode. The output produced is a steady d.c. indication of the light level and this is particularly useful in photometric applications.

The values of components shown give approximately 14 V/mW/cm^2 of irradiation. The values of R1 and R2 may be reduced for less sensitivity but must be kept equal. For values less than 100 kΩ, a less sophisticated amplifier such as the μA741 may be used. The 1000 pF capacitors may be increased in value to

Figure 8.13 Extremely sensitive light-operated relay using a light-dependent resistor

Figure 8.14   Analogue light level sensor

reduce any ripple from a.c. lighting or to control the response time accordingly.

8.6.4   INTERFACING LINEAR HALL EFFECT TRANSDUCERS (LOHETs) AND OPERATIONAL AMPLIFIERS

Figure 8.15[6] shows three methods of interfacing an LOHET with common single supply operational amplifier circuits. The operational amplifier characteristics limit the output voltage (V_0) equations at the high and low ends.

The circuits shown can be used with adjustable gain and adjustable offset, although the adjustments will not be completely independent. One method is to adjust the gain to the desired value with V_1 at approximately half the operational amplifier supply

308

$$V_o = (V_1 - V_{IN})\frac{R_2}{R_1} + V_1$$

(a)

$$V_o = V_{IN} \left(1 + \frac{R_2}{R_1}\right)$$

(b)

$$V_o = V_{IN}$$

(c)

Figure 8.15 Interfacing LOHET devices and single supply operational amplifiers. (a) Inverting; (b) non-inverting; (c) voltage follower voltage. Then adjust V_1 to give the exact offset at the V_0 required for the application.

8.7 Phase shifting circuit (all pass filter)

The circuit shown in Figure 8.16 uses an operational amplifier to generate an arbitrary phase shift. All frequencies within the closed loop bandwidth are passed at unity gain but with a phase shift which varies with frequency. If the resistor R' connected to the non-inverting input terminal is made variable, the circuit provides a convenient means of phase adjustment. A phase variation between 0 and almost 180° is possible; the output signal is available at the low output impedance output of the operational amplifier.

Figure 8.16 Operational amplifier phase shifter

8.8 Capacitance multipliers

Large value capacitors are rather expensive and bulky and designers generally try to avoid using them. In setting long time constant values, resistance values are often made large to allow the use of smaller value capacitors, but inevitably there are always design trade-offs such as increased offset and noise in operational amplifier circuits, and sometimes resistor values are fixed by other circuit conditions making a large capacitance value essential. In such cases it is still possible to avoid the use of a large capacitor by using a capacitance multiplier circuit to increase the effective value of a smaller capacitor. Capacitance multipliers can also be useful in creating an effectively variable capacitance from a fixed value capacitor.

Figure 8.17 Variable capacitance multiplier

The circuit illustrated in Figure 8.17 allows the effective capacitance between the input terminal and earth to be adjusted by simply varying the gain of the inverting amplifier stage A_2. Amplifier A_1 acts as a unity gain follower; its function is to isolate the capacitance formed by the circuit from the loading imposed by the inverting amplifier stage. The practical limit on the size of capacitance that can be created is determined by the fact that the capacitance multiplication achieved is almost the same as the gain of the inverting amplifier stage; thus the larger the capacitance multiplication, the smaller is the allowable input signal that can be tolerated without exceeding amplifier A_2 output voltage limitation. Choice of amplifier type to be used in the practical circuit is determined largely by signal frequency requirement. Amplifier bias currents and input offset voltage cause an offset voltage in the output of A_2 but this is not of major significance other than in its effect in limiting signal output sweep. The bias current of amplifier A_1 represents a leakage current of the synthesised capacitor, but it is not as is the usual case, a function of the applied voltage. Bias current continues to flow even with zero applied input signal, and if the synthesised capacitor is used say to perform a timing function the bias current causes an offset error rather than a scaling error caused by the leakage current of a conventional capacitor.

Another example of a capacitance multiplier circuit is shown in Figure 8.18. The operational amplifier is connected as a unity gain follower and neglecting offsets its output voltage at any instant is equal to the voltage across the capacitor C_1. This output voltage is fed back via resistor R_2 to the input end of resistor R_1 in a 'bootstrap' fashion and increases the effective capacitance value between the input terminal and earth. Circuit behaviour is represented by the performance equation and equivalent circuits given in Figure 8.18. Note that the multiplied capacitance has an effective resistance in series with it so that high Q capacitors cannot be realised and the circuit cannot be used for tuned filter applications. However, it can be used in timing applications and simple RC low pass filters where resistance is always connected in series with the capacitor.

In timing application the multiplied capacitance is connected in series with an external resistor to a d.c. supply voltage. The voltage across the actual capacitance C_1 rises exponentially but the time constant is determined by the multiplied capacitance value; also the voltage across C_1

Note with e_{in} open circuit
offsets continuously charge C_1

Figure 8.18 Capacitance multiplier

312

Figure 8.19 Timing circuit not requiring excessively large CR values

is available at the low impedance output terminal of the operational amplifier. A timing circuit using the principle is illustrated in Figure 8.19. The timing period is initiated by opening the switch when the voltage across capacitor C_1 rises exponentially governed by the time constant

$$C_{\text{eff}}\,[R_{\text{eff}} + R_3] \;=\; C_1 \left[1 + \frac{R_1}{R_2}\right]\,[R_2 \| R_1 + R_3]$$

The second amplifier in the circuit is used as a comparator to sense when the exponentially rising voltage reaches a reference value set by the potential divider $R_4 R_5$; when this voltage level is reached the output of A_2 switches from its positive to its negative saturation value. Long timing periods can be obtained with this circuit without the necessity for very large CR values. With the component values shown in the circuit the time delay is approximately 90 s.

8.9 Averaging

Providing an indication of the average value of a signal or signals is a signal processing operation which must sometimes be performed in instrumentation systems. Average value can have different meanings: it can refer to the arithmetic mean value of a series of signals, or it can be the time average of a signal or the time average of a sum of several signals. Time average can be a continuously running average or it can be an average value over some specified time interval.

8.9.1 ARITHMETIC MEAN

The arithmetic mean of a number of signals $e_1, e_2, \ldots e_n$, is simply found by adding together the signal values and dividing the resultant signal sum by the number of signals. This operation is readily performed by the inverting adder circuit of Figure 8.20. The coefficients which

$$e_0 = -\frac{1}{n}\left[e_1 + e_2 + e_3 + \ldots e_n \right]$$

(a)

$$e_0 = -\left[a_1 e_1 + a_2 e_2 + \ldots a_n e_n \right]$$

$$a_k = \frac{\| \text{sum } R_{1_1} \; R_{2_1} \ldots R_n}{R_k}$$

(b)

Figure 8.20 Forming an arithmetic mean (a) Output of summing amplifier gives arithmetic mean of input signals (b) Matching resistors used for weighted average

314

are used to multiply individual signals can be adjusted by choice of resistor value to give an output signal that is equal to a weighted average of input signals. The use of matching resistors in dividing networks connected as shown in Figure 8.20(b) ensures that the sum of the weighting coefficients is always unity. In a practical circuit due consideration must of course be given to bandwidth limitation and offset errors.

8.9.2 TIME AVERAGING

Running time averages are normally obtained using some form of first order low pass filter. Consider the arrangement in Figure 8.21 in which a signal which has high frequency fluctuation superimposed on a constant or slowly varying mean value is applied to a long time constant CR filter. The capacitor charges up and if the time constant CR is much greater than the period associated with the input signal fluctuations the capacitor voltage takes up a steady value equal to the mean value of the input signal, with a small fluctuation superimposed upon it. Long time constants without recourse to big CR values can be realised by making use of the capacitor multiplying principle as shown in Figure 8.21(b).

$$e_0 = \frac{1}{1+j\omega\tau}e_{in}$$
$$\tau = CR$$

(a)

$$e_0 = \frac{1}{1+j\omega\tau}e_{in}$$

$$\tau = C_1R_1 \left[1 + \frac{R_3}{R_2} + \frac{R_3}{R_1} \right]$$

(b)

Figure 8.21 Averaging filters (a) Simple RC averaging filter (b) Long time constant averaging filter uses capacitance multiplier

315

It is sometimes necessary to average out noise which is superimposed upon a signal which has sudden large variations in its mean value (e.g. a noisy square wave). A long time constant first order low pass filter is unsuitable because its output is unable to follow rapid changes in mean value. A modification to the circuit of Figure 8.21(b) which is shown in Figure 8.22 can sometimes be used in such applications. The back to back diodes connected across resistor R_1 bypass resistor R_1 when large changes in the mean level of the input signal occur and give the circuit the shorter time constant CR_3. Small noise fluctuations of amplitude less than the diode voltage drop are not passed by the diodes so that the circuit has a much larger time constant for the small noise fluctuations.

Figure 8.22   Averaging filter with time constant dependent upon signal amplitude

8.10 Precise diode circuits

An ideal diode is a device which exhibits zero resistance for applied voltages of one polarity and an infinite resistance to the opposite polarity. When used in a simple rectifier circuit as in Figure 8.23(b) it would completely block signals of one polarity and transmit perfectly those of the other. Real solid state diodes pass no appreciable current for small applied voltages and exhibit a nonlinear finite resistance when conducting; the voltage drop across a forward biased diode has a marked temperature dependence. These characteristics cause perform-

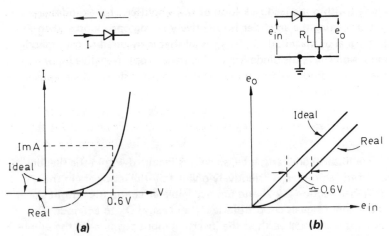

(a)

(b)

Figure 8.23   Real and ideal diodes (a) Real and ideal diode characteristics (b) Real
and ideal ½ wave rectifier

ance errors at low signal levels when a solid state diode is used in a
simple rectifier circuit. Diode deficiencies can be largely overcome by
combining them with operational amplifiers; the diode operational
amplifier circuit of Figure 8.24 produces a near ideal half wave rectifier
characteristic. In the circuit of Figure 8.24, diodes D_1 and D_2 are

Figure 8.24   Operational amplifier diode circuit performing ideal ½ wave
rectification

included within the feedback loop of the amplifier. If the diodes are non-conducting the amplifier is effectively acting open loop and an input signal of magnitude V_f/A_{OL} is all that is required to cause diode conduction (V_f is the diode forward voltage drop). Negative input signals cause diode D_1 to conduct and the output signal which appears at the cathode of D_1 is

$$e_o = -\frac{R_2}{R_1} e_{in}$$

The nonlinear diode resistance, since it is included within the feedback loop of the amplifier, is effectively divided by the loop gain in the circuit (see Section 2.2.2) and has negligible effect on the output signal. Positive input signals cut off diode D_1 and cause D_2 to conduct. This maintains the virtual earth at the inverting input terminal of the amplifier, and the output signal is zero since it is connected directly to this point via resistor R_2.

There are a variety of operational amplifier diode circuits which are used for precise diode action; some examples are given in the following sections. The main performance limitation in precise diode circuits arise as a result of amplifier slew rate; because of slew rate limitations the amplifier output voltage takes a finite time to overcome diode forward voltage drops and this restricts the frequency response of precise diode circuits.

8.10.1 ABSOLUTE VALUE CIRCUITS (FULL WAVE RECTIFIERS)

An absolute value circuit or full wave rectifier gives an output signal which is directly proportional to the instantaneous magnitude of the input signal applied to it; it converts bipolar signals into unipolar form. Absolute value circuits are used extensively in a.c. measurements where the initial signal processing usually consists of an absolute value conversion. They are used to fit bipolar inputs to single quadrant devices, e.g. in log-antilog computation circuits.

The basic operation performed by an operational amplifier precise full wave rectifier is that of switching gain polarity. The circuits are arranged so that when the polarity of the input signal changes so also does the overall gain polarity, thus maintaining a constant polarity output signal. There are several operational amplifier circuit configurations possible which provide full wave rectification. Some of the factors in making a choice between the different configurations are:

Figure 8.25 Absolute value circuits having input impedance set by input resistor values

319

input impedance requirements, a possible requirement of a summing capability and cost/performance as set by the number of close tolerance resistors required in the circuit.

Three absolute value circuits are shown in Figure 8.25; they all use an inverting amplifier configuration at their input and therefore have an input impedance determined by the input resistor values. Figure 8.25(a) is probably the most obvious approach to full wave rectification; it consists of the precise half wave rectifier circuit of Figure 8.24 added to a summing amplifier. Negative input signals are simply passed by the inverting summer and blocked by the half wave rectifier. Positive signals are inverted and passed by the half wave rectifier; they are multiplied by 2, summed with the input and inverted by amplifier A_2. The circuit requires accurate matching of two pairs of resistors plus the selection of a half value resistor.

The circuit of Figure 8.25(b) requires only two matched resistors. Two parallel signal paths exist between input and output; positive input signals are passed at unity gain by amplifier A_1 (via D_1) and blocked by amplifier A_2. Negative signals are inverted and passed by A_2 but because of diode D_2 used to prevent saturation of A_1 they also appear at the inverting input of A_1 and draw a current from the output of A_2 through the resistor R_1. The value of R_1 must be sufficiently large to minimise this current.

The circuit of Figure 8.25(c) is an absolute value circuit which allows summation of signals at its input simply by adding extra input resistors. Equal value resistors are used throughout the circuit. Positive input signals reverse bias D_2 and forward bias D_1 and the two inverting amplifiers are in cascade. Negative input signals forward bias D_2 and reverse bias D_1 and there are then two feedback paths to the inverting input A_1. Assuming ideal circuit performance

$$I_1 = I_2 + I_3$$

giving

$$\frac{e_{in}}{R} = -\frac{e_o}{3R} - \frac{e_2}{R}$$

But $e_2 = e_3 = e_o \, 2R/3R$ and substitution gives

$$e_o = -e_{in} \qquad\qquad \text{for } e_{in} \text{ negative}$$

The circuit shown in Figure 8.26 is a high input impedance absolute value circuit which uses a follower connected operational amplifier at its input. The circuit requires only two closely matched resistors. The

Figure 8.26 High input impedance absolute value circuit

action of the circuit is as follows: positive input signals cause D_1 to conduct and D_2 to block, the feedback loop is connected round amplifier A_1 via A_2 and the signal is passed to the output of A_2 at unity gain. Because two amplifiers are included within the feedback loop for positive input signals additional phase compensation may be required. This can be obtained by a capacitance connected in the position shown. Negative input signals cause D_2 to conduct and D_1 to block; amplifier A_1 then acts as a unity gain follower which passes the input signal to the point X. Amplifier A_2 acts as a unity gain inverter on this signal at X.

8.10.2 PRECISE LEVEL SELECTION

The circuit shown in Figure 8.27 is an arrangement which provides at its output the most positive (or least negative) of several input voltages. The amplifier with the most positive input has its D_1 diode switched on and forces its input signal on the output by precise follower action. All other amplifiers are bounded by conduction of their D_2 diode.

8.10.3 PEAK DETECTORS

It is sometimes necessary to measure the maximum positive excursion (peak value) or negative excursion (valley value) of a waveform over a given time period or to capture and hold some maximum value of a positive or negative pulse. A circuit which performs this function is

321

(a)

(b)

Figure 8.28 Peak detection (a) Simple peak detector (b) Precise diode peak detector

called a peak follower or peak detector. A basic peak detector circuit consists of a diode and a capacitor connected as shown in Figure 8.28(a). The capacitor is charged by the input signal through the diode but when the input signal falls the diode is reverse biased and the capacitor volt-

322

age retains a memory of the peak value of the input signal. The simple circuit has errors because of the diode forward voltage drop but the errors can be removed by replacing the diode with a precise diode circuit as shown in Figure 8.28(b). The circuit of Figure 8.28(b) is useful in applications not requiring a long hold time, for example, for measuring the peak value of a repetitive signal. In the hold mode, the voltage across the capacitor decays exponentially governed by the time constant

$$ C \, \frac{R \, R_L}{R + R_L} $$

In applications requiring an appreciable hold time the output voltage across the capacitor must be read out by some form of high input impedance buffer. Peak detector circuits employing f.e.t. input operational amplifiers in the follower mode as buffers are shown in Figure 8.29. In Figure 8.29(a) a two-diode arrangement is used to

(a)

(b)

Figure 8.29 Precise peak detector circuits (a) Low drift peak detector (b) Fast peak detector

323

reduce diode leakage current, and it is only the input error voltage of amplifier A_2 which appears across diode D_2 in the hold period. Circuits of this type in which two operational amplifiers are enclosed within a single feedback loop, normally require added frequency compensation; this has the effect of slowing down the rate at which the circuit responds to rapid changes in peak value.

Two separate feedback loops, one connected round each amplifier, are employed in the circuit of Figure 8.29(b). Amplifier A_1 acts both as a comparator and a unity gain follower. Its feedback loop is open for input signals less than the voltage stored on the capacitor, but when the input exceeds the capacitor voltage diode D_1 is brought into conduction and the amplifier then acts as a unity gain follower and causes the capacitor voltage to follow the input signal. If appreciable hold times are required both amplifiers should be f.e.t. input types to minimise capacitor leakage caused by amplifier bias current. Amplifier A_1 should be a type which retains its high input impedance in the saturated overload condition and should be capable of fast recovery from this condition; it must further be able to drive a capacitive load without serious reduction in phase margin.

The choice of capacitor values in a peak follower circuit is governed by the conflicting requirements of fast charging to allow a rapid acquisition of quickly changing input peaks, and long hold time. The smaller C the more rapid is its charging rate, but by the same reasoning the more rapidly will it discharge due to leakage during the hold period. One way of increasing capacitor charging current in order to obtain faster acquisition is to use a current booster at the output of the operational amplifier. A simple emitter follower booster can be used since only single polarity output currents are required in order to charge capacitor C.

Figure 8.30 Peak to peak detector

Positive peak detector circuits have been described but they can all be modified in order to detect negative peaks (valleys) by simply reversing diode directions. Peak to peak detectors can be implemented by connecting the output of a positive peak detector and a negative peak detector to a subtractor connected amplifier as shown in Figure 8.30.

8.11 Sample hold circuits

In signal processing it is sometimes necessary to retain a memory or hold the value that a signal has at a specified instant in time; a circuit used to perform this function is called a sample hold. A sample hold circuit has input and output terminals and control inputs to allow switching between the sample and hold mode. In the sample mode the output of the sample hold circuit is ideally equal to the input signal and follows or tracks variations in the input signal. When switched to the hold mode the output ideally remains constant at that value of the output signal which existed at the instant of switching. The phrase 'track hold' is often used when referring to a sample hold circuit which is in the sample mode for an appreciable time.

In principle the sample hold function can be performed by simply a switch and a capacitor as shown in Figure 8.31(a). Ideally when the switch is closed (sample mode) output and input signals are equal and the output follows or tracks the input. When the switch is opened, the output voltage remains constant at the value which it had at the instant the switch opened. A simple practical sample hold circuit is shown in Figure 8.31(b). It uses an f.e.t. switch and an operational amplifier unity gain follower to minimise capacitor discharge in the hold mode.

Practical sample hold circuits do not behave ideally; departures from the ideal sample hold function are in respect of speed and accuracy. When switched from the hold to sample mode a finite time is required for the output to become equal to the input signal within a specified accuracy; this time is referred to as the *acquisition time*. When switched from the sample to the hold mode there is a time delay between the instant at which the hold command signal is applied and the time the circuit actually goes into the hold state. This time delay is referred to as the *aperture time* and with fast changing input signals the held signal is in error to an extent determined by the aperture time. Performance accuracy in the sample mode is sometimes expressed in terms of the percentage gain error; the ideal sample hold is normally

Figure 8.31 The sample hold function (a) Principle of sample hold circuit (b) Simple practical sample hold circuit

designed to have unity gain (output signal = input signal). In the hold mode the output of a practical sample hold does not remain constant as it should; any loading of the hold capacitor causes the capacitor to discharge and the output drifts towards zero.

The choice of capacitor values for use in a sample hold is normally a compromise based upon the conflicting requirements of fast acquisition time and long hold time. In the sample mode the capacitor must charge up to the value of the input signal; the larger the capacitor value the longer it takes to charge. In the hold mode there is inevitably always some capacitor discharge current (amplifier bias current, switch leakage current, capacitor leakage current). The larger the capacitor value the longer the time taken for a capacitor leakage to cause appreciable capacitor discharge.

Ready built sample hold circuits have been available for some time in modular form; they are now available as comparatively inexpensive monolithic integrated circuits[7] (e.g. National Semiconductor LF 198/ 298/398). The user simply connects a hold capacitor externally. The choice between buying ready made and building from operational amplifiers and f.e.t. switches can only be made satisfactorily by the designer from a knowledge of his particular performance requirements; the choice then becomes a cost/performance decision.

As a help to the reader who does want to build his own sample hold circuit, some of the possible design approaches are now given. In the simple sample hold circuit of Figure 8.31(b) the storage capacitor is charged directly by the signal source through the f.e.t. switch. The capacitor loads the signal source. Capacitor charging rate when switched from hold to sample mode is determined by the time constant $C (R_s + R_t)$; R_s is the source resistance and R_t is the switch on resistance. An upper limit of the charging rate is set by the output current limit of the signal source. An alternative one-amplifier sample hold circuit is shown in Figure 8.32(a); it is essentially an integrator which is switched between the set and the hold mode. The circuit, unlike that of Figure 8.31(b), does provide some input isolation, in the form of resistor R_1. Its main deficiencies are its limited tracking bandwidth and comparatively slow acquisition time. Both tracking bandwidth and acquisition time are controlled by the time constant $C_1 R_1$.

The circuit for a two-operational amplifier high accuracy sample hold is given in Figure 8.32(b). Amplifier A_1 is follower connected and imposes negligible holding on the signal source. In the sample mode (S_1 closed, S_2 open) the feedback loop is closed round both amplifiers and the output is forced to track the input within the gain, common mode, offset errors and current output capability of amplifier A_1. Common mode and offset errors in the output follower A_2 are compensated by the action of the feedback loop. Extra frequency compensation in the form of $C_1 R_1$ is required to ensure closed loop stability in the sample mode and this inevitably slows down the circuit response. In the hold mode S_1 is open, isolating the hold capacitor, and S_2 is closed so as to complete the feedback loop round amplifier A_1 and prevent A_1 from going into saturation.

If speed is more important than high accuracy a circuit of the form shown in Figure 8.32(c) can be used. The two amplifiers in this circuit work independently; each has its own closed feedback loop and in the sample mode (both switches closed) the switches are enclosed within

Figure 8.32 Sample hold circuits (a) Simple integrating sample hold (b) High accuracy sample hold (c) Fast sample hold

A_1's feedback loop. The circuit is faster than that of Figure 8.32(b) because no additional frequency compensation is required. It is less accurate because of the summation of the offset and common mode errors of both amplifiers.

8.12 Circuit with switched gain polarity

The circuit configuration shown in Figure 8.33 allows the closed loop signal gain of an operational amplifier to be switched between + and − unity by a control voltage applied to the gate of the f.e.t. T_1. When T_1 is cut off both non-inverting and inverting input terminals are driven by the same input signal. Neglecting amplifier bias current and offset voltage the output of the amplifier must be equal to the input signal in order to maintain the amplifier input error voltage at zero; the gain is then + 1. If the 'on' resistance of the f.e.t. were zero, switching T_1 on would earth the non-inverting input and convert the circuit into an inverter. In practice when T_1 is on the finite value of the switch on resistance allows a small fraction of the input signal to be applied to

Figure 8.33   f.e.t. switches closed loop signal gain polarity

the non-inverting input terminal of the amplifier. The use of the matching 'on' resistance of f.e.t. T_2 and resistors proportioned as shown in Figure 8.33 balances the effect of the 'on' resistance of T_1 and gives

329

the operational amplifier a closed loop gain of -1. This is readily
verified as follows:

Assume that the 'on' resistance of the f.e.t.'s is R_{DS} and that the
operational amplifier behaves ideally. Thus:

$$V_- = V_+ = e_{in} \frac{R_{DS}}{R_{DS} + 2R_1}$$

and

$$V_- = e_{in} \frac{R_{DS} + R_1}{R_{DS} + 2R_1} + e_o \frac{R_1}{R_{DS} + 2R_1}$$

Giving

$$e_o \frac{R_1}{R_{DS} + 2R_1} = - e_{in} \left[\frac{R_{DS} + R_1}{R_{DS} + 2R_1} - \frac{R_{DS}}{R_{DS} + 2R_1} \right]$$

or

$$e_o = - e_{in}$$

D.C. offset error cannot be completely balanced out in the switched
gain circuit. The effective source resistance through which bias current
flows to the non-inverting input terminal changes as the gain polarity is
switched. The offset error due to bias current cannot therefore be
balanced out but it can be reduced to a very small value by use of an
f.e.t. input operational amplifier. Another practical consideration in
the circuit is that the effective input resistance of the circuit changes as
the gain polarity is switched: when the circuit has a gain of $+1$ the circuit
has the high input impedance characteristic of the follower configuration,
but when switched to a gain of -1 the circuit has the input resistance
of the inverter configuration which is set by the value of resistor R_1.

The switched gain polarity circuit can be made the basis of a phase
sensitive detector[8]. The signal to be detected is applied as an input
signal and a phase coherent squarewave reference signal of the same
frequency is used to switch the gain polarity.

8.13 Voltage to frequency conversion

A voltage to frequency converter is a system which produces a periodic
signal with frequency proportional to an analogue control voltage. The
waveform produced may be a squarewave, a pulse train, a triangular
wave or a sine wave. The voltage controlled waveform generator discussed in Section 7.4 acts as a voltage to frequency converter.

330

Inexpensive pulse train output voltage to frequency converters can be realised using two operational amplifiers, one acting as an integrator and the other as a regenerative comparator. The principle of action of such circuits is illustrated by the configuration in Figure 8.34.

Figure 8.34 Simple V to f converter

Starting at the time at which the comparator switches to its positive level V_o^+ the action of the circuit is as follows: D_1 is reverse biased and the output of the integrator falls linearly at a rate determined by the magnitude of the positive d.c. voltage e_{in}. When the integrator output reaches a level $-V_o^+ R_1/R_2$ the comparator switches to its negative output state, diode D_1 is forward biased and the integrator output is made to run up rapidly. When it reaches a level $-V_o^- R_1/R_2$ the comparator reverts to its positive output state and the cycle repeats. If the time taken for the integrator output to run up is made negligibly small compared to the run down time the frequency of oscillations becomes directly proportional to the input voltage e_{in}. An expression for the frequency of oscillation obtained by neglecting integrator reset time and comparator switching time is

$$f \cong \frac{R_2}{R_1 (V_o^+ - V_o^-) CR_1} e_{in}$$

In practice finite comparator switching time causes the frequency of oscillation to be less than that predicted by this equation. The lower limit for linear voltage to frequency conversion of the circuit is set by the integrator drift error due to amplifier bias current and input offset voltage (see Section 6.3.1). The upper conversion limit is determined by switching times; conversion becomes nonlinear at the higher frequencies when the integrator reset time becomes comparable to the run down period.

Simple voltage to frequency converters using the type of circuitry discussed above can be expected to provide a ± 1% accuracy over the 2 to 3 decades at the most. Greater conversion accuracy and wider dynamic range require the use of more sophisticated circuitry[9], or as an alternative to building a voltage to frequency converter ready built modules are available (see Analog Devices, Burr Brown or Teledyne Philbrick product guides).

8.14 Frequency to voltage conversion

Simple frequency to voltage conversion circuits operate by first converting the signal whose frequency is to be measured to a constant amplitude pulse train. The pulse train is then differentiated, rectified and averaged to give a d.c. indication of the frequency. A simple frequency to d.c. converter using this principle of operation is illustrated in Figure 8.35.

Figure 8.35   Frequency to d.c. voltage conversion

Amplifier A_1 acts as a zero reference comparator; its output is bounded by back to back zener diodes and it produces a constant amplitude pulse train with the same frequency as the input signal. Capacitor C_1 and diodes D_1, D_2, constitute a simple diode pump circuit. On each positive going part of the input pulse a quantity of charge $C_1 2V_z$ is transferred through D_2 to the summing point of amplifier A_2. The

332

charge pulses are averaged (averaging time constant = C_2R_2) to give an average current of $fC_1 2V_z$ through the feedback resistor R_2 and the amplifier develops an output voltage

$$e_o = C_1 2V_z R_2 f$$

The output voltage is directly proportional to the frequency of the input signal.

Our simple analysis has assumed that capacitor C_1 discharges completely, and it has neglected diode forward voltage drops. Temperature dependence of diode voltage may be expected to cause scaling factor changes but if $2V_z \gg \Delta V_d$ (where ΔV_d is the change in diode forward voltage due to temperature dependence) the effect is small. Temperature dependence of zener voltages also directly affects frequency to voltage scaling. The upper frequency limit of the circuit is determined by the comparator switching times.

8.15 Modulation

Modulation is a process whereby a characteristic of one signal, a so-called carrier signal, is varied in proportion governed by a second signal called the modulating signal. The voltage controlled waveform generator of Section 7.4 is an example of frequency modulation, the control signal to the generator acting as the modulating signal. Four quadrant multipliers (see Section 5.7) are particularly useful in a variety of modulator and demodulator applications. The reader will find these applications treated elsewhere[6]; this Section is confined to a treatment of some basic operational amplifier modulator circuits.

8.15.1 PULSE AMPLITUDE MODULATION

In a pulse amplitude modulated signal the carrier is a squarewave or pulse whose amplitude is varied by the modulating signal. The switched gain polarity circuit discussed in Section 8.12 can be used as a pulse amplitude modulator. A squarewave or pulse wave train acting as the carrier signal is applied to the gate of the control f.e.t. so as to switch the gain polarity. The modulating signal is applied as an input signal to the amplifier giving an output squarewave with amplitude proportional to the value of the input signal.

8.15.2 PULSE WIDTH MODULATION

A pulse width modulator produces a train of pulses that have widths proportional to the amplitude of a modulating signal. A combination of an operational integrator and a comparator may be used to produce a pulse width modulated signal; the principle of operation of such a system is illustrated by the circuit shown in Figure 8.36. The carrier

Figure 8.36   Pulse width modulator

signal, a fixed frequency squarewave of constant amplitude, is applied to the input of an a.c. coupled integrator. The integrator converts the carrier into a triangular wave which is applied to one input of the comparator; the modulating signal is applied to the other input of the comparator. The waveforms produced are as shown in the diagram.

REFERENCES

1. CLAYTON, G. B., *Linear Integrated Circuit Applications*, Macmillan (1975). *Linear I.C. Applications Handbook*, Tab Books (1977)
2. MIYARA, J., 'Measuring Air Flow using a Self Balanced Bridge', Analog Dialogue **5**, No. 1 (1971)

3. RS Components Ltd — Data Sheet 10322 (1990)
4. RS Components Ltd — Data Sheet 11072 (1991)
5. RS Components Ltd — Data Sheet 11088 (1991)
6. RS Components Ltd — Data Sheet 8852 (1990)
7. 'Sample-Hold Chip Delivers Precision', Electronics, **50**, No. 21, 121 (1977)
8. CLAYTON, G. B., *Experiments with Operational Amplifiers*, Macmillan (1975). *88 Practical Op Amp circuits you can build*, Tab Books (1977)
9. *Non Linear Circuits Handbook*, Analog Devices Engineering Staff (1974)

Exercises

8.1 A charge amplifier (Figure 8.4) has a feedback capacitor of value 100 pF, with a resistor of value 100 MΩ connected in parallel with it in order to prevent its continuous charging. The operational amplifier used in the circuit is internally frequency compensated and has unity gain frequency 10^6 Hz, bias current I_B = 10 pA and input offset voltage V_{io} = 2 mV. The charge amplifier is supplied by a transducer whose output impedance is capacitative and of value 900 pF. Find the upper and lower frequency of the 3 dB bandwidth limits and the output offset voltage of the circuit. What would be the effect on circuit performance of connecting the transducer to the amplifier by means of a cable of capacitance 200 pF? Sketch the Bode plots to illustrate your answers.

In order to avoid the use of a very large feedback resistor and yet still maintain the same low frequency bandwidth limit a 1 MΩ resistor and a resistive T network is used in place of the 100 MΩ resistor (see Figure 4.2). What effect will this have on the output offset voltage?

8.2 In the phase shifting circuit of Figure 8.16 R = 10 kΩ, C = 0.005 μF and a 1 MΩ potentiometer is used for R'. If the input signal has frequency 1 kHz, what phase variation is possible with the circuit? If R' is fixed at 1 MΩ sketch the phase variation as the frequency of the input signal is increased from zero upwards.

8.3 Component values C_1 = 0.1 μF, R_1 = 1 MΩ, R_2 = 100 Ω are used in the capacitance multiplier circuit of Figure 8.18. Find the effective capacitance and series resistance of the circuit thus formed. If the input point to the circuit is left open, what is the maximum drift rate in the output voltage of the amplifier that you might expect, assuming the operational amplifier has input offset V_{io} = 2 mV and bias current I_B = 50 pA?

8.4 The circuit of Exercise 8.5 is used as the basis for a timing circuit with additional components added to it as shown in Figure 8.19. If component values R_3 = 22 kΩ, R_4 = R_5 = 5.6 kΩ and supply voltage 15 V are used calculate the timing period, neglecting the effect of amplifier offsets. If account is taken of the bias current and input offset voltage of amplifier A_1 what effect does this have on your calculation? What effect on the time period do you expect to be produced by changes in the supply voltage? (Use equation 7.2.)

8.5 The following component values are used in the circuit of Figure 8.34: R = 100 kΩ, C = 0.1 μF, R_1 = 10 kΩ, R_2 = 20 kΩ, R_3 = 2 kΩ. What is the frequency of the oscillations for an input signal of 1 V? At what value of the input voltage may you expect a 5% departure from linearity of voltage to frequency conversion? Sketch circuit waveforms. Assume ideal amplifier characteristics and that the output limits of the comparator amplifier A_2 are ± 10 V.

8.6 A basic sample hold circuit consisting of an f.e.t switch and a unity gain buffer stage (Figure 8.31(b)) is supplied by a signal source of output resistance 250 Ω, the f.e.t. has an 'on' resistance of 50 Ω and the operational amplifier has a bias current I_B = 50 pA, C = 0.01 μF. Find the acquisition time to 1% for a 10 V change in output when switched between the hold and sample mode. Assume that the required initial capacitor charging current does not exceed the current output limit of the signal source. What must the current output capability of the source be for this assumption to be valid?

Neglecting switch and capacitor leakage find the output drift rate in the hold mode.

CHAPTER NINE

ACTIVE FILTERS

9.1 Introduction

In the context of this chapter, *filters* are electrical networks that
have been designed to pass alternating currents generated at only
certain frequencies and to block or attenuate all others. Filters have
a wide use in electrical and electronic engineering and are vital
elements in many telecommunications and instrumentation
systems where the separation of wanted from unwanted
signals—including noise—is essential to their success.

There are two generic types of filter: passive and active. The first
type comprises simple resistors, capacitors and inductors while the
second has the addition of active components, usually in the form
of operational amplifiers. Both of these types are sub-divided into
the four classes according to their use. These are *low-pass*,
high-pass, *band-pass* and *band-stop*. This chapter is mainly
concerned with active filters employing operational amplifiers, but
it may serve as a useful introduction for some readers if firstly a
brief examination is made of the passive type.

9.2 Passive filters

9.2.1 THE LOW-PASS FILTER

The circuit of a simple CR low-pass filter is shown in Figure 9.1.
This is essentially a potential divider comprising a resistance in
series with a capacitor. The output voltage, e_o, is taken from across
the capacitor and is related to the input voltage, e_i, by the equation:

$$e_o = -jX_c \, e_i/(R - jX_c)$$

Algebraic manipulation of this complex number equation shows
that the amplitude of e_o is given by the expression:

$$|e_o| = e_i X_c/\sqrt{(R^2 + X_c^2)}$$

Figure 9.1　First order, low-pass passive CR filter circuit

Even though e_i may be held constant over a range of input frequencies, the amplitude of e_o decreases as the frequency is increased. This is because the reactance of the capacitor, $X_c = 1/2\pi fC$, varies as the inverse of the frequency, f, and tends from an infinitely high value at zero frequency to zero at an infinitely high frequency. The circuit output is effectively shorted out at very high frequencies. Figure 9.2 shows the response curve for this circuit which is typical of the low-pass filter.

At low frequencies the output volts : input volts ratio remains sensibly level up to a frequency, f_c, at which a marked fall off starts. At about $2f_c$ the fall off (or roll off, as it is usually called) becomes linear at 20 dB per decade (which is the same as 6 dB per octave). The frequency f_c is known as the cut-off frequency and is taken as that frequency at which the reactance of the capacitor has the same magnitude as the resistance in the circuit. Also, f_c is the frequency at which the output voltage has fallen to $1/\sqrt{2}$ times its d.c. value to give half the d.c. power output. Simple calculations based on these facts show that the cut-off frequency is given by the equation:

$$f_c = 1/2\pi RC \text{ hertz}$$

For frequencies below f_c the circuit gain (output volts : input volts) is taken as being reasonably constant, while for frequencies higher than f_c the gain is regarded as being so low that the passage of these signals is effectively blocked. The circuit is known as a low-pass filter having a bandwidth extending from d.c. to f_c.

Because the response of the circuit depends upon frequency to the mathematical first order, the filter is known as a *first order filter*. (Also note that the circuit contains only a single component—the capacitor—the performance of which is frequency conscious.)

338

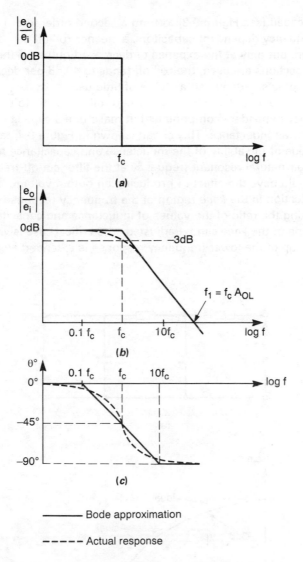

Figure 9.2 Low-pass filter response curves for (a) ideal magnitude; (b) actual magnitude with Bode approximation; (c) phase shift with Bode approximation

The ideal low-pass passive filter frequency response curve or *transfer function* would show no loss of gain for frequencies below f_c and zero output above f_c (see figure 9.2). Clearly, the first order low-pass filter achieves neither of these ideals. If two *CR* sections are cascaded (see Figure 9.3) to form a second order filter having two frequency dependent capacitors, a steeper roll-off can be obtained, but only at the expense of decreased output. If these two similar sections are used, the roll-off tends to 40 dB per decade but the output is so attenuated as to be of little use.

A better solution for achieving a steep roll-off is still to use two frequency-dependent components but make one a capacitor and the other an inductance. This circuit, shown in Figure 9.4, takes advantage of the ability of the inductance and capacitance to be near their natural resonant frequency at the filter cut-off frequency. This would have the effect of producing an output voltage magnification in the *knee* region of the frequency response curve. By varying the ratio of the values of inductance and capacitance, the shape of the knee can be adjusted. The critical case is where the flat top of the lower frequency response is extended along the

(a)

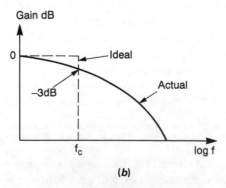

(b)

Figure 9.3   Second-order, high-pass passive CR filter circuit (a) and response curve (b)

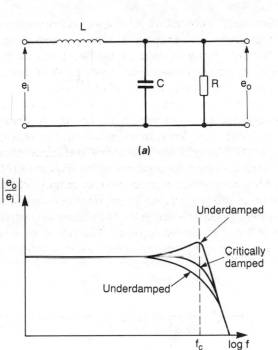

Figure 9.4 *Second-order, low-pass passive LCR filter circuit (a) and response (b)*

frequency scale before falling in a steeper roll-off yet without introducing the undesirable effects of underdamping or overdamping. These include output voltage oscillations before finally settling or having an excessively long response time to transient inputs. The combined high frequency effect of the high inductive reactance coupled with the low capacitive reactance eventually produces a second order filter linear roll-off dependent upon the inverse square of the frequency.

9.2.2   THE HIGH-PASS FILTER

To form a high-pass filter, the *CR* components of the low-pass filter are simply interchanged. Figure 9.5 shows the first order high-pass circuit and Figure 9.6 its frequency response curve. The gain roll-off

Figure 9.5   First-order, high-pass passive CR filter circuit

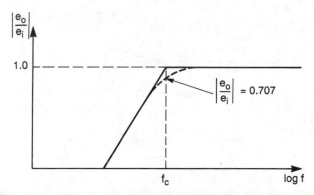

Figure 9.6   High-pass filter response curve showing the actual (dotted) response and the Bode approximation

is once again 20 dB per decade and the cut-off frequency is still given by the equation

$$f_c = 1/2\pi RC \text{ hertz}$$

At low input frequencies the capacitor has a high reactance and effectively rejects any input voltage. As the input frequency is increased the capacitor progressively lowers its reactance, allowing an increasing proportion of the input voltage to be developed across the resistor and appear at the circuit output. Frequencies below f_c are regarded as being in a stop-band; those above, as being in the circuit pass-band.

342

(a)

(b)

Figure 9.7 Second-order bandpass passive LCR filter circuit (a) and response (b)

9.2.3 THE BAND-PASS FILTER

A second order band pass filter can be obtained by using the series LCR circuit arrangement shown in Figure 9.7. At low input frequencies the capacitive reactance predominates and the circuit behaves as a simple series capacitor with a 6 dB per octave increasing response from d.c. As the frequency of the input signal approaches circuit resonance, there is a marked up-turn in the response curve to climax in a peak at:

$$f_0 = 1/2\pi\sqrt{(LC)} \text{ hertz}$$

Once the resonant frequency has been exceeded, the inductive reactance becomes increasingly dominant and the response falls away but not as sharply as was the build up from the low frequencies.

Thus, there are two frequencies where the response is 3 dB less than the peak and they are called the upper and lower cut-off frequencies, f_{cu} and f_{cl}. They are equally disposed about the resonant or centre frequency; the centre frequency is always taken as the geometric mean of the two.

$$f_o = \sqrt{(f_{cu}.f_{c1})}$$

The difference between f_{cu} and f_{cl} is taken as the bandwidth or pass-band, B, of the filter and together with the *goodness* or Q of the circuit is related to the centre frequency, f_o, by the following equation:

$$B = f_o/Q \text{ Hz}$$

The higher the Q of the circuit the smaller is its pass-band and the filter is said to be more *selective*.

If several such LCR circuits, each having a slightly different resonant frequency, are connected in series, the resulting circuit is a band-pass filter.

9.2.4 THE BAND-STOP FILTER

A second order band-stop filter can be obtained by using the parallel CL circuit arrangement shown in Figure 9.8. At low input frequencies the circuit is effectively a low-pass arrangement comprising only the L and the R. At the circuit resonant frequency, determined by $f = 1/2\pi\sqrt{(LC)}$, the parallel L and C presents an infinitely high impedance and the circuit output is zero. Once the resonant frequency has been exceeded, the inductive reactance continues to increase while that of the capacitor decreases, making the circuit perform more as if comprising only the C and the R in a simple high-pass filter arrangement.

Figure 9.8 Second-order bandstop passive LCR filter circuit (a) and response (b)

9.2.5 PASSIVE FILTER SUMMARY

The four basic frequency sensitive filter circuits described above can be cascaded using any mix of first and second order variants that is necessary to produce a desired response. The shape of filter response curves has been studied by many eminent people, some of whom have had their names credited to particular circuits which satisfy particular requirements. These names include Butterworth, Bessel, Chebyshev, Cauer and, together with other special filter circuits, they will be discussed later in this chapter.

Passive filter circuits contain various combinations of resistors, capacitors and inductors and in most cases suffer from several shortcomings. Mathematically, they are difficult to design; they are often pulled off frequency by the load current drawn from them; even in their pass-band they usually attenuate signals and are not easily tuned over a wide frequency range without changing their response characteristics. Further problems can be associated with the use of inductors. Not only are they expensive, bulky and heavy; they are prone to magnetic field radiation unless expensive shielding is used to prevent unwanted coupling.

9.3 Active Filters

9.3.1 THE CASE FOR ACTIVE FILTERS

The advent of the low-cost, integrated circuit operational amplifier has made it possible to overcome most of the problems associated with the passive filter circuit. Not only will the high input impedance and low output impedance of the operational amplifer effectively isolate the frequency sensitive filter network from the following load, it can also provide useful current or voltage gain. More significantly, the operational amplifier can be designed into a CR only circuit in such a way as to provide a filter response virtually identical with that of a passive inductive filter network. This means that the use of inductors in filters is now unnecessary. Unlike the inductor, the operational amplifier does not possess a magnetic field which stores energy, rather it is designed to behave mathematically in the same way as the whole passive circuit it replaces. Any additional circuit energy is obtained from the separate power source used by the operational amplifier.

9.3.2 NEGATIVE IMPEDANCE CONVERSION

The circuit shown in Figure 9.9 is designed to have an input impedance, Z_i, which appears to be the negative of the impedance Z.

$$i_i = \frac{e_x - e_o}{R} \tag{9.1}$$

Normal operational amplifier action causes

$$e_i = e_x = e_y, \text{ and } e_y = \frac{e_o Z}{R + Z}$$

Figure 9.9 Impedance convertor

Rearranging,

$$e_o = \frac{e_i(R + Z)}{Z} \tag{9.2}$$

Substituting (9.2) and $e_x = e_i$ into (9.1),

$$i_i = \frac{e_i - \dfrac{e_i(R + Z)}{Z}}{R} = \frac{e_i\left(1 - \dfrac{R + Z}{Z}\right)}{R}$$

$$\therefore \ i_i = \frac{e_i(Z - R - Z)}{ZR}$$

and rearranging,

$$Z_i = \frac{e_i}{i_i} = \frac{ZR}{Z - R - Z}$$

$$\text{or} \ \underline{\underline{Z_i = -Z}}$$

Suppose Z is a capacitor, C. Then $Z = -j/\omega C$ and so it follows that:

$$Z_i = -(-j/\omega C) = +j/\omega C$$

347

The $+j$ means that the current lags the voltage, that is, has an inductive reactance, jX_L, but where $X_L = 1/\omega C$. However, while the result is to produce an inductive effect, the 'inductive reactance' decreases with increasing frequency rather than increases as would the reactance of a true inductance.

9.3.3 IMPEDANCE GYRATION

A single negative impedance invertor is not capable of simulating the true action of an inductor. However, this effect can be achieved if a pair of negative impedance convertors are used. Such a circuit is shown in Figure 9.10 where the input impedance is $Z_i = R^2/Z$. Suppose that $Z = -jX_c$. then $Z_i = R^2/-jX_c = +j\omega CR^2$. Now the capacitor C is being made to act as if it were a true inductor of value $L = CR^2$.

Similarly, it can be shown that if Z were an inductive reactance of value jX_L, then the gyrator would make this appear to the preceding circuit as a true capacitance of $C = L/R^2$.

9.3.4 MAKING A SIMPLE ACTIVE FILTER

The response curve shown in Figure 9.4 for a passive second order low-pass LR filter can be simulated using only resistors and capacitors. A first attempt may include two cascaded first order, low-pass CR sections with the addition of an emitter follower. This has a high input impedance but low output impedance and so minimises any loading effects on the frequency sensitive CR sections. This circuit, which produces a highly damped response, is shown in Figure 9.11 (a).

Figure 9.11(b) illustrates a major design improvement by the introduction of positive energy feedback to the centre of CR section. This 'bootstrapping' has a maximum effect only near the cut-off frequency. At very low frequencies the normal gain enhancement of positive energy feedback is largely negated by the high reactance of $C1$ in the feedback path. In excess of the cut-off frequency, the low reactance of $C2$ allows the signal to leak to earth and attenuate the output accordingly. The values of the filter network capacitors and resistors can be selected to eliminate the damping problem of the previous circuit. While the product of the resistors and capacitors decides the cut-off frequency it is the ratio of the capacitors which affects the circuit response rate. Compared with the values of $C1$

Impedance gyrator

Figure 9.10 Analysis of the circuit at (a) by progressive simplification through (b) and (c) to (d) shows that the input impedance is proportional to the reciprocal of the load impedance, Z

and $C2$ for critical damping, a large $C1$ with a small $C2$ will produce an underdamped response while the reverse will cause overdamping.

Figure 9.11 Second-order active filter

9.4 Active filters using operational amplifiers

In practical active filters the emitter followers used above are
invariably replaced by operational amplifiers in the form of
integrated circuits. The frequency sensitive filtering networks are
either placed before the operational amplifier input terminals or in
the feedback circuits.

9.4.1 FIRST ORDER HIGH-PASS AND LOW-PASS FILTERS

Examples of simple first order high and low-pass active filters are
shown in Figure 9.12. As expected, the frequency selective

$$f_c = \frac{1}{2\pi CR} \text{ Hz}$$

$$\frac{e_o}{e_{i(jf)}} = \frac{1}{1 + j\frac{f}{f_c}}$$

$$f_c = \frac{1}{2\pi CR} \text{ Hz}$$

$$\frac{e_o}{e_{i(jf)}} = \frac{1}{1 - j\frac{f_c}{f}}$$

Figure 9.12 First order low and high pass active filters (a) First order low pass response (b) First order high pass response

351

resistor–capacitor circuit elements decide the frequency response. The cut-off frequency is $f_c = 1/2\pi CR$ at which the magnitude of the filter response is 3 dB less than that in the pass-band, and the higher frequency roll-off tends to 20 dB per decade. If a low value of f_c is required, a general purpose Bi-fet operational amplifier should be suitable. This will allow the use of large resistance values without introducing any appreciable bias current off-set error. Resistor values up to 10 MΩ may be used so avoiding the expense of a high value, close tolerance capacitor.

First order low-pass filters are often used to perform a running average of a signal having high frequency fluctuations superimposed upon a relatively slow mean variation; for this purpose it is simply necessary to make the filter time constant CR much greater than the period of the high frequency fluctuations.

A practical point to remember is that all operational amplifier active high-pass filters show a band-pass characteristic. This is because their response eventually falls at frequencies which exceed the closed loop bandwidth of the operational amplifier.

9.4.2 SECOND ORDER LOW-PASS AND HIGH-PASS FILTERS

Examples of simple second order low-pass and high-pass active filter circuits are shown in Figure 9.13. The second order filter response has a 40 dB per decade roll-off in the stop-band. The sharpness of the response curve knee depends upon the choice of values for the components forming the frequency sensitive element of the filter. In Figure 9.13, the components are proportioned to give a so-called Butterworth response (further details later in this chapter) and the cut-off frequency $f_c = 1/[2\sqrt{2}(\pi CR)]$ hertz.

9.5 Choosing the frequency response of the low-pass filter

Figure 9.2 shows the ideal shape for a low-pass filter. It has a perfectly flat (horizontal) response from zero frequency up to the cut-off frequency where a vertical fall then occurs. In practice this perfectly rectangular shape is unattainable. Depending upon the intended role of the filter, it can be designed to approximate to the ideal response in varying ways and these are mentioned briefly below.

$$f_c = \frac{1}{2\sqrt{2}(\pi CR)} \text{ Hz}$$

Low pass $\dfrac{e_o}{e_{i(jf)}} = \dfrac{1}{1-\left(\dfrac{f}{f_c}\right)^2 + \sqrt{2}\left(j\dfrac{f}{f_c}\right)}$

$$\left|\frac{e_o}{e_{i(jf)}}\right| = \frac{1}{\sqrt{\left\{\left[1-\left(\dfrac{f}{f_c}\right)^2\right]^2 + 2\left(\dfrac{f}{f_c}\right)^2\right\}}}$$

$$f_c = \frac{1}{2\sqrt{2}(\pi CR)} \text{ Hz}$$

High pass $\dfrac{e_o}{e_{i(j\omega)}} = \dfrac{1}{1-\left(\dfrac{f_c}{f}\right)^2 - \sqrt{2}\left(j\dfrac{f_c}{f}\right)}$

$$\left|\frac{e_o}{e_{i(j\omega)}}\right| = \frac{1}{\sqrt{\left\{\left[1-\left(\dfrac{f_c}{f}\right)^2\right]^2 + 2\left(\dfrac{f_c}{f}\right)^2\right\}}}$$

Figure 9.13 Second order low and high pass active filters

9.5.1 THE BUTTERWORTH LOW-PASS RESPONSE

This response requires that at zero frequency the circuit gain is flat
and remains as near flat as possible up to the designed cut-off
frequency. The higher the order of filter the more accurately does
its response approximate to this ideal, as illustrated by Figure 9.14.

Figure 9.14   Butterworth response

9.5.2   THE CHEBYSHEV LOW-PASS RESPONSE

The Chebyshev approximation is an attempt to overcome the
practical failure of the Butterworth response to maintain a truly flat
pass-band as the frequency of operation is increased up to the
cut-off frequency. The Chebyshev circuit is designed uniformly to
spread any deviation of gain over the pass-band in the form of
ripples as shown in Figure 9.15. Above the cut-off frequency, like
the Butterworth, the Chebyshev roll-off eventually tends to be
monotonic at $20n$ dB per decade where n is the order of the filter.
Even so, the second and third order Chebyshev filters tend to have
a less steep initial roll-off than their Butterworth counterparts,
whereas comparable fourth order and above filters show the
Chebyshev response to have the sharper knee.

9.5.3   THE CAUER (OR ELLIPTIC) LOW-PASS RESPONSE

Using a Butterworth or Chebyshev filter, a complete signal stop is
usually regarded as having been achieved when the filter
attenuation has reached a designed level. The frequency at which
this degree of attenuation first occurs is taken as the start of the
filter stop-band. However, while a continued increase in frequency
initially causes further signal attenuation, a practical limit is

Figure 9.15 Third-order Chebyshev response

reached. This is where, owing to unwanted leakage through stray reactances, further increasing the frequency can produce an unwanted output from the filter. The Cauer response is designed to cater for those applications where it is required that an infinite attenuation is achieved at a particular frequency and that for any higher frequencies a designed minimum attenuation is maintained.

Figure 9.16 shows the Cauer filter circuit diagram and the typical response curves it produces. The infinite attenuation is caused at the frequency, f_2, because at this frequency L2 and C2 are in resonance and present an infinite impedance to the signal flow.

9.5.4 THE BESSEL LOW-PASS RESPONSE

The above studies on the Butterworth, Chebyshev and Cauer filter responses have all emphasised the relative amplitudes of the filter input and output voltages. No mention has been made of the phase shift which occurs as the signal travels through the filter. In applications involving voice or other analogue transmissions, phase shift is not important and optimum amplitude responses are often obtained at the expense of phase shift. However, in the case of digital transmissions it can be important that the pulses are not distorted and linear-phase filters are often used.

Figure 9.17 shows the ideal linear relationship between the signal frequency and the resulting phase shift introduced by the filter. With regard to the signal transit time through the filter, signals of all frequencies should ideally suffer the same time delay and so any signals in phase at the input will still be in phase at the output. But

Figure 9.16 Cauer or elliptic, third-order filter circuit (a) and the response it produces (b)

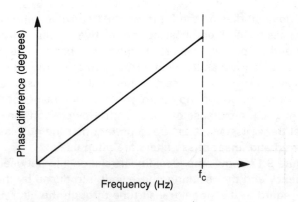

Figure 9.17 Linear relationship between frequency and phase shift in ideal low-pass filter

a signal of double the frequency of another will suffer twice its phase shift. This effect is shown at Figure 9.18.

The Bessel approximation is an attempt to produce such a linear-phase filter. The Bessel response circuit has the same appearance as the Butterworth and Chebyshev circuits and differs only in the component values necessary to produce the required constant transit time at all frequencies.

9.5.5 COMPARATIVE RESPONSES OF THE DIFFERENT LOW-PASS FILTERS

See Figure 9.19 for a summary of the various low-pass responses.

9.6 Choosing the frequency response of the high-pass filter

Figure 9.6 shows the ideal shape for a high-pass filter response curve. It has a zero output at low frequencies but continued frequency increase eventually causes the response to rise monotonically until just short of the −3 dB cut-off frequency which marks the start of the pass-band. At frequencies higher than this, the response in the pass-band levels at the maximum gain. But,

Figure 9.18 Time-related waveforms showing how a constant time delay of 0.25 s produces a 90° phase shift in a 1 Hz signal but only a 45° shift in a 0.5 Hz signal

Figure 9.19 Comparison of the different low-pass filter performance in their stopbands. (1) Cauer (elliptic); (2) Chebyshev; (3) Butterworth; (4) Bessel

because of practical component inadequacies and stray reactances becoming increasingly significant at the higher frequencies, the flat response of the passive filter circuit element does not extend to infinity and eventually declines. Additionally, in the case of the active filter, the inherent high frequency gain roll-off of the operational amplifier effectively makes any high-pass filter behave as a form of band-pass filter—but with the upper cut-off frequency being above the highest frequency to be passed.

The studies of the low-pass filtering transfer functions and response curves made in Section 9.5 can be readily modified to suit the high-pass conditions. Basically, the high-pass filter is a mirror image of its low-pass equivalent; the capacitors and resistors are simply interchanged. The mathematical process involved in this change is called *mathematical transformation by 1/f*. Figure 9.20 shows a graphical summary of the high-pass response curves.

9.7 Band-pass filters using the state variable technique

A bandpass filter characteristic can be obtained by cascading a high and a low pass filter, but when a highly selective (high Q) bandpass characteristic is required a different approach is necessary. Many examples of active bandpass filters will be found in the literature and in manufacturers' notes, but high Q bandpass filters, based upon a single operational amplifier, have a Q value which is very sensitive to

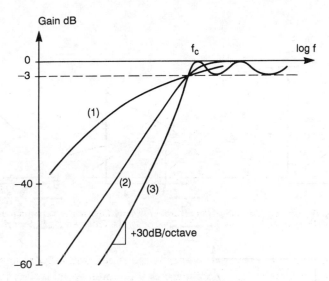

Figure 9.20 Example of high-pass filter (5th order) response curves. (1)
Overdamped; (2) flattest; (3) Chebyshev

component variation. The so called state variable filter approach[1],
which is based upon the use of analogue computer techniques
(Section 6.10), is less component sensitive and although it requires
the use of three operational amplifiers it is now economically attract-
ive because of the availability of inexpensive quad Bi-Fet operational
amplifiers (e.g. TI 084).

The circuit for a second order state variable filter is shown in
figure 9.21. It is particularly versatile in that it allows the simultaneous
realisation of high pass, low pass and bandpass characteristics at three
separate circuit points; if a quad amplifier is used the fourth amplifier
can, if required, be used to form a band reject characteristic. The
steady state sinusoidal response equation for the circuit of Figure
9.21 is now derived—operational amplifiers are assumed to behave
ideally. It is the action of the feedback loops which forces the
desired relationships between inputs and outputs; we derive the
bandpass relationship which is exhibited between the input signal
and the output of amplifier A_2 (e_{bp}). The relationship between e_{bp}
and the other output signals is readily found if it is remembered
that the action of an integrator is to multiply by

$$-\frac{1}{j\omega T}$$

where $T = CR$ is the integrator time constant.

Figure 9.21 Amplifiers – use Bi-fets eg. LF 356 TI080 singles or TL 084 quad

Amplifier A_1 is connected as an adder-subtractor. It sums the input signal with the output of A_2 and subtracts the output of A_3, thus:

$$e_{o_1} = \frac{e_{bp}}{-\dfrac{1}{j\omega T_1}} = \left[e_i \frac{R_4}{R_3 + R_4} + \frac{e_{bp}R_3}{R_3 + R_4} \right] \left[1 + \frac{R_6}{R_5} \right]$$

$$- \left[-\frac{1}{j\omega T_2} \; e_{bp} \; \frac{R_6}{R_5} \right]$$

where $T_1 = C_1 R_1$ and $T_2 = C_2 R_2$

Algebraic manipulation yields

$$\frac{e_{bp}}{e_{i(j\omega)}} = \frac{-\dfrac{1}{T_1} \dfrac{\left[1 + \dfrac{R_6}{R_5} \right]}{\left[1 + \dfrac{R_3}{R_4} \right]} \, j\omega}{\dfrac{1}{T_1 T_2} \dfrac{R_6}{R_5} + \dfrac{j\omega}{T_1} \dfrac{\left[1 + \dfrac{R_6}{R_5} \right]}{\left[1 + \dfrac{R_4}{R_3} \right]} - \omega^2} \tag{9.3}$$

This is a second order bandpass response which can be put in the more general form

$$\frac{e_{bp}}{e_{i(j\omega)}} = - \frac{A_{o(bp)}}{1 + jQ \left[\dfrac{\omega}{\omega_o} - \dfrac{\omega_o}{\omega} \right]} \tag{9.4}$$

With response constants related to circuit parameters by

$$A_{obp} = \frac{R_4}{R_3}, \; \omega_o = \sqrt{\left(\frac{R_6}{R_5 C_1 R_1 C_2 R_2} \right)}$$

$$Q = \sqrt{\left(\frac{C_1 R_1 R_6}{C_2 R_2 R_5} \right)} \frac{1 + \dfrac{R_4}{R_3}}{1 + \dfrac{R_6}{R_5}} \tag{9.5}$$

In practice it is convenient to make $R_5 = R_6$, $C_1 = C_2$, $R_1 = R_2$. The centre frequency ω_o can then be tuned without altering the Q by simultaneously changing R_1 and R_2. The Q value can be varied by changing R_4 without altering the centre frequency.

The response at the low pass and high pass output is readily derived by the substitution

$$e_{lp} = -\frac{1}{j\omega T_2} e_{bp} \quad \text{and} \quad e_{hp} = -j\omega T_1 e_{bp}$$

in equation 9.3 yielding after some algebraic manipulation equations of the form

Second order low pass

$$\frac{e_{lp}}{e_{i(j\omega)}} = \frac{A_{o(lp)}}{1 + 2\zeta j \dfrac{\omega}{\omega_o} - \left(\dfrac{\omega}{\omega_o}\right)^2} \tag{9.6}$$

With
$$A_{o(lp)} = \frac{1 + \dfrac{R_5}{R_6}}{1 + \dfrac{R_3}{R_4}}, \quad \omega_o = \sqrt{\left(\frac{R_6}{R_5 C_1 R_1 C_2 R_2}\right)}$$

and
$$\zeta = \frac{1}{2} \frac{1 + \dfrac{R_6}{R_5}}{1 + \dfrac{R_4}{R_3}} \sqrt{\left(\frac{C_2 R_2 R_5}{C_1 R_1 R_6}\right)}$$

Second order high pass

$$\frac{e_{hp}}{e_{i(j\omega)}} = \frac{A_{o(hp)}}{1 - 2\zeta j \dfrac{\omega_o}{\omega} - \left(\dfrac{\omega_o}{\omega}\right)^2} \tag{9.7}$$

With
$$A_{o(hp)} = \frac{1 + \dfrac{R_6}{R_5}}{1 + \dfrac{R_3}{R_4}}, \quad \omega_o = \sqrt{\left(\frac{R_6}{R_5 C_1 R_1 C_2 R_2}\right)}$$

and
$$\zeta = \frac{1}{2} \frac{1 + \dfrac{R_6}{R_5}}{1 + \dfrac{R_4}{R_3}} \bigg/ \sqrt{\left(\frac{C_2 R_2 R_5}{C_1 R_1 R_6}\right)}$$

9.8 Band reject filter (notch filter)

An active band reject filter based upon a modified twin tee network[2] is given in Figure 9.22. With the components proportioned as shown the performance equation is governed by the relationship

$$\frac{e_o}{e_{i(j\omega)}} = \frac{Q\left[j\,\dfrac{\omega}{\omega_o} - \dfrac{\omega_o}{\omega}\right]}{1 + jQ\left[\dfrac{\omega}{\omega_o} - \dfrac{\omega_o}{\omega}\right]}$$ (9.8)

where
$$\omega_o = \frac{1}{CR} \text{ and } Q = \frac{1}{4(1-m)}$$

The circuit allows the adjustment of Q by means of a single potentiometer. In practice the depth of the notch obtainable with the filter is very much dependent upon component matching and high Q circuits are very component sensitive.

Figure 9.22 High Q band reject filter (notch filter)

A band reject filter which is less sensitive to component tolerance than Figure 9.22 can be realised with the state variable filter of the previous section. If a quad operational amplifier is used for the state variable filter, the fourth amplifier can be used to sum the input signal with the band pass output giving a band reject response; thus the output of amplifier A_4 in Figure 9.21 is

$$e_{br} = - \left[e_i \frac{R_9}{R_8} + e_{bp} \frac{R_9}{R_7} \right]$$

Substituting for e_{bp} from equation 9.4 gives

$$\frac{e_{br}}{e_{i(j\omega)}} = - \left[\frac{R_9}{R_8} - \frac{A_o \dfrac{R_9}{R_7}}{1 + jQ \left[\dfrac{\omega}{\omega_o} - \dfrac{\omega_o}{\omega} \right]} \right]$$

If resistors are proportional so that $A_o = R_4/R_3 = R_7/R_8$ the response becomes

$$\frac{e_{br}}{e_{i(j\omega)}} = - \frac{R_9}{R_8} \left[\frac{jQ \left[\dfrac{\omega}{\omega_o} - \dfrac{\omega_o}{\omega} \right]}{1 + jQ \left[\dfrac{\omega}{\omega_o} - \dfrac{\omega_o}{\omega} \right]} \right] \qquad (9.9)$$

Note that Q and ω_o have the values for the bandpass function given previously (equation 9.5).

9.9 Filter design

The preceding paragraphs have given an insight into the different shapes of filter response curve which may be obtained by the careful selection of the order of filter required together with the correct component values. The mathematical prediction of a particular response using manual methods becomes lengthy, tedious and error prone as the filter order increases. The recent proliferation of personal computers has made these design calculations a less onerous task, but even more important is the availability of ready-made designs for which tables of 'normalised' frequency against component values are published and which can be used to design a filter having a particular cut-off frequency and input impedance. The tabulated 'normalised' figures are 'scaled' to give practical component values.

Suppose we consider one of the simplest active filter circuits, the single pole low-pass filter, a typical circuit for which is shown in Figure 9.23(a) (the feedback resistor is included for DC off-set purposes). It is shown in Section 9.2.1 that the cut-off frequency for this CR circuit is given by:-

$$f_c = 1/2\pi CR \text{ Hz}$$

or $\omega_c = 1/CR$ radians/second

If the circuit were required to have an impedance level of 1 ohm and a cut-off angular frequency of 1 rad/s, then the capacitor would need a value of 1 farad. The circuit would be said to have been 'normalised' to 1 ohm and 1 rad/s and is shown in Figure 9.23(b). With these values for resistance and capacitance the circuit is not of great use but it can be 'scaled' to determine the values of resistance and capacitance which give a particular cut-off frequency and impedance level.

Suppose we wish to raise the impedance level from the normalised 1 ohm to 500 ohms. The rule for this is to raise all the circuit impedances by a factor of 500. This means that all the resistances must be multiplied by 500 but that all the capacitances will require their values to be divided by 500, since capacitive reactance is inversely proportional to frequency.

(*Rule No 1*—To increase the impedance level, multiply the resistors and divide the capacitors by the scaling factor.)

Further, suppose that we now wish to increase the cut-off frequency from 1 rad/s to 50 rad/s without changing the newly adjusted impedance level. The requirement now is for the circuit time constant to be reduced in that same ratio, that is, by 1/50. This means that the product of CR must be reduced by 1/50 without altering the fixed value of R at 500 ohms. Therefore, from the relationship, $\omega = 1/CR$ and knowing that ω must be 50 rad/s and that R is newly fixed at 500 ohms, C becomes $1/(50 \times 500)$ which is 40 F.

(*Rule No 2*—To increase the cut-off frequency, divide either the resistors or the capacitors, by the scaling factor.)

Figure 9.23 Simple first-order, low-pass active filter. (a) Typical circuit. (b) Circuit normalised to 1 ohm impedance level and cut-off frequency of 1 rad/s. (c) Circuit scaled to change the impedance level to 500 ohms and cut-off frequency to 50 rad/s (7.96 Hz)

When the circuits comprise more than a single resistor with a single capacitor, as is the case with the higher order filters, the same basic rules still apply. But remember, for multi-section filters, the ratios of the frequency sensitive capacitor and resistor pairs must remain unchanged if the overall filter frequency response is not to change. Also, if the frequency of one section is altered then all sections must be changed to the same frequency.

366

It was shown in Figure 9.12(a) that the closed loop gain for this single pole low-pass filter is:

$$\frac{e_o}{e_i(jf)} = \frac{1}{1 + jf/f_c} = \frac{1}{1 + j\omega/\omega_c}$$

Expressed in polar form, this becomes,

$$\frac{e_o}{e_i} = \frac{1}{\sqrt{1 + \left(\dfrac{\omega}{\omega_c}\right)^2} \angle \tan^{-1}\dfrac{\omega}{\omega_c}}$$

$$\text{or } \frac{e_o}{e_i} = \frac{1}{\sqrt{1 + \left(\dfrac{\omega}{\omega_c}\right)^2}} \angle -\tan^{-1}\dfrac{\omega}{\omega_c}$$

This expression can be used to calculate the circuit gain for varying values of ω, these being made a known fraction of ω_c. The table of data together with the plotted response curve are shown in Figure 9.24.

9.9.2 SALLEN–KEY SECOND ORDER ACTIVE FILTERS

There are many circuit configurations which operate successfully as second order filters but perhaps any dissertation on active filters would be incomplete without at least a mention of the circuits jointly attributable to Sallen and Key. There are two basic Sallen–Key designs: the unity-gain filter and the equal-component filter.

While these circuits are relatively simple to construct, in order that they operate as expected, the various component values must have a definite relationship which is a function of the circuit Q-factor. Typical Sallen–Key second order low-pass filter circuits of the two types mentioned are shown in Figure 9.25. It is important to note that both these circuits have constant gains; one being unity, the other 2 : 1.

An advantage of the unity-gain circuit is that it requires the minimum number of components; even the feedback resistor is not necessary in some circuits and may be omitted. However, the unity-gain circuit does not lend itself to easy conversion to a high-pass or bandpass filter by simply interchanging the circuit positions of some components. On the other hand, the equal-component circuit

367

ω (rad/sec)	$\left\| \dfrac{e_o}{e_i} \right\|$	Phase angle (degrees)
$0.0625\,\omega_c$	0.998	$-\ 3.58$
$0.100\,\omega_c$	0.995	$-\ 5.71$
$0.125\,\omega_c$	0.992	$-\ 7.13$
$0.25\,\omega_c$	0.970	-14.0
$0.5\,\omega_c$	0.894	-26.6
ω_c	0.707	-45.0
$2\,\omega_c$	0.447	-63.4
$4\,\omega_c$	0.243	-76.0
$8\,\omega_c$	0.124	-82.9
$10\,\omega_c$	0.100	-84.3
$16\,\omega_c$	0.062	-86.4
$100\,\omega_c$	0.010	-89.4

(a)

(b)

Figure 9.24 Response curve and data table for a first-order, low-pass filter. (a) Table of data for closed loop magnitude and phase; (b) response curve plotted from data at (a)

(a)

(b)

Figure 9.25 Second-order Sallen–Key, low-pass active filters. (a) Typical unity-gain
circuit; (b) typical equal-component circuit

requires more components but has the advantage of being simple
to convert to a high-pass filter by interchanging the frequency
determining resistors and capacitors.

REFERENCES

1. KERWIN, W. J., HUELSMAN, L. P. and NECOMB, R. W., 'State Variable Synthesis
    for Insensitive Integrated Circuit Transfer Functions, *I.E.E.E.J. Solid-State Circ.*, Sept.
    1967, 87
2. FARRER, T. E., *Elect. Eng.* 139, no. 470, 219.

369

Exercises

9.1 Show that for the circuit at Figure 9.1, the cut-off frequency, f_c, is given by the expression:

$$f_c = 1/2\pi RC \text{ Hz}$$

9.2 Also for the circuit at Figure 9.1, show that the magnitude of the voltage output, $|e_o|$, and its phase angle, ϕ, are given respectively by the equations:

$$|e_o| = e_i X_c / \sqrt{(R^2 + X_c^2)} \text{ and } \phi = \tan^{-1}(X_c/R)$$

9.3 Plot a graph of gain, $|e_o/e_i|$, against frequency for a simple first order, passive, low pass, CR filter and use this graph to show that:
(a) at f_c the gain is 3 dB less than its d.c. value
and
(b) At frequencies higher than $2f_c$ the output declines at 20 dB/decade.

9.4 A second order, low-pass filter with a Butterworth response and a 3 dB cut-off frequency of 20 Hz is required. The filter circuit of Figure 9.13 is to be used with close tolerance capacitors of value 0.001 μF and 0.002 μF. What values of resistors are required and what will be the output offset voltage if the operational amplifier has an input offset voltage $V_{io} = 2$ mV and the bias current $I_B = 50$ pA?

9.5 A high-pass, active filter is required and the circuit to be used is the one shown in Figure 9.13. If both of the capacitors are to have the same value of 0.001 μF and the 3 dB cut-off frequency is to be 1125 Hz, calculate the necessary values of each of the two resistors.

9.6 If the high-pass, active filter circuit of Figure 9.13 has a 3 dB cut off frequency of 1125 Hz and its input is connected to a 100 mV variable frequency signal, estimate the voltage output from the filter at (a) 100 Hz, (b) 1 kHz and (c) 2 kHz.

9.7　It is required to produce a unity-gain, low-pass Sallen–Key active filter based upon the circuit shown in figure 9.25(a). If the filter input resistance is to be 20 kΩ and the 3 dB cut-off frequency 1 kHz, suggest suitable values for the capacitors and resistors.

9.8　Component values $R_3 = R_5 = R_6 = 10$ kΩ, $C_1 = C_2 = 0.001$ μF, $R_1 = R_2 = R_4 = 1$ MΩ are used in the state variable filter of Figure 9.14. Calculate the constants of the equation (equation 9.9) relating the output of the amplifier A_2 to the input signal. Sketch the response on a dB against log f plot. In order to obtain a bandpass response, a fourth operational amplifier is used to sum the input signal with the output of A_2. Suggest suitable values for the components to be connected to this fourth amplifier. If the output voltage limits of the operational amplifiers are ± 10 V, calculate the maximum allowable input signal amplitude at the rejection frequency. Explain this limitation. Suggest modifications to component values in order to give the bandpass response a centre frequency of 200 Hz and a Q-factor of 100.

CHAPTER TEN

PRACTICAL CONSIDERATIONS

The principal fields of application for operational amplifers have
been outlined in the preceding Chapters and more applications will
be found in Appendix A.1. In some circuits, specific amplifiers and
component values have been given but this is simply for the
convenience of the reader, for in general applications are not
confined to particular amplifier types and specific component
values. The amplifier and component values suited to a particular
application are dictated by the performance requirements of that
application. This Chapter describes how to set about choosing an
amplifier and discusses some of the more important practical
points which should be considered when designing and using
operational amplifier circuits.

10.1 Amplifier selection and design specification

There is a wide variety of different operational amplifiers from
which to choose and the newcomer to operational amplifier circuits
is advised to start his or her work on practical circuits using an
internally compensated general purpose device. The performance
capabilities of a good general purpose operational amplifier are
normally adequate for the majority of operational amplifier
applications. Amplifiers of the '741' type have been used in industry
for many years and are still available. However, the ever expanding
range of newer operational amplifier designs using improved
technologies has made the '741' obsolete in many applications.
Readers wishing to familiarise themselves with some of the many
operational amplifiers now available commercially should refer to
Appendix A5.

Once the reader has gained experience in using general purpose
operational amplifiers he or she will be in a better position to start
designing his own operational amplifier circuits and might require
the use of more specialised amplifier types. Choice of amplifier type
and other design decisions are made easier by a systematic
approach to the design process. To choose an amplifier or

amplifiers for a given application the designer must first of all clearly spell out the nature of the application, its objective and all factors which might conceivably influence the attainment of those objectives. The designer must further make quite sure that he understands the meaning of published amplifier specifications and that he is able to translate them into terms meaningful to his design requirements (a thorough study of Chapter 2).

Essential to the proper formulation of any operational amplifier design is a knowledge of the following:

The nature of the signal source. Is the source a voltage or current source? What is the source impedance? What is the expected amplitude range of the input signal? What are the expected time/frequency characteristics of the input signal?

The nature of the load. What is the load impedance? What output voltage and current are required? Remember that it is always possible to increase the output current capability of an amplifier by the addition of a suitable booster amplifier (see Section 10.9).

The required accuracy. With what degree of accuracy is it required to amplify or manipulate the input signal? Accuracy must of course be defined in terms meaningful to the application with respect to band-width, d.c. offset and other parameters. It is important to make a realistic estimate of accuracy requirements. The operational amplifier circuit may represent a sub-system of a complete measurement or instrumentation system, and accuracy must be related to the overall system accuracy which is aimed for. A mistake often made by beginners is either to start a design with little more than the haziest idea of performance errors or to specify a far greater accuracy than is really required.

The environmental conditions. What is the maximum range of temperature, time and supply voltage over which the circuit must operate to the required accuracy without readjustment? In assessing the relevant factors it is necessary to consider the total environment in which the circuit is required to operate; the physical environment, for example temperature, humidity, mechanical vibration, the presence of near by sources of interference noise etc., and the human environment, for example, the degree of skill to be expected from the operators of the finished design.

Having carefully considered the details of an application specification, the designer must then decide as to how best to meet it, the decision normally being made from a cost/performance stand-point; 'best' is regarded as that which achieves a desired performance specification at minimum cost. In making a realistic cost estimate the designer should look carefully for hidden costs—many factors other than the price of an amplifier, components, or complete circuit modules can contribute to the overall cost of implementing an application. Cost relates to the external requirements necessary to make a device compatible with other elements of a complete system in which the device may form just a small part, and to the time factor involved in any setting up procedure. For example, there may be a specific circuit function that can be performed by a low cost general purpose operational amplifier but which requires the use of trim potentiometers and an adjustment procedure which must be performed by a skilled operator. Under such circumstances there could well be an overall cost advantage in using a higher performance more expensive amplifier if the adjustment requirement were thereby eliminated. Savings would be in the price of the potentiometer (perhaps a precision one), space savings, and skilled operator time savings. In addition, there could well be an added bonus in the superior performance provided by the more expensive amplifier.

10.2 Selection processes

Amplifier performance parameters influence the behaviour of an amplifier in a specific application. They control the performance errors which determine the ability of a particular amplifier to meet a desired accuracy requirement as set out in an application specification. In most applications all amplifier parameters are not of equal importance and the ability to recognise the important performance limiting specifications simplifies amplifier selection. A general purpose amplifier, if it can meet desired performance requirements, is likely to be the best choice of amplifier type reckoned on a cost/performance basis. Where the use of a general purpose operational amplifier is not possible, it is generally because of limitations encountered in two areas; d.c. offset and drift performance and/or bandwidth and slew rate requirements.

10.2.1 D.C. AND LOW FREQUENCY APPLICATIONS

Many operational amplifier applications are concerned with slowly varying signals in which a knowledge of the d.c. level of the signal is important. In such applications it is the amplifier offset voltage, bias current
and drift parameters which largely influence the final selection of an
amplifier.

The non-inverting feedback configuration, where performance
requirements allow its use, is usually best for processing the signal from
voltage sources. A high input impedance which minimises loading errors
is obtained without the use of high value resistors and it is the amplifier
input offset voltage and drift which are of prime importance in determining d.c. errors. Non-inverting applications require a consideration of
common mode errors and limits.

Voltage summing applications with isolation between signal sources
require the use of the inverting feedback configuration, and the resistor
values required to minimise loading errors normally make both amplifier
bias current and offset voltage important in determining d.c. errors.
Inverting configurations do not of course require a consideration of
common mode errors and limitations.

Applications in which the input signals are essentially currents applied
by current sources invariably use the inverting amplifier configuration.
In such applications it is usually amplifier bias current and its drift
which dominate the d.c. errors.

In examining offset and drift specifications the designer must ask
himself how much offset and drift error can be tolerated. The question
is related to the input signal level and the required accuracy. For
example, to amplify or otherwise manipulate a d.c. input signal of 1 V
with an accuracy of 0.1% the offset error must be 1 mV or less. Note
that the offset error is a combination of the effects of amplifier input
offset voltage and bias current (see Section 2.10.4); values are made up
of initial values plus drift. Initial offsets can be balanced out by a suitable trimming arrangement (see Section 10.6); errors are then due
to drift. This of course assumes that other sources of error such as
input loading, noise and gain errors have already been assessed.

10.2.2 WIDE BAND APPLICATIONS

Applications in which d.c. levels are not of interest can use a capacitor
to block out d.c. offset. Amplifier offset and drift specifications can

then largely be ignored, except in so far as a large d.c. output offset might restrict the dynamic swing of alternating output signals.

Some significant points relating to the selection of an amplifier for amplifying or manipulating continuous sinusoidal, complex or random wave forms are as follows:

1. What closed loop bandwidth is required?
   Closed loop bandwidth is determined by the intersection of the open loop and $1/\beta$ frequency response plots (see Section 2.5).

2. What loop gain βA_{OL} is required?
   The available loop gain at a particular frequency or over a range of frequencies is often more important than closed loop bandwidth in an application. Closed loop gain stability, output impedance and nonlinearity all depend upon loop gain.
   Closed loop gain stability

$$\frac{\Delta A_{CL}}{A_{CL}} \cong \frac{\Delta A_{OL}}{A_{OL}} \frac{1}{\beta A_{OL}} \tag{10.1}$$

$\Delta A_{OL}/A_{OL}$ is the open loop gain stability which is dependent upon temperature and power supply voltages.
Closed loop output impedance (see Section 2.2.2)

$$Z_{oCL} \cong \frac{Z_{oCL}}{\beta A_{OL}} \tag{10.2}$$

Closed loop distortion (nonlinearity)

$$D_{CL} \cong \frac{D_{OL}}{A_{OL}} \tag{10.3}$$

The open loop input/output transfer curve for an amplifier may exhibit nonlinearity but the effects of such nonlinearity on the closed loop behaviour are reduced to an extent dependent upon the magnitude of the loop gain.

A loop gain of 100 (40 dB) is normally adequate for most applications but remember that loop gain decreases with increase in frequency (see Section 2.5) and this makes it difficult to obtain large loop gain at high frequencies. For this reason it may be necessary to use an operational amplifier with a 10 MHz unity gain bandwidth in order to achieve adequate loop gain over a 10 kHz bandwidth, or in high gain wide band applications it may be necessary or more economical to use two amplifiers in cascade each at lower gain.

3. What power response and/or slew rate are required (see Section 2.9)? An amplifier should be selected whose slew rate exceeds the maximum expected rate of change of the output signal. In applications in which signals are sinusoidal, a wide small signal bandwidth is not much help if the available output signal is only a fraction of a volt when a much larger output signal is required. An amplifier should be chosen whose power bandwidth is not exceeded at the highest operating frequency.

In applications such as pulse amplifiers and sample holds used in A/D and D/A converters, the transient response of the wide band amplifier is generally more important than gain bandwidth characteristics. Slewing rate, overload recovery and settling time are the amplifier specifications which normally govern the choice of amplifier type.

10.3 Attention to external circuit details

Having taken care to select an amplifier which, according to its published specifications, should reduce the errors in an application to allowable limits, it is important not to degrade performance by improper attention to external circuit details. Care is required in order to make use of the full capabilities of those amplifier types which reach state of the art limits in specialised aspects of performance. Neglect of possible error sources external to the amplifier nullifies the attention to detail that has gone into the design of the amplifier, and of course wastes the extra money that has been paid for the high performance amplifier. The performance offered by the better present day general purpose designs approaches that of the so-called high performance amplifiers of not so many years ago, but gross inattention to external circuit conditions could prevent the capabilities of these newer devices from being realised.

Some practical external circuit details requiring attention are now outlined. The experienced electronics designer will already be aware of the importance of the physical layout of a circuit and of the quirks and idiosyncrasies of practical electronic components. Operational amplifiers are being increasingly used by non-electronics specialists (with great effect) and it is these readers who may perhaps benefit most from the practical hints and tips that are to be given.

Operational amplifier circuits, in common with most other electronic circuits, give best results if some care is taken over the physical arrangement of the components external to the amplifier. A neat circuit layout which minimises the effects of stray capacitance should be sought. The following are some points worth observing:

Low resistance and low inductance earth and power supply leads should be used. Power supplies should be RF bypassed to earth. Bypass capacitors should be connected at or as near as possible to the device socket; values of bypass capacitors in the range 0.01 to 0.1 μF are normally satisfactory. Proper attention should be paid to frequency compensation of the amplifier. Input and output leads should be kept as short as possible and shielded if required. It is advisable to use one common tie point for all earth connections and this should be near to the amplifier. Finally, when a circuit is completed waveforms should be monitored with an oscilloscope to verify correct operation.

10.4 Avoiding unwanted signals

A practical operational amplifier circuit is prone to disturbances that are not in any way suggested by its circuit diagram. If the sources of such disturbances are identified and understood their ill effects, if not completely avoidable, can be considerably reduced by the use of appropriate circuit techniques. The smaller the input signal to be measured, the greater attention to detail must be paid. Particular care is required in the measurement of very small currents at high impedance levels.

10.4.1 EARTH LOOPS

In an electronic system separate earthing of individual units in the system can create large area circuit loops. Stray magnetic fields at power line frequencies can induce currents in such loops thus injecting unwanted signals into the system. In amplifier applications if both signal source and amplifier are separately earthed as shown in Figure 10.1 an earth loop is created; the obvious remedy is to earth the system at one point only, usually at the amplifier input. If this is not possible a differential input circuit configuration should be used in order to reject the unwanted pick up signal (see Section 4.4).

Another earth loop error which can occur when measuring small signals is when the power supply current or load current is allowed to

Figure 10.1 Separate earthing of source and amplifier creates earth loop

flow through the input signal return connection. If this happens a voltage is developed which is applied to the input of the amplifier in series with the signal source; this results in an error at the output. The proper connection to avoid this effect is shown in Figure 10.2; signal return and load return should be connected to power supply common as close to the amplifier pins as possible.

Figure 10.2 Proper connections to avoid earth loops

10.4.2 THERMOCOUPLE EFFECTS

A possible source of error which may be encountered in the external connections to an amplifier is thermocouple voltages which are generated when a difference of temperature exists between the junction of dissimilar metals. Thermocouple voltages can cause an offset voltage drift comparable in magnitude to the specified offset voltage drift of very low voltage drift amplifiers. Careful attention to inter-connection layout serves to minimise possible temperature differential between junctions and reduces the effect.

379

Unwanted alternating signals can be introduced into the input of an amplifier via capacitive coupling or via inductive coupling. Take for example the inverting amplifier shown in Figure 10.3. If $e_s = 100$ mV and $R_1 = 1$ MΩ it only needs a leakage coupling of 10^{11} Ω from a 200 V a.c. supply to introduce an unwanted noise signal equivalent to 2% of the input signal. A capacitance as small as 0.03 pF will provide this at 50 Hz.

Capacitively coupled signals can be minimised by surrounding the input circuitry of the amplifier, or preferably the whole of the amplifier, with an earthed electrostatic shield. The shield should be ferromagnetic as well as conductive when the unwanted signal is introduced by inductive coupling. High permeability magnetic shielding is best for shielding away signals under 100 Hz but high conductivity shielding (aluminium or copper) is more effective for frequencies above 100 Hz.

D.C. leakage paths
Unless particular care is taken to avoid d.c. leakage paths when using low bias current amplifiers the overall circuit performance is often limited by leakages in capacitors, diodes, analogue switches or printed circuit boards rather than by the operational amplifier itself.

Printed circuit boards must be thoroughly clean, solder fluxes should be removed with trichloroethylene or alcohol and boards blown dry with compressed air. The fluxes may be good enough insulators at low circuit impedance levels but can cause gross errors in low current high impedance circuits and erratic behaviour as the temperature is changed. Even the leakage of properly cleaned boards can be trouble-

Figure 10.3 Unwanted signal can be picked up at amplifier input because of stray coupling impedance

some at elevated temperatures. At 125°C the leakage resistance between adjacent runs on a clean, high quality epoxy glass board (0.05 in separation, parallel for 1 in) may be no more than 10^{11} Ω. The leakage becomes worse if the board becomes contaminated, indicating the desirability of coating boards with epoxy or silicon rubber after cleaning to prevent such contamination from taking place.

Guarding

Guards can be used to interrupt the leakage paths to the input terminals of an operational amplifier. The standard pin configuration used with most integrated circuit operational amplifiers has the input pin adjacent to pins at the power supply voltage. A guard in the form of a circuit board trace forming a conducting ring surrounding the input pins can be used to prevent leakage currents from the pin at the supply potential passing to the input pins. A suitable board layout for input guarding of an 8-lead TO5 device is shown in figure 10.4. A 10-lead pin circle is used; the leads of the integrated circuits are formed so that the holes adjacent to the input pins are vacant when the integrated circuit is inserted into the board. Note that if the integrated circuit leads or the leads of other components connected to the input of the amplifier go through the board it may be necessary to guard both sides of the board.

Guard

Figure 10.4 Printed circuit board input guard for eight-lead amplifier

In order to be effective the potential of the guard must at all times be constrained to be equal to that of the input terminals. Proper guard connections for the common feedback configurations are shown in Figure 10.5. In circuit configurations in which the input terminals are close to earth potential, e.g. inverters and integrators, a guard is simply connected to earth. In follower configurations in which the input terminals are above earth potential the guard is maintained at the same potential as the input terminals by driving it by a signal derived from the amplifier output. A low impedance drive should be used calling for relatively low input, and feedback resistors as shown by R_1 and R_2 in Figure 10.5(c).

Figure 10.5 Guard connections (a) Inverter (b) Unity gain follower
(c) Follower with gain

Another consideration in high impedance low level measurement besides insulation problems is the capacitance of cable used for input shielding; this should be minimised by using as short a length of cable as possible. In high input impedance follower applications, cable capacitance lowers the effective input impedance; it forms a long time constant CR low pass filter with the high value signal source resistance, which restricts bandwidth and increases rise time. If the cable shield is driven (rather than connecting it to earth) by connecting it to the guard drive as in Figure 10.5(c), cable capacitance is uncharged and its effect in increasing input capacity is reduced. It is possible to decrease input capacitance in the follower with gain circuit by connecting a capacitor C_p as shown in Figure 10.6 as a positive feedback path from output to input.

A further point about input cable and input leads in high input impedance circuits is that they should be kept as mechanically rigid as

$$C_p = C_s \frac{R_1}{R_2}$$

For neutralisation

Figure 10.6 Neutralising input capacitance in follower with gain

possible. Movement causes capacitor changes which in turn cause
changes in charge and a spurious signal flow.
Note:

$$I = \frac{dQ}{dt} = \frac{d}{dt}(CV) = C\frac{dV}{dt} + V\frac{dC}{dt}$$

This term
causes
problems

10.5 Ensure closed loop stability

The comparative simplicity of the external circuitry used in many
operational amplifier applications should not allow the designer to
forget the underlying principles responsible for precise circuit operation.
Precise performance is obtained as a result of negative feedback. An
operational amplifier circuit is essentially a feedback system and like
other feedback systems it can, if suitable precautions are not observed,
become unstable and oscillate as a result of excess phase shift in the
feedback loop. The factors governing closed loop stability and the
methods used for frequency compensating amplifiers in order to obtain
closed loop stability were discussed in detail in Chapter 2. As a general
rule when using externally frequency compensated amplifiers, it is best
to follow the manufacturer's recommended method of frequency compensation for a particular amplifier, but remember that if speed and
bandwidth are not a design limitation increased phase margin against

instability can generally be obtained by increasing the size of compensating capacitors.

External circuit influences that have perhaps not been accounted for can decrease stability phase margin and it is always as well to check closed loop stability when evaluating a new design. A transient response test gives a quick practical evaluation of closed loop phase margin (see Section 3.8.2).

Some external factors which can adversely affect closed loop stability are : load capacitance (stray wiring capacitance or an actual capacitance at the amplifier output), capacitance at the inverting input of the amplifier, a large resistance at the non-inverting input terminal, supply voltage not adequately bypassed.

10.5.1 EFFECT OF LOAD CAPACITANCE

Capacitance between the output terminal of an operational amplifier and earth forms a first order lag network with the amplifier output resistance and introduces an extra break in the amplifier open loop frequency response. If this break occurs at a frequency before that at which the loop gain becomes unity (or near to it) the extra phase lag introduced by the added break decreases the stability phase margin and can cause an otherwise stable closed loop configuration to become unstable.

The output resistance of an operational amplifier is generally low so that most amplifiers will tolerate quite a few pF at the output, but when load capacitance reaches 100 pF or more it may be necessary to take appropriate steps to ensure closed loop stability. It does not take much load capacitance (driving a coaxial cable for instance) to make some 741's oscillate when used at unity closed loop gain.

Compensation for a capacitance load can be achieved using the circuit arrangements shown in Figure 10.7. Resistor R_3 (value of order 100 Ω or so) is used to isolate the capacitance load from the output, and a feedback capacitor C_f is connected directly from the amplifier output to the phase inverting input terminal. The value of C_f should be chosen so that its capacitive reactance at the unity gain cross over frequency is no more than 1/10 of the resistance of R_2. At high frequencies feedback is predominantly via C_f making the high frequency value of $1/\beta$ approach the limiting value of unity. This means that when using this capacitance load isolation scheme the amplifier must be frequency compensated for unity gain operation regardless of the closed loop signal gain.

Figure 10.7 Preventing instability due to capacitive loading

10.5.2 EFFECT OF INPUT CAPACITANCE

Capacity between the inverting input terminal of an amplifier and earth decreases the feedback fraction at high frequencies and introduces a phase lag into the feedback loop, thus reducing phase margin and leading to possible instability (see Section 2.6). The stability problem is most conveniently examined in terms of Bode plots as shown in Figure 10.8. The break in $1/\beta$ occurs at the frequency $1/(2\pi C_i R_1 \parallel R_2)$, and if this frequency is less than the frequency at which the loop gain becomes unity (the $1/\beta$ and A_{OL} intersection) phase margin is reduced. Clearly the effects of stray capacitance at the input are likely to be significant in applications using large values for input and feedback resistors.

A simple method of compensating for the effect of input capacity is to connect a phase lead capacitor C_f in parallel with the feedback resistor R_2. This causes $1/\beta$ to break back at the frequency $1/(2\pi C_f R_2)$, the leading phase shift thus introduced cancelling the lag due to C_s. The value of the capacitor C_f should be chosen so that the frequency $1/(2\pi C_f R_2)$ is at least an octave before the $1/\beta$ and A_{OL} intersection frequency. The closed loop signal bandwidth is set at $1/(2\pi C_f R_2)$ by the value used for C_f.

10.5.3 EFFECT OF A LARGE RESISTANCE AT THE NON-INVERTING INPUT TERMINAL

A large resistor value connected to the non-inverting input terminal such as might be required for bias current compensation can lead to closed loop instability. The resistor forms a potential divider with the differential input impedance of the amplifier, whose frequency dependence introduces a phase lag in the feedback signal at the higher frequencies. The effect is remedied by bypassing the bias current compensating resistor with a capacitor to earth.

$$\text{Without } C_f \quad \frac{1}{\beta_{(j\omega)}} = \left[1 + \frac{R_2}{R_1}\right]\left[1 + j\omega\, C_i\, R_1 /\!/ R_2\right]$$

$$\text{With } C_f \quad \frac{1}{\beta_{(j\omega)}} = \left[1 + R_2\right]\left[\frac{1 + j\omega\,(C_1 + C_f)\,R_1 /\!/ R_2}{1 + j\omega\, C_f R_2}\right]$$

Figure 10.8 Capacitance at the input can cause instability

Any impedance that the individual stages of a multistage operational amplifier have in common allows interaction between the stages and represents a possible source of regenerative feedback and instability. The power supplies represent a source of common impedance. Normally power supplies have a very low impedance but the inductance of the supply leads, particularly if long leads are required, can present an appreciable impedance at the high frequencies at which the transistor stages in the operational amplifier still have appreciable gain. A simple cure for instability caused by this effect is to bypass the positive and negative supply terminals of the amplifier to earth with capacitors of value at least 0.01 μF.

Different amplifier types vary in the amount of inductance they can tolerate in the supply leads without requiring supply bypass capacitors. Some amplifier types, when connected singly, do not require bypass capacitors to prevent oscillation, but when a large number of the same amplifier type are run from common power supply lines oscillations occur if bypass capacitors are not used.

Supply bypassing and decoupling are particularly important when current amplifiers are used in conjunction with operational amplifiers to boost the output current capabilities of the operational amplifier (see Section 10.9). Current boosters can feed a considerable signal back into the supply lines and bypass capacitor values should be increased accordingly.

10.6 Offset nulling techniques

A requirement for zero d.c. output voltage with zero input signal exists in most operational amplifier applications but d.c. offsets arise at the output of an amplifier as a result of the effects of amplifier input offset voltage and bias current (see Section 2.10.4). There are a variety of circuit techniques which may be used to balance out the effects of initial offset; choice of the most suitable technique depends upon the nature of the application, inverting, non-inverting, differential, the circuit impedances involved and the effect, if any, of the technique used on the offset temperature drift. In some applications a small initial offset may be tolerable, if so it might be possible and cheaper to choose an amplifier type which will set the resultant input offset voltage within the allowable limit. Savings will be in the cost of the balance potentiometer which will not be required, and of course in time, since no adjustment procedure will be required.

In applications requiring zero initial offset some form of offset balancing must be employed. Most operational amplifiers have provision for adjustment of V_{io} with a single trim potentiometer connected directly to the amplifier, but what is perhaps not generally realised is that published drift specifications do not normally apply to the amplifier when it is connected to this offset potentiometer. Nulling the input offset voltage with the recommended trim potentiometer can induce additonal voltage drift with temperature. In applications with large source resistance the offset due to bias current should never be balanced with the amplifier's internal trim potentiometer. Internal offset balancing can cause considerable variations in the gain characteristics of many integrated circuit operational amplifiers (see Section 3.7.1).

External offset balancing techniques do not affect the amplifier's internal circuit conditions and consequently cannot introduce any extra drift. In external offset balancing an adjustable d.c. signal is added directly to one or other of the amplifier's input terminals. There are several ways of doing this and in many cases there is little to choose between them. Some of the more commonly used offset balancing techniques are now given.

Input offset error voltage in the inverting amplifier configuration can be balanced by supplying a small additional current to the amplifier summing point using the arrangement shown in Figure 10.9(a). The

Figure 10.9 Offset balancing—inverting configuration (a) Current bias
(b) Voltage bias

388

balancing current is adjustable in the range $\pm V_s/R_3$; it produces an equivalent input offset voltage range $\pm V_s/R_3 (R_1 \| R_2)$, which is adjusted to be equal and opposite to the total input offset voltage E_{os} equation 2.39). If a fine offset balance is to be possible the total range should not be significantly greater than the expected maximum value of E_{os} and this normally dictates that R_3 should be made several thousand times greater than the parallel combination of R_1 and R_2. In applications using large values of R_1 and R_2 the resistance value required for R_3 becomes excessive and offset balancing is more readily achieved by adding a small adjustable voltage to the non-inverting input terminal of the amplifier, using the balancing circuit of Figure 10.9(b)

Offset range $\pm V_s \dfrac{R_3}{R_4} \left[R_3 \ll R_4 \right]$

Offset range $\pm V_s \dfrac{R_2}{R_1}$

(a)

(b)

Figure 10.10 Offset balancing—follower configuration (a) Follower (b) Unity gain follower

Offset balancing with a small adjustable voltage effectively in series with the input is also used in the non-inverting amplifier configuration as shown in the circuit of Figure 10.10(a). In this circuit the correction voltage is developed across the resistor R_3; the balancing adjustment alters the gain but provided $R_3 \ll R_1$ the effect is not significant.

A balancing arrangement suitable for a unity gain follower is shown in Figure 10.10(b). The value of resistor R_1 should be made very much greater than R_2 but this still leaves the circuit with a gain slightly greater than unity. For example, with typical values say $R_1 = 2$ MΩ, $R_2 = 2$ kΩ, the gain error is 0.1%.

389

Figure 10.11 *Offset balancing a differential amplifier*

The arrangement illustrated in Figure 10.11 provides a method of offset balancing for a differential amplifier configuration. The offset correction is applied by the small bias voltage developed across the resistor R_4. A disadvantage of this method is that unless $R_5 \gg R_4$ the offset nulling procedure will degrade the c.m.r.r. of the circuit because of the resistive unbalance introduced (see Section 4.4).

10.6.1 OFFSET BALANCING FOR VARYING SOURCE RESISTANCE

In applications in which the value of the source resistances connected to amplifier input terminals are changed, e.g. variable gain and switched gain circuits (see Section 4.2), the equivalent input offset error also changes because of its dependence upon amplifier bias current. Such circuits require separate balancing of the input offset voltage and bias current contributions to the total input offset error voltage. This can be done by using a combination of a current bias and a voltage bias, as shown in Figures 10.10(a) and (b). The voltage balance is adjusted first to give zero output voltage when the inverting input terminal is shorted to earth by means of a small resistance. The short is removed and the current balance is then adjusted to make the output again zero. Alternatively and more conveniently a low bias current f.e.t. input amplifier is used in order to make the bias current offset contribution negligibly small.

390

10.6.2 OFFSET BALANCING WITH DRIFT COMPENSATION

A variety of circuit techniques have been devised to balance amplifier bias currents and offset voltage and at the same time to provide compensation for the temperature drift of these parameters[1]. Most of these techniques require extensive external circuitry and although they can substantially improve the drift performance of general purpose amplifiers they require considerable care and time spent in adjustment procedures. The techniques are not described here, for if you want a really low drift amplifier it will probably cost you less in the long run to buy one rather than attempt to modify say a '741'.

Figure 10.12   Offset balancing with bias current drift compensation

The arrangement shown in Figure 10.12 is a simple offset balancing technique which provides a measure of compensation against the bias current temperature drift encountered in bipolar input operational amplifiers. The technique is easy to apply. The total equivalent input offset error voltage for the circuit is

$$E_{os} = \pm \ V_{io} + I_{b-} \ R_1 \| R_2 - I_{b+} R_3$$

R_3 is adjusted to make the output voltage zero when the input signal is zero; the adjustment makes E_{os} zero. In applications with source impedance values such that amplifier bias current is a major source of offset this technique provides a marked improvement in the temperature drift performance. In the case of bipolar input operational amplifiers the currents I_{b-} and I_{b+} tend to track well with temperature. The method is only applicable for fixed values of input and feedback resistors and readjustment is required if the values of these components are changed.

391

10.7 Importance of external passive components

The versatility of the operational amplifier approach to instrumentation arises from the ability to set performance characteristics precisely by means of a small number of passive circuit elements connected externally to the amplifier. The designer should not forget that the resistor and capacitor symbols which he uses so freely in circuit diagrams represent idealisations. Practical components have deficiencies: component magnitudes all have a tolerance factor and they exhibit a temperature dependence; a pure resistance, capacitance or inductance is not to be had in the form of a practical component. In many cases the ultimate limit to accuracy and stability in an operational amplifier application is determined not by the amplifier and its power supply but by the characteristics of the external components connected to it. The tolerance on the values of the components featured in a closed loop gain expression obviously sets a limit on closed loop accuracy, but the designer should look very closely at other characteristics of the passive components which he intends to use in a critical application.

10.7.1 RESISTORS

Resistors are the most frequently used external passive components in operational amplifier circuits. There are many different kinds of resistors available, differing widely in performance characteristics and cost. In the case of fixed value resistors, following initial tolerance the largest component of error introduced by resistors is that due to temperature coefficient of resistance (TCR). Other sources of error are leakage (particularly in high value resistors), humidity effects, drift with time and voltage coefficients. Resistors have some series inductance (particularly wire wound types), stray capacitance across their terminals and also, dependent upon type, exhibit several kinds of noise generation effects.

The chief types of fixed value discrete resistors used are, in ascending order of temperature stability (decreasing TCR), carbon composition, carbon film (cracked carbon), metal film and wire wound. Carbon film resistors are superior to carbon composition; they are marginally more expensive but the small extra cost is usually well worth while. They make an excellent choice for experiment and not so critical applications. In critical, gain setting and filter applications metal film resistors are normally used. Metal film resistors are available with TCR in the range

100 ppm/°C to 25 ppm/°C. Carbon film resistors can be usefully employed in trimming and padding, metal film and wire wound resistors when the greater TCR of the carbon resistor causes only a small percentage change in the overall value. Wire wound resistors provide the greatest initial accuracy and stability; their noise level approaches the theoretical minimum, but they are expensive and for a given resistance value they have a higher shunt capacitance and series inductance than other resistor types.

As an alternative to the use of discrete resistors integrated circuit technology has produced resistor networks which have come into increasing use as their cost comes down and their properties are better understood. Low cost networks that replace discrete resistors are now available.

Potentiometers

Potentiometers are another important component used in operational amplifier circuits. Potentiometers are essentially resistive dividers and as such must be considered to suffer from the same deficiencies inherent in fixed value resistors. In addition the effects of linearity error, end resistance, resolution and stray capacitance and inductance effects which vary with potentiometer setting all require consideration. Potentiometers together with several fixed value thin film resistors are commercially available in dual in line packages and provide a convenient means for implementing operational amplifier offset balancing circuits.

10.7.3 CAPACITORS

Although theoretically the pF's or μF's provided by any capacitor are the same, in practice the freedom from capacitor deficiencies and the package in which the capacitance is provided vary widely according to the dielectric used. A number of factors therefore have to be considered in assessing the most suitable capacitor dielectric for a particular application.

A choice of capacitor will normally be dependent upon a required combination of the following parameters: capacity/size, voltage and/or current rating, tolerance/stability/temperature coefficient, frequency, power factor/insulation resistance/Q environmental conditions, shape, finish and cost.

A further consideration not contained in the above list is dielectric absorption which results in the capacitor failing to discharge itself fully. This feature also prevents the capacitor from charging to a predicted voltage immediately. Dielectric absorption can be a particular nuisance in hold type circuits and at higher frequencies it causes an apparent loss of capacitance. The higher grade dielectrics exhibit low absorption characteristics with polystyrene being very good in this respect.

Capacitors are important components in operational amplifier applications such as integrators, track hold circuits and active filters; they are also important in frequency compensating techniques. The designer should look very carefully at the characteristics of any capacitor he intends to use in a performance limiting location.

10.8 Avoiding fault conditions

Most modern operational amplifiers incorporate internal circuit protection against accidental overloads but it is important to examine those overloads which they will not tolerate. An attempt should be made to avoid or protect against any unusual circuit condition or set of conditions which might occur in a particular application to cause amplifier limits to be exceeded and to lead to amplifier failure.

First, rather an elementary point, do not make the wrong pin connections; in particular, amplifiers do not take kindly to having power supplies applied to the wrong pins or reverse power supplies. Find out what are the maximum safe differential input voltage, common mode voltage and output voltage limits. Remember that these limits only apply for the amplifier connected to rated values of power supplies. Quite small values of input voltage applied before the power supplies are switched on can cause damage.

It might seem, at first sight, that nothing can go wrong if an amplifier's input terminals are internally protected up to power supply voltage levels and if its output is short circuit proof to earth or supply line. However, even with such an internally protected amplifier, under certain external circuit conditions, faults can arise which might lead to device destruction.

Look out for voltages retained on capacitors that can apply input signals when the power supplies are switched off or for conditions in which capacitor voltages can cause input voltage limits to be exceeded. Also, guard against conditions which might possibly allow the amplifier's output to be connected to a voltage higher than the supply voltage because, say, of an inductive load.

An example of a capacitor discharge that can cause an amplifier

Figure 10.13 An amplifier input limit can be exceeded by transient capacitor discharge

input limit to be exceeded is shown in Figure 10.13. The amplifier is connected as an integrator. The integrating capacitor, assumed charged up to the positive output limit of the amplifier, is to be discharged to the negative limit of the amplifier by connecting the amplifier output to the negative supply. Closing the discharge switch in Figure 10.13 will almost certainly blow out the amplifier for it in effect puts a transient − 30 V on the inverting input terminal. The inverting input terminal is made negative with respect to the negative supply voltage and the capacitor discharge current is supplied by the low impedance negative supply line. Fortunately it is not difficult to prevent the condition envisaged in Figure 10.13 from destroying the amplifier. It is merely necessary to limit the capacitor discharge current to a safe value of a few mA, and a resistor (say 10 kΩ) connected directly to the amplifier's input terminal will normally provide the necessary protection.

Diode clamps can be used to protect against input overloads (see Section 2.1) and diodes in series with power supply leads can be used to guard against inadvertent power supply reversal. Rather more elaborate circuits are required to guard against transient supply over voltages[2].

The output circuitry of most amplifiers is internally protected by some means of output current limiting to prevent excessive current being drawn because of an accidental short to either earth or power supply lines. In such amplifiers damage to the output stage can still be caused because of voltage breakdown if the output terminal of the amplifier is brought to a potential higher than that of the supply line (say because of an inductive load). The zener diode output clamp shown in Figure 10.14 can be used to provide protection against such

Figure 10.14 Zener diodes protect against output overvoltage

output over voltage. The output signal from the amplifier is taken from the side of R_s remote from the amplifier. R_s is included within any feedback loop; its value need be just sufficient to prevent excessive current being passed through the zeners as a result of the output point of the circuit being connected to a point at a higher potential than that of the amplifier supply voltages.

10.9 Modifying an amplifier's output capability

It should be remembered that operational amplifiers are normally designed for high voltage gain and economy of power consumption and are not intended to deliver an appreciable amount of output power. Typically operational amplifiers work off ± 15 V power supplies to give a rated output voltage of ± 10 V with output currents limited to something of the order of ± 5 mA. Operation with reduced power supply voltages is normally possible but results in a corresponding reduction of output capabilities.

Applications occur, for example if an operational amplifier is to be used to drive some form of electro-mechanical device, when a greater than normal output voltage and/or current swing is required. The answer to this problem, like the answer to many others, is largely dictated by economic considerations. By paying more money you can get an amplifier with greater than normal output capabilities. Alternatively by expending time and effort and using additional circuitry you can modify the output characteristics of a lower cost amplifier to fit the requirements of your application.

10.9.1 OUTPUT CURRENT BOOSTING

The output current capabilities of an operational amplifier can be increased by connecting a unity voltage gain current amplifier directly

To limit output current

Feedback

+V$_S$

To feedback loop

Output

−V$_S$

(a)

+V$_S$

To limit output current

Feedback

To feedback loop

Output

To limit output current

(b)

Figure 10.15   Output current (a) simple emitter follower current booster (b) Simple output current booster for dual polarity output current

to the amplifier's output terminal. The current amplifier is normally included within any feedback loop in which the amplifier is used so that any nonlinearities in the current amplifier are reduced in proportions governed by the loop gain in the circuit. Provided that the bandwidth of the current amplifier extends beyond the unity loop gain frequency the inclusion of the current amplifier in the feedback loop does not alter the closed loop stability phase margin.

In applications requiring only a single polarity output current and moderate current gain a single transistor emitter follower serves as a satisfactory current amplifier. In order to protect the transistor against excessive short circuit current it is advisable to include a resistor in series with the collector of the transistor as shown in Figure

397

10.15(a). The collector resistor inevitably imposes some restriction on the positive output voltage swing. A pnp transistor and reversed supply voltages can be used for negative output currents.

Single emitter follower boosters are not really suitable for dual polarity output currents for they consume an excessive amount of standby power and their dynamic voltage range is rather limited. Unity voltage gain current amplifiers providing dual polarity current output are available as single units for use as operational amplifier boosters (e.g. Analog Devices Type B100, National Semiconductor Type LH0002) or they can be made up using discrete components. A simple circuit for dual polarity output currents is shown in Figure 10.15(b). Its dynamic output voltage swing is largely determined by the magnitude of the collector resistors which are included to limit output short circuit current. The simple circuit is only suitable for low frequency applications at which the voltage gain of the operational amplifier can satisfactorily overcome the crossover distortion (dead zone as the output of the amplifier passes through zero) of the complementary pair. At the higher frequencies the finite time taken by the amplifier output voltage to switch between the output transistors introduces an output distortion which cannot be neglected.

The current booster circuit shown in Figure 10.16 eliminates crossover distortion by using transistor T_3 to establish class AB biasing of the complementary output pair. The action of T_3 is to hold the voltage across the 1 kΩ potentiometer at a multiple of the base emitter voltage

Figure 10.16   Class AB biasing of current amplifier eliminates crossover distortion

Figure 10.17 current booster gives ± 10 V across 10 Ω

of T_3 determined by the potentiometer setting. Larger output currents are obtained in the circuit of Figure 10.17 by adding further current gain in the form of Darlington pairs; transistors T_6 and T_7 are used to set an output current limit. If the bias regulator transistor T_3 is mounted on the same heat sink as the output transistors T_1 and T_2 thermal feedback ensures bias stability. The circuit gives ± 10 V across a 10 Ω load.

Note that offset drift in current boosters is of little significance because it is divided by the open loop gain of the operational amplifier when it is referred to the input of the composite amplifier.

10.9.2 OUTPUT VOLTAGE BOOSTING

Operational amplifier types are available which give a greater than normal output voltage swing (e.g. Analog Devices AD 171J/K, output voltage swing ± 140 V, Burr Brown 3582J, output voltage swing ± 145 V, National LM 144, output voltage swing ± 30 V), but by adding an output voltage gain stage employing high voltage transistors an ordinary general purpose amplifier can be adapted for higher than normal output voltage swings. A simple circuit of the type shown in Figure 10.18 can be used if only single polarity output signals are required. Dual polarity output voltage boosters normally employ some

Figure 10.18   Simple single polarity output voltage: 0–100 V

form of complementary transistor circuit. An example of a complementary transistor output voltage booster is shown in Figure 10.19, current limiting is provided both by emitter resistors and diodes D_3, D_4. Note that this voltage gain stage is phase inverting so that the overall feedback circuit around the composite amplifier must be returned to the non-phase inverting input terminal of the operational amplifier.

400

Figure 10.19   Voltage booster with class AB biasing and output current limit

Voltage boosters normally have their gain defined by a local feedback path connected around them. They are then connected to the
output of the operational amplifier and like current boosters are included within any feedback path in which the amplifier is used. The closed
loop stability of an operational amplifier voltage booster circuit
requires particular attention; unity gain frequency compensation of
the operational amplifier does not ensure closed loop stability of the
cascaded arrangement and the operational amplifier normally requires
greater than unity gain frequency compensation. The effect can be
examined in terms of the appropriate Bode plots.

It will be remembered (Section 2.7) that unity gain frequency
compensation of an operational amplifier involves using a value of
frequency compensating capacitor which reduces the open loop gain
down to unity before the second break in the amplifier open loop
response. Assuming that, as it should be, the bandwidth of the voltage
booster is greater than this second break frequency it is necessary to
use the operational amplifier with a value of frequency compensating
capacitor which reduces the gain of the composite amplifier down to
unity before this second break frequency in the operational amplifier

401

Figure 10.20 Operational amplifier must be used with greater than unity gain frequency compensation when followed by a voltage booster

response. Thus, if the voltage booster has a gain of 10 (20 dB) the operational amplifier requires a value of frequency compensating capacitor which is ten times greater than that which is required for unity gain operation of the operational amplifier used alone. See Figure 10.20.

10.9.3 OUTPUT LEVEL BIASING

It is sometimes necessary to develop the output signal of an amplifier about a reference level other than zero and this can be done without the use of an output voltage booster. An output zero reference level which is in fact considerably greater than an amplifier's normal output

rating can be obtained by the appropriate addition of a fixed d.c. bias
to the output of the amplifier.

In the circuit shown in Figure 10.21 the zener diode serves to hold
the amplifier output voltage within its rated limits and the zener breakdown voltage should be selected accordingly. The output voltage which
is applied to the load appears about the fixed positive level $E_{ref}R_2/R_3$.

Figure 10.21 Output level biasing with zener included within feedback loop

If the output to the load is to be developed about a fixed negative
reference level the zener should be inverted and the opposite polarity
used for E_{ref} and the zener bias supply. The output current rating of
the operational amplifier must be such that it can supply the load
current, the feedback and the zener bias current. Note that since the
zener is connected within the feedback loop the effects of its internal
impedance and any drift in zener voltage are divided by the loop gain
when they appear at the output which is applied to the load.

10.10 Speeding up a low drift amplifier

Amplifiers designed for low drift and d.c. stability are generally somewhat limited in their bandwidth and slewing rate, whereas amplifiers
in which slewing rate and wide bandwidth are optimised tend to exhibit
considerable offset voltage drift with both time and temperature. It is
possible to combine a low drift amplifier with a high speed amplifier in
such a way that the composite amplifier has the d.c. stability of the
low drift amplifier combined with the fast response of the high speed
amplifier.

Figure 10.22 Composite connections used to speed up a low drift amplifier in inverting applications

A composite amplifier connection is shown in Figure 10.22. The low drift amplifier A_1 is connected as an integrator so that its closed loop signal gain rolls off at 20 dB/decade reaching unity gain (zero dB) at the frequency $1/(2\pi CR)$. The output of amplifier A_1 is connected to the non-inverting input terminal of the wide band amplifier A_2, diodes D_1 and D_2 are included to prevent a possible latch-up condition which can exist in the composite amplifier if the output limits of the low drift amplifier exceed the common mode range of the wide band amplifier.

The open loop gain of the composite amplifier is

$$A_{2(jf)} \left[1 + A_1 \frac{1}{1 + j \dfrac{fA_1}{f_o}} \right]$$

$A_{2(jf)}$ is the frequency dependent open loop gain of the wideband amplifier.

A_1 is the zero frequency open loop gain of the low drift amplifier.

$f_o = 1/(2\pi CR)$ is the frequency at which the signal gain of the integrator connected low drift amplifier is unity.

A choice of component values CR which makes $f_o = f_1/A_2$ gives the composite amplifier a single 20 dB/decade roll off down to unity gain. A_2 is the zero frequency value of the open loop gain of the wideband amplifier and f_1 is the unity gain frequency of the wideband amplifier. It is assumed that the wideband amplifier has a unity gain compensated open loop response. At frequencies higher than f_o there is negligible signal transmission through the integrator connected low drift amplifier and at these frequencies the gain of the composite amplifier is provided by the direct path to the inverting input terminal of the wideband amplifier. D.C. offset and drift in the wideband amplifier when it is referred to the input of the composite amplifier is effectively divided by the open loop gain of the low drift amplifier, and the d.c. offset and drift characteristics of the composite amplifier are essentially those of the low drift amplifier. The integrator connected low drift amplifier senses and integrates any offset voltage from earth present at the inverting input terminal of the composite amplifier and applies an off-set compensating voltage to the non-inverting input of the fast amplifier. Note that if A_1 is an externally frequency compensated type, an alternative to integrator connection is to use A_1 with a much greater than normal frequency compensating capacitor so that its open loop unity gain frequency is set at the value f_o.

The interconnection technique of Figure 10.22 is essentially that present in feed forward frequency compensation (see Section 2.7.2) and it is also the same principle which underlies the low drift characteristics of chopper stabilised operational amplifiers.

10.11 Single power supply operation for operational amplifiers

Operational amplifiers are generally designed to operate from symmetrical positive and negative power supplies and in most of the circuits given in this book the connection of such twin supplies to the amplifier has been assumed. The use of dual supplies permits output signal voltages which can swing both positive and negative with respect to the potential of the power supply common terminal, which point is normally connected to earth. Applications not requiring a response down to zero frequency or in which only single polarity output signals are of interest can be implemented with an amplifier powered by a single voltage supply.

The problem of single power supply operation is simply that of maintaining the d.c. voltage levels between particular circuit points at their proper values. Most amplifiers have three reference levels when operated from dual supplies; these are: $+ V$, earth and $- V$ (V is the value of the supply voltage). For single supply operation these reference levels can be maintained by using $2V$, V and earth obtained from a resistive divider network or split zener biasing system. The negative supply terminal of the amplifier is connected to earth, the positive supply terminal is connected to $2V$, and the differential input terminals which are normally at earth potential when using dual supplies are biased up to V.

Connections for single power supply operation of an a.c. inverter and an a.c. follower (see Section 4.7) are shown in Figure 10.23. The d.c. blocking capacitors at input and output determine the low frequency bandwidth. Their presence means that amplifier offsets and their temperature drift are of no great importance except in so far as any output voltage offset reduces the dynamic output voltage swing capability of the amplifier. If necessary offset due to bias current can be minimised by connecting resistors of equal magnitude to the two input terminals of the amplifier. The resistors are returned to the V bias supply when using single power supply operation. The input impedance of the a.c. follower can be increased using the boot strapping technique previously mentioned in Section 4.7.

In single power supply application requiring a response down to zero frequency blocking capacitors cannot be used and the design problem usually centres on ensuring that the potential of the input terminals of the amplifier (measured with respect to the voltage midway between the supply pins) remains within the allowable common mode range of

Figure 10.23 Single power supply operation of a.c. amplifiers (a) Inverting a.c. amplifier
with single supply operation
(b) a.c. follower with gain–single supply operation

the amplifier. Single power supply operation of course allows only
single polarity output signals.

There are amplifier types available whose common mode range
extends down to the negative supply rail and which are specifically
designed with single power supply operation in mind, e.g. National

Semiconductors LM 124/224/324, Motorola MC 3503/3403; the Bi-Mos operational amplifiers from RCA, e.g. CA3140, allows single supply operation. The one type of application made possible by amplifiers specifically designed for single supply operation which cannot be implemented directly with general purpose operational amplifiers is shown in Figure 10.24 but the connection really does not work very well with the output near earth potential. Operational amplifiers designed specifically for single power supply operation can be used with dual or referenced supplies but they do not normally perform as well as most general purpose types.

Figure 10.24   Connections possible with operational amplifier designed for single power supply operation

Figure 10.25   Single power supply operation of general purpose operational amplifier with a second amplifier used to supply a reference

A method of referencing a general purpose operational amplifier for single power supply operation is shown in Figure 10.25. The common mode range at the input terminals is

$$V_{ref} \pm \text{(Common mode range of amplifier)} \times \frac{R_1 + R_2}{R_2}$$

which for low gains can include earth or even negative common mode signals. Either input terminal may be earthed and a signal which is positive or negative with respect to earth applied to the other. The output can swing both positive and negative with respect to the reference level and the reference output can be easily connected to other operational amplifier stages.

A recent innovation in the field of biasing the single-supply operational amplifier has been the introduction, by Texas Instruments, of their 'virtual ground generator'. This device is capable of overcoming many of the problems encountered by the existing biasing methods discussed above. More information on this new device is given in Section 10.12.3.

10.12 New technologies

Some of the data to be found in Appendix A5 refer to two new developments which are worthy of a brief explanation at this juncture.

10.12.1 COMPLEMENTARY BIPOLAR (EXCALIBUR) TECHNOLOGY.[3,4]

There are thousands of different types of integrated circuit operational amplifiers available commercially and the number is increasing almost daily. The many manufacturers of these different types are ever striving to produce the best devices by giving them improved d.c. precision, faster a.c. performance and lower power consumption.

One of the problem areas concerning the improved performance of operational amplifiers is the relative slow response of the PNP transistor compared with that of the NPN. (The reason is the low mobility of the majority carrier 'holes' in the PNP type compared with that of the electrons in the NPN type.) Many operational amplifiers require to use the PNP configuration in their signal paths, particularly in differential pair inputs and emitter follower outputs, but the PNP tends to be slow, having a typical f_T of only 5 MHz compared with 150 MHz for the NPN transistor. One solution to the problem of slow complementary bipolar circuits appears to be to increase the speed of the NPN element and this is the approach of several semiconductor manufacturers.

Conventionally, a vertical NPN transistor will have a P-type substrate and an N-type epitaxial layer. The theory of achieving faster PNP transistors is simply to reverse this design and have a substrate of N-type silicon. However, while this method successfully increases the NPN speed, it is only at the expense of reducing the speed of the complementary PNP fabricated in the same IC.

In contrast to this method, one manufacturer, Texas Instruments, has developed a manufacturing process for a fast vertical PNP device structure which also retains the speed of the NPN devices. Texas Instruments call this their Excalibur process and it uses a deeply submerged N-region as a 'fake substrate' in which a buried P-region becomes the collector. The fast PNP Excalibur transistor can be integrated directly into the signal path without fear of limiting the bandwidth or slew rate and very often with the added benefit of requiring less supply current than its predecessors.

An example of this new range of operational amplifier is the TLE2021/2/4 family which are low power, precision operational amplifiers in single, dual and quad packages respectively. They were designed to replace the standard OP21 or LT1013 amplifiers. Typically, the TLE2021 achieves a unity-gain bandwidth in excess of 2 MHz and a slew rate of 0.9 μs, all for a supply current of under 200 μA and while still holding an input voltage offset of less than 100 μV. Generally, the newer devices are faster and more stable.

10.12.2 'CHOPPER' STABILISATION[5]

The TLC265X family of cmos technology operational amplifiers offers enhanced d.c. performance using a technique known as chopper stabilisation. The chopper operational amplifier is designed continuously to undertake self-calibration to provide an ultra low offset voltage which is extremely time and temperature stable and at the same time increase the CMRR and reduce the 1/f noise content.

Figure 10.26 shows a typical chopper stabilised operational amplifier. Basically, the enhanced performance is the result of using

410

Figure 10.26 Chopper stabilised operational amplifier

two operational amplifiers—a nulling and a main amplifier—
together with an oscillator, switches and two external or internal
capacitors to create a system that behaves as a single amplifier.
With this approach, the TLC2652 operational amplifier achieves a
sub-microvolt input offset voltage, a sub-microvolt input noise
voltage and offset variations with temperature in the nV/°C range.

The on-chip control logic produces two dominant clock phases: a
nulling phase and an amplifying phase. During the nulling phase,
switch 'A' is closed, shorting the nulling amplifier inputs together
and allowing the nulling amplifier to reduce its own input offset
voltage by feeding its output signal back to an inverting input node.
Simultaneously, the external capacitor, CXA, stores the nulling
potential to allow the offset voltage of the amplifier to remain
nulled during the amplifying phase.

During the amplifying phase, switch 'B' is closed, so connecting
the output of the nulling amplifier to a non-inverting input on the
main amplifier. In this configuration, the input offset voltage of the
main amplifier is nulled. Also, the external capacitor, CXB, stores
the nulling potential to allow the offset of the main amplifier to
remain nulled during the next phase.

This continuous chopping process allows offset voltage nulling
during variations in time and temperature, and over both the
common mode input voltage range and the power supply voltage
range. Additionally, because the low frequency signal path is
through both the null and the main amplifiers, an extremely high
gain is obtained.

The low frequency noise output of the chopper operational amplifier depends upon the magnitude of the component noise prior to chopping and the capability of the circuit to reduce this noise while chopping. The low frequency noise is reduced by increasing the chopping frequency. Limiting the input signal frequencies to less than half the chopping frequency reduces the effects of intermodulation and aliasing.

10.12.3 THE 'VIRTUAL GROUND GENERATOR'[6]

In Section 10.11 the problems associated with the correct biasing of an operational amplifier powered from a single supply rail were discussed. For example, the proper biasing of a single power supply, inverting operational amplifier requires that the non-inverting terminal is biased at half the supply voltage. This will then allow the amplified signal at the output to swing evenly about half the supply voltage, as a virtual ground, without suffering distortion unless over-driven. Figure 10.27 illustrates the circuit requirement for this condition.

Some of the traditional methods of providing bias (see Figure 10.25 for one method) have the disadvantages of a high standing power dissipation (in the bias resistors) together with a poor input regulation because of any supply voltage variations being automatically passed on to biasing network. The use of additional buffer amplifiers and Zener diode voltage references help to reduce but do not eliminate these disadvantages.

An important step in achieving the proper biasing of single supply operational amplifiers has been the introduction of the Excalibur technology device TLE2425 which can act as a 'virtual

Figure 10.27 Operational amplifier properly biased for single supply applications

412

Figure 10.28 Use of Excalibur virtual ground generator for the biasing of single supply operational amplifiers

ground generator' around which the output of a single rail operational amplifier can be made to swing. The device is designed to operate with +5 V systems and produces a virtual ground voltage of +2.5 V. (Variants of the TLE2425 which will operate at higher voltages are under development.)

The TLE2425 comprises a precise band gap reference and high performance operational amplifier housed in a single monolithic analogue integrated circuit. The TO-226AA packaging provides a single three-lead component which readily lends itself to the retrospective modification of older type biasing arrangements. With respect to power line variations, it is capable of providing 5 μV/V regulation and while supplying ± 20 mA of load current, the output regulation varies only 8 μV/mA. Over a bandwidth from d.c. to 10 kHz, the output impedance is less than 6 μΩ and the device consumes only 0.9 mW.

The TLE2425 can also be used in other operational amplifier applications such as generating power rails (of 5 V, 10 V and 15 V, each with a sustained accuracy to within less than 1%), or as an ideal current source accurate to within 0.75% at room temperature.

10.13 Comparison of operational amplifiers[5]

10.13.1 D.C. CONSIDERATIONS

Table 10.1(a) and (b) show typical d.c. performance figures and highlight some of the features and benefits for the three basic types of operational amplifier technology: bipolar, bifet and cmos.

Table 10.1(a) D.C. comparison of operational amplifiers

D.C. parameter	Type of operational amplifier		
	Bi polar	Bifet	Cmos
Input offset voltage	10 μV–7 mV	500 μV–15 mV	200 μV–10 mV
Input offset voltage drift	0.1–10 μV/°C	5–40 μV/°C	1–10 μV/°C
Input bias current	100–50 000 pA	1–100 pA	1–10 pA
Input bias current drift	Fairly stable with temperature change	Doubles for every 10°C increase	Doubles for every 10°C increase

Notes:

Features	Benefits

Bipolar

- Very low offset and drift — Allow low level signal conditioning
- Stable bias current — Remains low at high temperature
- High voltage gain — Ensures accurate amplification
- Lowest voltage noise — Wide dynamic range

Bifet

- Very low input bias and noise current — Matches high impedance circuits
- Good a.c. performance — Useful for combined a.c. and d.c. applications

Cmos

- Very low input bias and noise currents — Matches high impedance circuits
- Single supply operations — Can be operated from battery or 5 V supply
- 'Chopper' stabilising techniques are used to overcome the large input offset drift prevalent in standard MOS devices. (See Section 10.12.2) — Can be used to amplify very small signals and provide a wide dynamic range

Comparing d.c. error parameter values for the different types of operational amplifiers shows that modern bipolar devices designed using the most modern processes and techniques have a performance far exceeding that of the old industry standard bipolar

Table 10.1(b)   D.C. comparison of operational amplifiers

Process	Bipolar		Excalibur	Bifet	LinCMOS		
Device No.	μA741	TLE2021B	TLE2027A	TL051A	TLC2201A	TLC2652A	Units
Input offset voltage	6000	100	25	800	200	1	μV
Input bias current	500	50	35	0.2	0.005	0.01	nA
CMRR	70	100	117	75	90	120	dB
Differential voltage amplification	86	120	140	94	104	135	dB
Input offset voltage drift	(10)	(2)	1	25	(0.5)	(0.03)	μV/°C

Note: The above figures are worst case values at 25°C except those in parentheses, which are for typical values.

operational amplifiers such as the μA741, LM301, LM324 and the like. The offset, its drift and the input bias current have been greatly reduced while the open loop gain has been increased dramatically. The TLE2021A has been designed for low power precision circuits and the TLE2027 for high precision circuits which still require low noise or high bandwidth. The new Excalibur type technology has improved the offset stability with time over one hundred fold and drift figures of 6 nV/month are now typical for the TLE202X family.

Bifet operational amplifiers normally have higher input offset voltage and drift compared with bipolar devices. The TL051 represents a second generation bifet operational amplifier with improved offset and stability compared with its predecessors. Although the input bias current of fet designs is insignificant when compared with that of bipolars, the fet bias current doubles for every 10°C temperature increase. Some bipolar designs actually have lower bias currents at higher temperatures than do fets.

Silicon gate cmos technologies such as LinCMOS have reduced the problem of unstable offsets in cmos designs. The TLC2201, designed using LinCMOS, is an example of the new breed of cmos devices. It offers extremely low and stable offsets while simultaneously featuring the high input impedance and low noise current found in the best of jfet devices. For high order d.c. precision, the chopper stabilised operational amplifiers, such as the TLC2652, provide the lowest input offset and drift.

Operational amplifiers require both a.c. and d.c. accuracy. While input offset voltages and bias currents are relevant in d.c. applications other parameters must be considered for a.c. applications.

The basic structure of the operational amplifier is designed to perform an amplification function. Unfortunately, this amplification of the signal is usually accompanied by a phase shift which can lead to instability of the device. To prevent instability, a Miller compensation capacitor is used but only at the expense of the slew rate and gain of the operational amplifier—ultimately defining the unity-gain bandwidth. These a.c. performance parameters are determined by the process technology used in manufacture or the design techniques employed. For example, the bandwidth of a device could be improved by increasing the supply current or by changing the technology.

Bipolar operational amplifiers were the first to be developed. These offer good gain and bandwidths but their slew rate for a given bandwidth is slow. This is a limitation of the bipolar technology process (high transconductance, g_m) and is not easily designed out.

PNP transistors are normally used somewhere in the operational amplifier design. Because they are usually much slower than their NPN counterparts they can severely affect the a.c. performance of the operational amplifier. The complementary bipolar technology (see Section 10.12.1) provides PNP transistors with bandwidths approaching those of NPN transistors—this enables the design of much faster operational amplifiers without having to compromise parameters such as supply current.

Bifet operational amplifiers are designed using a combination of bipolar and jfet structures. P-channel jfets (with much lower transconductance than have bipolar transistors) are used in the input stage and the remaining circuit is designed using bipolar transistors. This combination has produced operational amplifiers with significantly higher slew rates than purely bipolar designs.

Cmos technologies such as LinCMOS have a similar performance to designs using bifets, but are of particular benefit for low power or single supply applications.

The slew rate and bandwidth of an operational amplifier are related. For example, for a single pole bipolar operational amplifier the slew rate is approximately 0.32 × the unity-gain bandwidth. Devices with additional poles in their transfer function or processed using an alternative technology will have a different relationship.

Table 10.2 shows how the performance of a device is dependent upon the technology used and the supply current. Old designs such as the LM324 using very basic integrated circuit technologies do not compete with devices, such as the TLE2021, that have been designed using a complementary bipolar process. The TLE2027 has multiple poles, so although its bandwidth is high its slew rate is lower than would normally be expected. Bifets have high slew rates for a particular bandwidth without having particularly high supply currents but, unfortunately, they have poorer and less stable input offset voltages.

Table 10.2 Comparison of operational amplifier a.c. performance

Technology	Device No.	Bandwidth (MHz)	Slewrate (V/μs)	Supply current (μA)
Standard bipolar	μA741	1.0	0.9	1700
	LM324	0.6	0.2	250
Complementary	TLE2021	2.0	0.9	200
bipolar	TLE2027	15.0	1.7	3800
	TLE2061	2.1	3.4	280
Bifets	TLO51	3.1	18.0	2700
	TLO31	1.1	2.9	200
Cmos	TLC272	1.7	2.9	1000
(LinCMOS)	TLC27M2	0.5	0.4	105
	TLC27L2	0.14	0.04	15

The TLC27X series of LinCMOS devices highlight how the supply current affects a.c. performance. Each device is essentially the same but the internal bias currents have been altered, making the supply currents different in each case. This also affects the gain which is higher for the low bias current version. Unlike bifets, LinCMOS devices can operate from a single 5 V supply.

10.13.3 NOISE CONSIDERATIONS

Table 10.3 shows a comparison of the noise performance between the various operational amplifier technologies.

Bipolar devices offer the lowest voltage noise among those commercially available. The voltage noise from a bipolar input stage in the flatband is dominated by thermal noise from the base

Table 10.3 Comparison of operational amplifier noise parameters

Process	Bipolar	Excalibur	Bifet	LinCMOS		
Parameter	NE5534	TLE2027	TLO51	TLC1078	TLC2201	Units
Noise voltage	3.5	2.5	18.0	68.0	8.0	nV/√Hz
Noise current	400	400	10.0	0.5	0.6	pA/√Hz
1/f corner	100	3.0	100	30.0	50.0	Hz

spread resistance and the emitter small signal resistance. The TLE2027 or the decompensated TLE2037 represent such operational amplifiers optimised for low voltage noise.

However, while bipolar operational amplifiers are the best for low noise voltage, the shot noise arising from the input bias current flowing through the input stage can be significant. Therefore, in order to reduce this current noise, bias current cancellation circuits are sometimes used.

Fet input operational amplifiers have an input noise current given by the shot noise arising from the gate current. This is very low at temperatures around 25°C compared with the base current in bipolar inputs. Consequently, fet input operational amplifiers have negligible current noise and provide a superior noise performance with high impedance sources.

A fet input stage will have higher voltage noise and also a higher 1/f corner frequency than will a bipolar input stage.

The noise sources of a mosfet input device are similar to those of the junction fet. The gate current is negligible and the input current is reduced to leakage current by the input protection network.

A common disadvantage of operational amplifiers having mosfet inputs is their relatively high voltage noise and high 1/f frequency. However, an exception is the TLC2201 which offers a current noise level similar to the very best fet input operational amplifiers and a voltage noise comparable to many bipolar designs. Further, an offset of only 200 μV maximum coupled with a very low drift with time and temperature change, plastic packaging, operation from a single 5 V supply plus rail-to-rail output swing make this a very versatile device.

In summary, bipolar input stages give the lowest voltage noise and lowest 1/f corner frequency and are well suited for interfacing with low impedance sources. Jfet and cmos input stages have negligible input current noise, allowing them to be used with extremely high source impedances. Noise current is related to the input bias current and it will increase by $\sqrt{2}$ for every 10°C rise in temperature.

10.13.4 POWER SUPPLY CONSIDERATIONS

Operational amplifiers are required to operate with many different supplies. Circuits may be battery powered and required to run off a single 2 V supply or they may have a ±22 V or more supply in an instrumentation application. The available power affects the choice of operational amplifier.

Single supply operational amplifiers will normally need to operate from a 5 V supply and will typically require an input common mode range down to the negative rail and an output that swings near to ground. It is unlikely that a system running from ±15 V will need such performance.

An operational amplifier with a common mode range down to the negative rail can easily be designed using a bipolar process. PNP transistors will ensure that the output can swing down to the negative rail, or below, without causing problems. A good output swing is not so easy to achieve. Many bipolar devices are optimised for dual supply operation, that is, capable of sinking and sourcing current—this therefore means that the output will not normally swing to the negative rail. If a device is optimised for single supply operation, as is the LM324, then when it is operated with dual supplies its output suffers from crossover distortion. Bipolar devices that are suitable for both single and dual supplies are uncommon.

Bifets have been designed for dual supply operation. Their common mode input range reaches (and sometimes exceeds) the positive rail potential while the output will normally swing to within 3 V of each supply voltage. They operate from a wide range of supplies (±3.5 V to ± 22 V for the newer designs) and are optimised for a.c. performance.

Cmos devices, like the LinCMOS, have been specifically designed for single supply operation. Their input range exceeds the negative supply and the output can swing within microvolts of the ground potential. The positive output swing typically exceeds 3.5 V for the standard TLC27X series and newer devices such as the TLC2202 have an output that will swing to +4.7 V with a 5 V supply. Unlike bipolar designs, cmos devices with push-pull outputs can also perform well with both single and dual supplies. However, a limitation in some dual supply applications is that their maximum operating voltage is typically only 18 V.

A common problem with many devices is that they suffer from 'phase inversion'. This happens if the input swings beyond the power rail potential, causing the output to change state and swing to the opposite rail. Devices such as the LM324 and the older bifets are known to have this problem. The newer products like the TLE 2021 family have been designed to avoid this.

REFERENCES

1. 'Drift Compensation Techniques for Integrated D.C. Amplifier', National Semiconductor Linear Applications, **1**, A.N.31
2. GRAEME, J. G., 'Protect Op Amps from Overload', *Electron Design*, May 10 1973
3. 'Excalibur Technology', *Electronic Engineering*, Oct. 1990
4. Texas Instruments, Technical Note 1990
5. Texas Instruments, European Design Seminar 1990
6. Texas Instruments, TLE 2425 Virtual Ground Generator 1991

Exercises

10.1 A photo diode used to detect variations in the intensity of the light emitted by a modulated light source produces a photo current with a peak to peak variation of 0.01 μA. This current is converted into a voltage by means of an operational amplifier current to voltage convertor (Figure 4.10). What stray coupling capacitance between the inverting input terminal of the amplifier and a 250 V, 50 Hz supply line is sufficient to introduce an unwanted output signal with an amplitude 10% of that of the desired signal?

10.2 An operational amplifier is used as a follower with a gain of 11, and it has an input capacity of 50 pF between the non-inverting input terminal and earth. How would you reduce the capacitive loading imposed on the signal source? (See Figure 10.6.)

10.3 An internally frequency compensated operational amplifier has a unity gain frequency 4×10^6 Hz. When connected as a unity gain follower its small signal step response shows no overshoot. If the amplifier is connected as a unity gain inverter estimate the magnitude of the input and feedback resistors for which a capacitance of 10 pF between the inverting input terminal and earth will cause the phase margin in the circuit to be less than 30°. If resistors of this magnitude must be used how would you overcome the lightly damped response? (See Section 10.5.2.)

10.4 Component values $R_1 = 5$ kΩ, $R_2 = 20$ Ω, are used in the circuit of Figure 10.9(a). The operational amplifier has a maximum input offset voltage $V_{io} = 10$ mV and a maximum bias current $I_B = 0.5$ μA; $\pm$ 15 V supplies are used. Find a suitable value for the bias current supply resistor R_3.

10.5 The following components are used in the circuit of Figure 10.11: $R_1 = 10$ kΩ, $R_2 = 100$ kΩ, $R_4 = 1$ kΩ. Find suitable values for resistors R_3 and R_5 assuming the amplifier has a worst case input offset voltage $V_{io} = 2$ mV and input different current $I_{io} = 5$ nA, and that $\pm$ 15 V supplies are used.

10.6 Component values $R_1 = 10$ kΩ, $R_2 = 20$ kΩ, and a + 30 V supply are used in the circuit of Figure 10.25. The rated common mode range of the amplifier when worked off $\pm$ 15 V supplies is $\pm$ 12 V. What is the allowable common mode range at the input terminals of the circuit? If the gain of the circuit is increased (by increasing R_2) what is the maximum gain for which connecting the non-inverting input of the circuit to earth will not exceed the common mode range of the amplifier?

APPENDIX A1
OPERATIONAL AMPLIFIER APPLICATIONS AND CIRCUIT IDEAS

The circuits given in this Appendix represent extensions or modifications to the circuits given in the main body of the text. The reader conversant with the factors controlling accuracy and performance limitations (Chapters 2 and 9) should be able to use them as a basis for practical designs. Most circuits will function with a general purpose operational amplifier (use a Bi-Fet say) but the amplifier type used will inevitably govern performance limits. In all cases care should be taken to ensure that applied signals do not exceed allowable amplifier limits.

SCALING CIRCUITS

$$e_0 = \frac{R_2}{R_1}\left[e_3 - e_1\right] + \frac{R_2}{R_3}\left[e_4 - e_2\right]$$

Figure A1.1   Adder-subtractor

$$e_{o_2} = \left[e_2 - e_1\right]\frac{R_2}{R_1}$$

Input common mode
$$\text{Range} = n \times \begin{bmatrix} \text{amplifier} \\ \text{output limit} \end{bmatrix}$$

Figure A1.2   Differential input amplifier configuration with large common mode range

422

Figure A1.3   Current difference to voltage conversion with variable scaling factor

SIGNAL SOURCES

Positive voltage reference

Negative voltage reference

Figure A1.4   Voltage references

423

Figure A1.5   Regulated voltage supply

Figure A1.6   Square wave generator with voltage control of pulse width
(See Section 7.2.1)

Figure A1.7   Square and triangular wave generator (See Section 7.2.1)

Figure A1.8   Positive ramp generator (See Section 7.2.1)

425

Figure A1.9 Square and triangular wave generator with voltage control of frequency using a switched gain polarity amplifier (See Sections 7.4.3 and 8.12)

Figure A1.10 Two-phase square and triangular waveform generator with waveforms in quadrature (See Section 7.4.1)

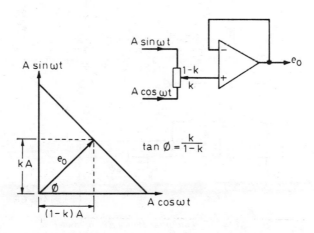

Figure A1.11   Sine, cosine and wave quadrature oscillator using phase shifter

Figure A1.12   Adjustable phase circuit for use with quadrature oscillator

Figure A1.13   Phase shift oscillator with single resistor frequency control and zener amplitude stabilisation

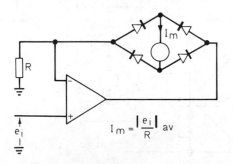

Figure A1.14   High input resistance a.c. voltmeter

$$I_m = \left| \frac{e_1 - e_2}{R} \right|_{av}$$

Figure A1.15 Differential input, high input resistance a.c. voltmeter

$$I_m = |i|_{av}$$

Figure A1.16 Average reading a.c. current meter

$$I_m = |i|_{av} \left[1 + \frac{R_2}{R_1}\right]$$

Figure A1.17 Average reading a.c. current meter with current amplification

$$V_O = V_{in}\frac{C_1}{C_2} + \underbrace{\frac{I_b}{C_2}\,t \pm V_{io}}_{error}$$

Figure A1.18 Measurement of high d.c. voltage with low reading voltmeter

$$V_x = \frac{E_{ref}}{R} R_x$$

$$V_0 = (1+k)V_x$$

$$= (1+k)\frac{E_{ref}}{R} R_x$$

$I = \frac{E_{ref}}{R}$

R₁ 10kΩ	kR₁ 1MΩ

R

R	kR	
100Ω	10kΩ	0.1 – 1Ω
1kΩ	100kΩ	1 – 10Ω
10kΩ	1MΩ	10 – 100Ω
100kΩ	10MΩ	100Ω – 1kΩ

E_{ref} 10V

Figure A1.19 Resistance measurement, earthed resistor

E_{ref} — R₁ — Zener diode feedback bound

V_{in} — C — For frequency stability

V_0 — V_z

$\frac{-E_{ref}}{CR_1}$

$\frac{dV_{in}}{dt}$

Figure A1.20 Rate comparator

Figure A1.21   Simple window comparator

Absolute value
circuit

Figure A1.22   Window comparator with control of window level and window width

432

Figure A1.23 Two-amplifier regenerative comparator with feedback bound and summing capability

Figure A1.24 Precise clipping circuit

$$I_{out} = \frac{V_{in} - E_{ref}}{R}$$

for $V_{in} > E_{ref}$

Figure A1.25   Ideal diode with current output

$e_0 = R |I_{in}|$

Figure A1.26   Single amplifier absolute value circuit for current input

$$e_O = \left| \frac{e_i}{k} \right|$$

$$kR \quad (1 \ k)R$$

$$e_i$$

Figure A1.27 High input impedance absolute value circuit with variable gain

APPENDIX A2

GAIN PEAKING/DAMPING FACTOR/PHASE MARGIN

The closed loop gain peaking and lightly damped transient response exhibited by closed loop configurations having an inadequate stability phase margin is in many cases due to the phase shift in the loop gain introduced by two break frequencies, one of which is remote (more than a decade away) from the other. Bode plots for commonly encountered situations are shown in figure A2.1 and in both the cases considered the frequency dependance of the loop gain can be expressed by the relationship

Figure A2.1   Bode plots showing frequency dependence of loop gain

$$\beta A_{OL\,(jf)} \cong - j \, |\beta A_{OL\,(o)}| \frac{f_{c_1}}{f} \frac{1}{1 + j \dfrac{f}{f_{c_2}}} \qquad (A2.1)$$

$(f_{c_2} \gg 10 f_{c_1})$

The closed loop signal gain of an operational amplifier feedback circuit can be expressed in the form (see Section 2.3.1)

$$A_{CL\,(f)} = A_{CL\,(o)} \left[\frac{1}{1 + \dfrac{1}{\beta A_{OL\,(jf)}}} \right]$$

$A_{CL\,(o)}$ is the ideal frequency independent closed loop signal gain. Substitution for $\beta A_{OL\,(jf)}$ and rearrangement give

$$A_{CL\,(jf)} = \frac{A_{CL\,(o)}}{1 + j \dfrac{f}{|\beta A_{OL\,(o)}| \, f_{c_1}} - \dfrac{f^2}{|\beta A_{OL\,(o)}| \, f_{c_1} f_{c_2}}} \qquad (A2.2)$$

Equation A.2 represents the closed loop sinusoidal response.

The more general closed loop transfer function is obtained in terms of the amplex variable s by the substitution, $s = jf$, $s^2 = -f^2$ giving

$$A_{CL\,(s)} = \frac{A_{CL\,(o)}}{1 + \dfrac{s}{|\beta A_{OL\,(o)}| f_{c_1}} + \dfrac{s^2}{|\beta A_{OL\,(o)}| f_{c_1} f_{c_2}}} \qquad (A2.3)$$

Equation A.3 represents a second order transfer function. Comparison with the general second order function

$$T_{(s)} = \frac{1}{1 + 2 \dfrac{\zeta}{\omega_o} s + \dfrac{s^2}{\omega_o^2}}$$

gives the relationships between the damping factor ζ, natural frequency f_o and amplifier parameters as

$$\zeta = \frac{\sqrt{f_{c_2}}}{2 \sqrt{(|\beta A_{OL\,(o)}| f_{c_1})}} \qquad (A2.4)$$

and

$$f_o = \sqrt{(|\beta A_{OL\,(o)}| f_{c_1} f_{c_2})} \qquad (A2.5)$$

Damping factor and phase margin

At the frequency f_1 at which the $1/\beta$ and the open loop gain frequency plots intersect the magnitude of the loop gain is unity. Equation A2.1 gives the magnitude as

$$|\beta A_{(jf)}| = 1 = |\beta A_{OL(o)}| \frac{f_{c_1}}{f_1} \frac{1}{\sqrt{\left[1 + \left(\dfrac{f_1}{f_{c_2}}\right)^2\right]}} \qquad \text{(A2.6)}$$
$$\text{at } f = f_1$$

Combining equations A2.4 and A2.6 gives

$$\zeta = \frac{1}{2\sqrt{\left\{\dfrac{f_1}{f_{c_2}}\left[1 + \left(\dfrac{f_1}{f_{c_2}}\right)^2\right]^{\frac{1}{2}}\right\}}} \qquad \text{(A2.7)}$$

Phase margin is related to f_1 and the break frequency f_{c_2} by the relationships

$$\frac{f_1}{f_{c_2}} = \frac{1}{\tan\theta_m} \; ; \; \left[1 + \left(\frac{f_1}{f_{c_2}}\right)^2\right]^{\frac{1}{2}} = \frac{1}{\sin\theta_m} \qquad \text{(A2.8)}$$

$(f_{c_2} \gg f_{c_1})$

Substitution in equation A2.7 gives the relationship between damping factor and phase margin as

$$\zeta = \frac{1}{2\sqrt{\left(\dfrac{\cos\theta_m}{\sin^2\theta_m}\right)}} \qquad \text{(A2.9)}$$

Gain peaking

The magnitude of the closed loop signal gain is, from equation A2.2

$$|A_{CL(jf)}| = \frac{A_{CL(o)}}{\sqrt{\left\{\left[1 - \left(\dfrac{f}{f_o}\right)^2\right]^2 + \left(2\zeta\dfrac{f}{f_o}\right)^2\right\}}} \qquad \text{(A2.10)}$$

where ζ and f_o are determined by equations A2.4 and A2.5.

The magnitude peaks for $\zeta < 1/\sqrt{2}$ and the frequency at which the gain peak occurs can be found by differentiating equation A.10 with respect to f and equating to zero. This gives the frequency at which the gain peak occurs as

438

$$f_p = f_o \sqrt{(1 - 2\zeta^2)} \quad \text{(For } \zeta < 1/\sqrt{2}\text{)} \tag{A2.11}$$

Substituting this value of f_p in equation A2.10 gives

$$|A_{CL(jf)}|_{\text{at peak}} = \frac{A_{CL(o)}}{2\zeta\sqrt{(1 - \zeta^2)}} \tag{A2.12}$$

The extent of the magnitude peaking may be expressed as

$$P_{\text{(dB of peaking)}} = 20 \log_{10} \frac{1}{2\zeta\sqrt{(1 - \zeta^2)}} \tag{A2.13}$$

The relationship between gain peaking and phase margin may be obtained by substituting the value of ζ from equation A2.9, thus

$$P_{\text{(dB of peaking)}} = 20 \log_{10} \frac{\dfrac{2\cos\theta m}{\sin^2\theta m}}{\sqrt{\left(4\dfrac{\cos\theta m}{\sin^2\theta m} - 1\right)}} \tag{A2.14}$$

APPENDIX A3

EFFECT OF RESISTOR TOLERANCE ON C.M.R.R. OF ONE AMPLIFIER DIFFERENTIAL CIRCUIT

In Figure A.29 the amplifier is assumed ideal, resistors have tolerance $100x\%$ and worst case c.m.r.r. is considered. An input common mode signal e_{cm} gives rise to an output signal

Figure A3.1   c.m.r.r. due to resistor tolerance with worst case distribution

$$
e_{o_{cm}} = e_{cm} \left[\frac{R_2(1-x)}{R_2(1-x) + R_1(1+x)} \cdot \frac{R_1(1-x) + R_2(1+x)}{R_1(1-x)} \right.
$$
$$
\left. - \frac{R_2(1+x)}{R_1(1-x)} \right]
$$
$$
= e_{cm} \frac{R_2}{R_1} \left[\frac{R_1(1-x) + R_2(1+x)}{R_2(1-x) + R_1(1+x)} - \frac{1+x}{1-x} \right]
$$
$$
= e_{cm} \frac{R_2}{R_1} \cdot \frac{R_1 \, 4x}{R_2(1-x)^2 + R_1(1-x^2)}
$$
$$
\cong e_{cm} \frac{R_2}{R_1} \frac{4x \, R_1}{R_2 + R_1}
$$

Thus common mode gain

$$
\frac{e_{o_{cm}}}{e_{cm}} = \frac{R_2}{R_1} \frac{4x \, R_1}{R_2 + R_1}
$$

440

and $\quad$ c.m.r.r.$_{(R)} = \dfrac{\text{differential gain}}{\text{common mode gain}} \cong \dfrac{\dfrac{R_2}{R_1}}{\dfrac{R_2}{R_1} \dfrac{4x\, R_1}{R_2 + R_1}}$

$$\text{c.m.r.r.}_{(R)} \cong \frac{1 + \dfrac{R_2}{R_1}}{4x} \qquad\qquad (A3.1)$$

C.M.R.R. of one amplifier differential circuit due to non-infinite c.m.r.r. of operational amplifier

Common mode signal applied to operational amplifier

$$= e_{cm} \frac{R_2}{R_1 + R_2} \qquad \text{(see Figure A3.2)}$$

Non-infinite c.m.r.r. of an operational amplifier is represented by an equivalent input error signal $e_{\epsilon cm}$ applied directly to the input terminal of the operational amplifier

Figure A3.2 *c.m.r.r. of circuit due to non-infinite c.m.r.r. of amplifier*

$$e_{\epsilon_{cm}} = \frac{e_{cm} \dfrac{R_2}{R_1 + R_2}}{\text{c.m.r.r.}_{(A)}}$$

$e_{\epsilon_{cm}}$ gives an output signal

$$e_{o_{cm}} = e_{\epsilon_{cm}} \left(1 + \frac{R_2}{R_1}\right)$$

Thus, common mode gain of circuit

$$= \frac{e_{o_{cm}}}{e_{cm}} = \frac{\dfrac{R_2}{R_1}}{\text{c.m.r.r.}_{(A)}}$$

and c.m.r.r. of the circuit = differential gain/common mode gain = c.m.r.r.$_{(A)}$.

Overall c.m.r.r. due to resistor mismatch and non-infinite c.m.r.r. of operational amplifier

Effects of resistor tolerance and c.m.r.r.$_{(A)}$ are represented by separate input error generators (Figure A.31). Output signal $e_{o_{cm}}$ due to input signal e_{cm} is

Figure A3.3

$$e_{o_{cm}} = e_{cm} \left[\frac{1}{\text{c.m.r.r.}_{(R)}} \pm \frac{1}{\text{c.m.r.r.}_{(A)}}\right] A_{\text{diff}}$$

$$\text{Overall c.m.r.r.} = \frac{\text{differential gain}}{\text{common mode gain}} = \frac{A_{\text{diff}}}{\left[\dfrac{1}{\text{c.m.r.r.}_{(R)}} \pm \dfrac{1}{\text{c.m.r.r.}_{(A)}}\right] A_{\text{diff}}}$$

$$= \frac{\text{c.m.r.r.}_{(R)} \times \text{c.m.r.r.}_{(A)}}{\text{c.m.r.r.}_{(A)} \pm \text{c.m.r.r.}_{(R)}} \tag{A3.2}$$

APPENDIX A4

INSTRUMENTATION TRANSDUCERS[1]

A4.1 Introduction

Because the 'front end' of an instrumentation system frequently consists of one or more sensors, the purpose of this appendix is to give the reader a brief introduction to these devices which often provide the input signal to operational amplifiers.

Most instrumentation systems comprise three basic sections: one for sensing the measurand, the next for conditioning the sensed signal and finally one for displaying or recording the conditioned signal. Figure A4.1 shows the block diagram of this arrangement.

The sensing element, known as a sensor or transducer, simply converts one form of energy into another. In this appendix all the transducers considered produce an electrical output when stimulated. However, the transduced electrical output may be of insufficient power and require amplification or other modification before it can be displayed or recorded. The necessary amplification, shaping, mixing or other such processing is undertaken in the signal conditioning section using the techniques variously described elsewhere in this book. Finally, the conditioned signal is recorded or displayed.

A simple example of the above basic system is that of a tank containing a hot liquid, the temperature of which needs to be monitored and recorded continually. The temperature sensing element could be a thermocouple (described later), the electrical output of which is conditioned by an operational amplifier to raise it to the necessary power level required to drive a chart recorder.

The remainder of this appendix will describe some of the more popular transducers used with instrumentation systems and will not consider signal conditioning or display and recording techniques.

Figure A4.1   Block diagram of basic instrumentation system

A4.2 Resistance strain gauges

These are devices which when subjected to mechanical strain are deformed, within their elastic limit, and change their ohmic resistance. The strain gauge therefore is most suitable for detecting and measuring small mechanical displacements. The strain gauge is firmly secured to the test piece which when strained under load causes the attached strain gauge also to distort. The strain gauge is usually part of an initially balanced resistive bridge network (see Section 8.1.1) and the accompanying change of gauge resistance causes an imbalance and an output signal from the bridge indicative of the amount of strain in the workpiece.

Figure A4.2 shows typical foil type strain gauges used for general engineering strain analysis.

The strain gauge measuring grid is manufactured from a copper nickel alloy which has a low and controllable temperature coefficient. The actual form of the metal grid, which changes its resistance under strain, is accurately produced by photo-etching techniques. A thermoplastic film is used to encapsulate the grid which helps to protect the gauge from mechanical and environmental damage. It also acts as a medium to transmit the strain from the test piece to the gauge material.

The principle of operation of this device is based on the fact that the resistance of an electrical conductor changes with a ratio of $\delta R/R$ if a stress is applied such that its length changes by a factor $\delta L/L$. This is where δR is the change in resistance from the unstressed value R and δL is the corresponding change in the unstressed length L.

Figure A4.2 Typical foil strain gauges (From RS Components Data Sheet No. 8155, page 1, issued November 1987)

The change in resistance is brought about mainly by the change in physical size of the conductor and, because of changes in its physical structure, an alteration in the conductivity of the material.

Copper nickel alloy is commonly used in the construction of strain gauges because the resistance change of the foil is virtually proportional to the applied strain, i.e.

$$\delta R/R = K . E$$

where K is a constant known as the gauge factor and E is the applied strain.

Therefore

$$\delta R/R = \text{gauge factor} \times \delta L/L$$

The change in resistance of the strain gauge can thus be utilised accurately to measure strain when connected to an appropriate measuring and indicating circuit.

Strain gauges are available commercially in a wide variety of preferred sizes and specifications. However, a typical specification for a small foil type strain gauge is as follows:

Measurable strain	2 to 4% maximum
Thermal output at 20–160°C	± 2 μstrain/°C*
at 160–180°C	± 5 μstrain/°C*
Gauge factor change with temperature	± 0.015%/°C max.
Gauge resistance	120 Ω
Gauge resistance tolerance	± 0.5%
Fatigue life	$\geqslant 10^5$ reversals at 1000 μstrain*
Foil material	copper nickel alloy
Temperature range	−30°C to +80°C
Gauge length	8 mm
Gauge width	2 mm
Gauge factor	2.1
Base length (single types)	13.0 mm
Base width (single types)	4.0 mm
Base diameter (rosettes)	21.0 mm

* 1 μstrain is equivalent to an extension of 0.0001%.

While the strain gauge is basically a displacement type transducer, because strain is caused by force the gauge can readily be adapted to measure force, torque, weight, acceleration and many other quantities.

It should be noted that because the resistance of strain gauges is affected by changes in temperature, they are often used in pairs; one in each of the balancing limbs of the measuring bridge circuits. Only one gauge is fixed to the test piece to act as the sensor, the other, being connected into the balancing limb of the bridge, is alongside but not fixed to the work piece and is purely for temperature compensation purposes. Figure A4.3 illustrates this connection.

Also, because the strain gauge resistance is quite low, typically 120 Ω, the remote connection of a sensor gauge away from the instrumentation bridge circuit can cause problems. This is because the resistance of the long connecting leads may have significant resistance compared with that of the sensor gauge itself. The problem can be overcome using three connecting leads to the remote sensor gauge as shown in Figure A4.4. The extra lead is connected so as effectively to place two equal resistance connecting leads in series with each of the bridge limbs a.c. and d.c. without disturbing the electrical balance.

Figure A4.3 Strain gauge bridge circuit with temperature compensation included

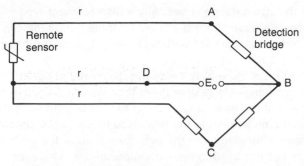

Ratio arms A.D. and D.C. both effectively
contain a total connector resistance of 2r

Figure A4.4   Three-wire compensating lead connection

A4.3 Platinum resistance temperature detectors

Like strain gauges, these are passive transducers which change
their resistance when stimulated. They are also known as resistance
thermometers because they suffer a change of resistance with
change of temperature. The change of resistance can be detected
using similar bridge circuits and operational amplifier conditioning
circuits as are used with strain gauges. The voltage output from
these circuits is calibrated to indicate temperature.

The resistive element is made from platinum because not only
does this metal exhibit a near linear variation of resistance with
temperature change, it also shows a large (38.5%) change of
resistance for a 100°C temperature change. The platinum
temperature sensing element is usually sheathed for protection and
may be mounted in a variety of probes, some being hand-held.
Figure A4.5 shows a typical platinum resistance element.

The platinum thermometers can be very accurate and typically
are manufactured to conform to BS 1904 Grade 2 and DIN 43 760. A
typical specification is as follows:

Resistance at 0°C	$100 \pm 0.1 \ \Omega$
Temperature coefficient	$0.385 \ \Omega/°C$
Maximum temperature	500°C
Minimum temperature	−50°C
Resistance tolerance at 0°C	$\pm 0.2 \ \Omega (\pm 0.3°C)$
at 500°C	$\pm 0.8 \ \Omega (\pm 2.4°C)$

Figure A4.5 Platinum resistance thermometer element (RS Components). (From RS catalogue, March–June 1991, page 952. RS stock No. 158–238)

Platinum resistance thermometers can be manufactured to give very accurate and long-term stable readings. They are often employed as laboratory temperature standards and where accurate temperature control is required. However, they can be fragile and have a slow response time of up to a second.

A4.4 Thermistors

These are devices which may have positive or negative temperature coefficients (p.t.c. or n.t.c.). They are manufactured from semiconductor materials and are often packaged as small discs, some 5–12 mm in diameter, or at the tip of a probe. Figure A4.6 shows a typical general purpose disc thermistor (n.t.c.).

The principle of operation of the thermistor is that its resistance changes with temperature. However, unlike the platinum resistance thermometer, the resistance : temperature relationship of the thermistor is very non-linear and its upper working temperature is usually much lower. The thermistor is quite robust, especially in disc form, and its small size and sensitivity make it very suitable for the temperature control of ovens, deep freezers, rooms, process control, temperature compensation, high temperature protection, high current protection and the like. But, because of its non-linearity and fairly wide tolerances, care should be taken to check the calibration of circuits after a thermistor change has been made. A typical specification is as follows:

Figure A4.6   General purpose disc thermistor (RS components). From RS catalogue, March–June 1991, page 956. RS stock No. 256–073)

Resistance at 25°	10 kΩ
at +125°C	260 Ω
Maximum temperature range	−30°C to +125°C
Maximum dissipation	900 mW
Thermal time constant	30 s

A4.5 Pressure transducers

There are several designs available and the purpose of each is to convert a fluid pressure into an analogous electrical signal. Some designs have a diaphragm which is moved by the measurand pressure and this movement is translated into a change of resistance, inductance or capacitance. Other pressure transducers use the piezo-resistive effect and one of the newest designs is shown in Figure A4.7. This new design incorporates advanced manufacturing techniques which includes laser trimmed bridge resistors for close tolerance on null and sensitivity. The sensing element is a 0.1 inch square silicon chip with integral sensing diaphragm and four piezo resistors. When pressure is applied to the diaphragm it is caused to flex, changing the resistance, which results in an output voltage proportional to pressure when a suitable excitation voltage is applied to the device. The sensing resistors are connected as a four-active element bridge for best linearity and sensitivity. A typical technical specification for one of these transducers is as follows:

450

Pressure range	0–30 p.s.i.
Full scale output	79 mV
Sensitivity/p.s.i.	2.63 mV
Excitation	10 V d.c.
Overpressure	60 p.s.i. max.

A4.6 Thermocouples

These are active transducers which convert the difference between two temperatures into a proportional voltage. The thermocouple principle is based upon the Seebeck effect which is simply the generation of a voltage by a heated junction of dissimilar metals. Figure A4.8 shows how a couple of two such metal junctions, at different temperatures, can be connected to cause a meter to indicate the voltage potential between the two junctions. One junction, the 'cold' or 'reference' junction, is usually held at 0°C (although room temperature suffices for some applications) while the other junction is used as a 'temperature sensor'.

The metals used include nickel, chromium, iron, platinum, rhodium, aluminium, constantan, manganese and silicon. Most

Figure A4.8 Principle of thermocouple action

$E_0 = (e_1 - e_2)$ and $(e_1 - e_2) \, \alpha \, (T_1 - T_2)$

thermocouple manufacturers use different pairs of these metals, or their alloys, to produce a selection of small, robust devices capable of measuring temperatures ranging from $-230°C$ to $+1300°C$. To indicate their designed temperature ranges and other characteristics, thermocouples are usually classified as being Type J, K, N or T. However, because the thermocouple sensor junction is placed in physical contact with the measurand, for temperatures above $+1300°C$, which is higher than the freezing point of most metals, optical pyrometer temperature measurement becomes more appropriate.

The device shown at Figure A4.9 comprises two meters of thermocouple wire insulated with varnish-impregnated glass fibre sleeving having an overall diameter of 1.5 mm. The hot junction tip is welded in an argon atmosphere to eliminate any oxidisation of the junction. It has an operating temperature range of $-50°C$ to $+400°C$. Its small size and flexibility make it suitable for temperature measurement in confined places such as inside electronic assemblies.

Figure A4.10 shows a high quality hand-held probe. This can be used not only for general purpose temperature measurements in the range $-100°C$ to $+600°C$ by, say, immersion in liquids, but also because of its sharpened stainless steel probe tip and robust construction it can penetrate solids for internal temperature measurements. This makes it ideal for use in the food industry for checking the temperature of frozen foods, or for general measurement of below-surface temperatures of soil, grain, and powders.

Figure A4.9 Welded tip glass fibre insulated wire thermocouple (RS Components). (From RS catalogue, March–June 1991, page 944. RS stock No. 151–192)

Figure A4.10 Insertion probe thermocouple (RS Components). (From RS catalogue, March–June 1991, page 946. RS stock No. 256–506)

A4.7 Linear variable differential transformers (LVDT)

Figure A4.11 shows a typical miniature a.c. energised LVDT and is one of a range of the most common forms of displacement transducer. LVDTs typically comprise three coils wound in-line on an insulating hollow former inside which is a movable nickel iron core. The centre of the three coils is energised by an alternating current and, with the movable core in the centre position, induces equal emfs across the other two coils which effectively form the secondary windings of a transformer. Movement of the core disturbs the balance of the magnetic coupling between the three coils. By comparison of the now unbalanced secondary coil output voltages, the magnitude and direction of the core movement can be determined.

The displacement to be measured is applied to the movable core and because it has no direct sliding contact is virtually friction-free. This gives the LVDT an advantage over the resistive potentiometric transducer which is sometimes used in similar applications. The LVDT can be used to detect movements in the range 0.5–25 mm.

A4.8 Capacitive transducers

These are passive displacement transducers which also require an a.c. excitation. A basic capacitor comprises a pair of parallel metal

Figure A4.11 Miniature LVDT (RS Components). (From RS catalogue, March–June 1991, page 964. RS stock No. 646–527)

plates between which is either a space or a solid dielectric material in which energy is stored when a voltage is applied between the plates. The capacitance, in farads, of a device is a measure of its ability to store this energy and depends upon the area of the plates, their spacing and the nature of the dielectric. The mechanical variation of any one of these three parameters will cause a sympathetic variation in the capacitance of the device. Since the spacing between the plates of a capacitor is usually less than 1.0 mm, a very detectable 10% variation in capacitance requires a change in the plate spacing of less than 100 microns. This sensitivity makes the capacitive transducer one of the most suitable sensors for the measurement of small displacements.

A4.9 Tachometers

These are used for the measurement of shaft angular velocity and are available in two basic types:

Pulse tachometers. These have a toothed ferromagnetic disc which must be coupled to and rotated by the shaft, the speed of which is to be measured. The ferromagnetic disc may be manufactured with only a single tooth-like protrusion which is arranged to fall close to a pick-up head once in each revolution of the disc. The pick-up head comprises a permanent magnet around which a coil is wound. The passage of the tooth through the magnetic field causes a distorting movement of the field and this flux movement induces an emf across the pick-up coil. The number of pulses counted or the average d.c. produced in a given time from a train of these pulses is indicative of the shaft speed.

Tachogenerators. These are really no more than either d.c. machines which produce a direct voltage proportional to their angular velocity or a.c. alternators which have a direct relationship between their speed of rotation and their output frequency.

A4.10 Electromagnetic flowmeters

Figure A4.12 shows the typical layout of an electromagnetic flowmeter as used for measuring the rate of flow of a wide range of

Figure A4.12 Principle of electromagnetic flowmeter

liquids. The basic principle involved is that of a current being induced into a conductor which is moving through a magnetic flux. In this case the moving conductor is the fluid itself and the magnetic flux is produced by external excitation. The fluid must have a resistance per cm^3 of less than 10 MΩ to permit the generation of a satisfactory signal (domestic tap water has a resistance per cm^3 of about 50 kΩ). The moving fluid is contained within a smooth-bore plastic pipe into which two pick-up electrodes are inserted and which collect the current generated, it being proportional to the rate of fluid flow.

Electromagnetic flow meters are available commercially with diameters ranging from 3 mm to 2000 mm. They have been used successfully to measure the flow of tap water, sea water, mercury, blood, chemicals and, because of the use of smooth-bored pipes, slurries and liquids containing solids.

A4.11 Hall effect transducers

Figure A4.13 shows how a current-carrying conductor situated in a perpendicular magnetic field experiences a transverse voltage which is proportional to the product of the current and the magnetic field flux density. The voltage so established is present in

456

Figure A4.13   The Hall effect

all conductors but is of particular significance in semiconductor materials. It can be shown that the current, I, the flux density, B, and the voltage generated, E, are related by the expression:

$$E = -R_H(I \times B)$$

R_H is known as the Hall coefficient and is given by $1/no$ where n is the number of charge carriers per unit volume which constitute the current and e is the charge on the carriers.

The Hall effect voltage in semiconductors in the presence of a magnetic field is used to produce 'bounce-free' switching. A switched transistor is turned on by a Hall effect device under the influence of a magnetic field exceeding a designed 'operate' strength. However, the Hall effect device does not 'release' until the magnetic field strength is reduced to a level below the operate level. It is this hysteresis effect that produces the 'bounce-free' switching.

The Hall effect vane switch transducer, as shown at Figure A4.14, makes use of bounce-free switching in its operation of transducing the presence of a ferrous metal into an electrical signal. A Hall effect sensor and magnet are housed in a pcb mounting package which will detect the presence of a ferrous metal vane passing through the gap between the sensor and the magnet. The device, which operates from 5 V direct (7 mA quiescent current), features two independent TTL compatible outputs capable of sinking up to 4 mA each or 8 mA combined. The switching time is less than 3 μs and the operating frequency can be up to 100 kHz. The device is

Figure A4.14   Hall effect vane switch (RS Components). (From RS catalogue, March–June 1991, page 967. RS stock No. 309–492)

useful in many position or counting operations, particularly in dusty or high ambient light environments, where an optical switch would be unsuitable.

The miniature linear Hall effect IC shown in Figure A4.15 is a magnetic field sensor in a moulded 4-pin dil plastic package less than 8 mm square. This device features a differential output stage. One output increases linearly in voltage, while the other decreases, for a linear increase in magnetic flux density over a ± 40 mT range. Typically, the output voltage varies linearly between 1.0 v and 3.0 v with a sensitivity of 1.0 mV/Gauss. Typical applications for this

Figure A4.15   Linear Hall effect IC (RS Components). (From RS catalogue, March–June 1991, page 967. RS stock No. 304–267)

device include the investigation of magnetic fields in the vicinity of transformers and cables and as current sensors with high isolation and in linear feedback elements in analogue control systems. The sensor is immune from damage by high values of flux density.

Figure A4.16 shows a Hall effect current transducer. The cable carrying the current to be measured is passed through the hole in the instrument casing and the magnetic field stimulates the device which produces an output calibrated in amps.

A4.12 Opto transducers

Opto transducers are devices which change one or more of their electrical characteristics when struck by light. The light is not necessarily visible to the human eye; it may be infra-red. Outlined below are brief details of but a small selection of the many opto devices available commercially.

Figure A4.16   Current transducer (RS Components). (From RS Catalogue, March–June 1991, page 968. RS stock No. 257–414)

The *light dependent resistor* (LDR) uses a strip of cadmium sulphide which may be illuminated by light passing through a clear window in the casing of the device. Figure A4.17 shows a typical LDR. The resistance of this particular device can vary from about 500 Ω in bright sunlight to 1.0 MΩ in darkness. It can be used at mains voltages (320 V d.c. or a.c. peak), it is cheap and sensitive but can have a slow response time of 120 ms. Typical applications for the LDR are the automatic control of public lighting, in intruder sensing devices and reflective smoke alarms.

All silicon diode junctions are affected by incident light and the *photodiode* is little more than a conventional silicon diode placed in a casing which is fitted with a window to allow the diode junction to be illuminated. The leakage current of the diode is very small but this increases when the junction is struck by light. The photodiode is operated in a reversed bias mode and in series with a load resistor through which the light dependent leakage current flows. The voltage drop across the load resistor is analogous to the intensity of the light striking the photodiode. Compared with the LDR, the photodidode is similarly packaged, will not operate with such high supply voltages, is not as sensitive to light stimulation but its response is much faster, being in the order of a few microseconds. Typical applications for photodiodes are in fast response a.c. circuits, in infra-red beam switching and with photographic flash circuits.

The *phototransistor* operates in much the same way as the photodiode, the base-collector junction being effectively reverse

Figure A4.17   Light dependent resistor (RS Components). (From RS catalogue, March–June 1991, page 605. RS stock No. 651–507)

biased and stimulated by light. However, the amplifying effect of the transistor makes the sensitivity of this device more than ten times that of the photodiode. But it cannot operate at such high frequencies; typically up to 200 kHz rather than the 500 MHz of the photodiode.

Optical shaft encoders are now available for sensing shaft position or angular velocity. Typically, these devices are 50 mm long and 50 mm wide and 50 mm in diameter and contain a light source beamed through a perforated rotating disc and detected by a light sensor. Rotation of the input shaft causes the energised encoder to produce an output comprising a number of TTL compatible pulses for each complete revolution. The device typically requires a d.c. excitation of 5–30 V and, depending upon the specification chosen, will provide resolutions of 100, 1250, 2000 or 2500 pulses per revolution. Typical applications are in machine tool control, robotics and position sensors for feedback on mechanical valve openings.

REFERENCE

1. R. S. Components Ltd, Catalogue, March–June 1991

APPENDIX A5

SPECIMEN DATA SHEETS FOR OPERATIONAL AMPLIFIERS

This Appendix contains specimens of the technical literature which operational amplifier manufacturers readily make available to electronic equipment designers. The following data sheets are reproduced just as they were received from the suppliers.

A5.1 Texas Instruments operational amplifier selection guide.

A5.2 Texas Instruments operational amplifier cross-reference guide

A5.3 Texas Instruments operational amplifier glossary of terms

A5.4 Motorola single bipolar operational amplifier Type MC1741/MC1741C.

A5.5 National Semiconductor dual JFET input operational amplifier Type LF412A/LF412.

A5.6 Texas Instruments advanced LinCMOS precision chopper-stabilised operational amplifier Type TLC2652/TLC2652A.

A5.7 Texas Instruments Excalibur high-speed, low-power precision quad operational amplifiers Type TLE2024/TLE2024A/TLE 2024.

APPENDIX A5.1

noncompensated, single

military temperature range (Values specified for $T_A = 25\,^\circ C$)

DESCRIPTION	SUPPLY VOLTAGE (V)		V_{IO} (mV)	I_{IB} (nA)	A_{VD} (V/mV)	B_1 (MHz)	SR (V/μs)	TYPE	PACKAGES	PAGE NO.
	MIN	MAX	MAX	MAX	MIN	TYP	TYP			
High Performance, Bipolar	±5	±22	2	75	50	1	0.5	LM101A	FK,JG,U,W	3-9
High Performance, Low Bias Current, Bipolar	±2	±20	2	2	50	1	0.3	LM108	L	3-13
High Performance, Low Bias Current, Bipolar	±2	±20	0.5	2	80	1	0.3	LM108A	L	3-13
BIFET, Low Power	±3.5	±18	6	0.2	4	1	3.5	TL060M	JG	3-47
BIFET, General Purpose	±3.5	±18	6	0.2	25	3	13	TL080M	JG	3-57
General Purpose,Precision Input, Bipolar	±9	±18	2	200	Typ 45	1	0.3	uA709AM	J,JG,U,W	–
General Purpose, Bipolar	±9	±18	5	500	Typ 45	1	0.3	uA709M	J,JG,U,W	–
General Purpose, Bipolar	±2	±22	5	500	50	1	0.5	uA748M	JG,U	3-108

industrial temperature range (Values specified for $T_A = 25\,^\circ C$)

DESCRIPTION	SUPPLY VOLTAGE (V)		V_{IO} (mV)	I_{IB} (nA)	A_{VD} (V/mV)	B_1 (MHz)	SR (V/μs)	TYPE	PACKAGES	PAGE NO.
	MIN	MAX	MAX	MAX	MIN	TYP	TYP			
High Performance, Bipolar	±5	±22	2	75	50	1	0.5	LM201A	D,JG,P	3-9
BIFET, Low Power	±3.5	±18	6	0.2	4	1	3.5	TL060I	D,JG,P	3-47
BIFET, Low Noise	±3.5	±18	6	200	50	3	13	TL070I	D,JG,P	3-52
BIFET, Low Power	±3.5	±18	6	400	25	3	13	TL080I	D,JG,P	3-57

noncompensated, single

commercial temperature range (Values specified for $T_A = 25\,^\circ C$)

DESCRIPTION	SUPPLY VOLTAGE (V)		V_{IO} (mV)	I_{IB} (nA)	A_{VD} (V/mV)	B_1 (MHz)	SR (V/μs)	TYPE	PACKAGES	PAGE NO.
	MIN	MAX	MAX	MAX	MIN	TYP	TYP			
High Performance	±5	±18	7.5	250	15	1	7.5	LM301A	D,JG,P	3-11
High Performance	±2	±18	7.5	2	25	1	0.3	LM308	D,JG,P	3-13
High Performance	±2	±18	0.5	2	80	1	0.3	LM308A	D,JG,P	3-13
BIFET, Low Power	±3.5	±18	6	0.2	4	1	3.5	TL060AC	D,JG,P	3-47
BIFET, Low Power	±3.5	±18	3	0.2	4	1	3.5	TL060BC	D,JG,P	3-47
BIFET, Low Power	±3.5	±18	15	0.4	3	1	3.5	TL060C	D,JG,P	3-47
BIFET, Low Noise	±3.5	±18	6	0.2	50	3	13	TL070AC	D,JG,P	3-52
BIFET, Low Noise	±3.5	±18	10	0.2	25	3	13	TL070C	D,JG,P	3-52
BIFET, General Purpose	±3.5	±18	6	0.2	50	3	13	TL080AC	D,JG,P	3-57
BIFET, General Purpose	±3.5	±18	15	0.4	25	3	13	TL080C	D,JG,P	3-57
General Purpose, Bipolar	±9	±18	7.5	1500	15	1	0.3	uA709C	D,JG,P	–
General Purpose, Bipolar	±2	±18	6	500	20	1	0.5	uA748C	D,JG,P	3-108

internally compensated, single

military temperature range

(Values specified for $T_A = 25\,°C$)

DESCRIPTION	SUPPLY VOLTAGE (V)		V_{IO} (mV) MAX	I_{IB} (nA) MAX	A_{VD} (V/mV) MIN	B_1 (MHz) TYP	SR (V/μs) TYP	TYPE	PACKAGES	PAGE NO.
	MIN	MAX								
High Performance	±5	±20	2	75	50	1	0.5	LM107	J,JG,U,W	3-10
Precision	±5	±22	0.015	2	450	0.8	0.25	LT1001AM	JG,L	3-18
Precision	±5	±22	0.06	4	400	0.8	0.25	LT1001M	JG,L	3-18
Low Noise, High Speed, Precision Input	±2.5	±22	0.025	35	7000	8	2.5	LT1007AM	JG,L	3-19
Low Noise, High Speed, Precision Input	±2.5	±22	0.060	55	5000	8	2.5	LT1007M	JG,P	3-19
Ultra Precision	±2.5	±20	0.035	0.1	300	0.8	2	LT1012M	L	3-21
Low Noise, High Performance	±4.5	±16	0.12	150	300	25	15	LT1028AM	D,JG,L,P	3-23
Low Noise, High Performance	±4.5	±20	0.18	300	200	25	15	LT1028M	D,JG,L,P	3-23
Low Noise, High Speed, Noncompensated, $A_{VL} \geq 5$	±2.5	±22	0.025	35	7000	60	15	LT1037AM	JG,L	3-24
Low Noise, High Speed, Noncompensated, $A_{VL} \geq 5$	±2.5	±22	0.060	55	5000	60	15	LT1037M	JG,L	3-24
Chopper-Stabilized	±1.9	±8	0.005	0.03	1000	1.2	4	LTC1052M	J,JG,L	3-25
Low Noise, High Speed	±3.5	±22	0.025	40	1000	8	2.8	OP-27A	JG,L	3-32
Low Noise, High Speed	±3.5	±22	0.1	80	700	8	2.8	OP-27C	JG,L	3-32
Low Noise, High Speed, Noncompensated, $A_{VL} \geq 5$	±4	±22	0.025	40	1000	40	17	OP-37A	JG,L	3-35
Low Noise, High Speed, Noncompensated, $A_{VL} \geq 5$	±4	±22	0.1	80	700	40	17	OP-37C	JG,L	3-35
Low Noise, High Performance	±3	±22	2	800	50	10	13	SE5534	FK,JG	3-28
Low Noise, High Performance	±3	±22	2	800	50	10	13	SE5534A	FK,JG	3-28
BIFET, Low Power, Precision	±3.5	±18	0.8	0.2	5	1.1	2.9	TL031AM	FK,JG,L	3-40
BIFET, Low-Power, Precision	±3.5	±18	1.5	0.2	5	1.1	2.9	TL031M	FK,JG,L	3-40
BIFET, Precision	±3.5	±18	0.8	0.2	50	3.1	20	TL051AM	FK,JG,L	3-44
BIFET, Precision	±3.5	±18	1.5	0.2	50	3.1	20	TL051M	FK,JG,L	3-44
BIFET, Low Power	±1.5	±18	6	0.2	4	1	3.5	TL061M	FK,JG,U	3-48
BIFET, Adjustable, Low-Power	±1.2	±18	6	0.2	4	1	3.5	TL066M	FK,JG	3-51
BIFET, Low Noise	±3.5	±18	6	0.2	35	3	13	TL071M	FK,JG	3-53
BIFET, General Purpose	±3.5	±18	6	0.2	25	3	13	TL081M	FK,JG	3-58
BIFET, Low V_{IO}	±3.5	±18	3	0.4	50	3	13	TL088M	JG,U	3-64
LinCMOS, Programmable, Low Bias	4	18	10	Typ 0.007	50	0.11	0.04	TLC271M	FK,JG	3-80
LinCMOS, Programmable, Medium Bias	4	18	10	Typ 0.007	25	0.64	0.56	TLC271M	FK,JG	3-81
LinCMOS, Programmable, High Bias	4	18	10	Typ 0.007	10	2.2	4.6	TLC271M	FK,JG	3-79
LinCMOS, Low Noise, Precision	4.6	16	0.2	Typ 0.001	400	1.9	2.7	TLC2201AM	D,FK,JG,L,P	3-96

internally compensated, single

military temperature range (Values specified for $T_A = 25\,°C$)

DESCRIPTION	SUPPLY VOLTAGE (V)		V_{IO} (mV)	I_{IB} (nA)	A_{VD} (V/mV)	B_1 (MHz)	SR (V/μs)	TYPE	PACKAGES	PAGE NO.
	MIN	MAX	MAX	MAX	MIN	TYP	TYP			
LinCMOS, Low Noise, Precision, 100% Noise Tested	4.6	16	0.2	Typ 0.001	400	1.9	2.7	TLC2201BM	D,FK,JG,L,P	3-96
LinCMOS, Low Noise, Precision	4.6	16	0.5	Typ 0.001	400	1.9	2.7	TLC2201M	D,FK,JG,L,P	3-96
LinCMOS, Precision, Chopper Stabilized	3.0	16	0.001	Typ 0.004	5600	1.9	2.8	TLC2652AM	D,FK,J,JG, L,N,P	3-97
LinCMOS, Precision Chopper Stabilized	3.8	16	0.003	Typ 0.004	1000	1.9	2.8	TLC2652M	D,FK,J,JG, L,N,P	3-97
LinCMOS, Low Noise, Precision, Chopper Stabilized	4.6	16	0.01	Typ 0.05	5600	1.9	2	TLC2654AM	D,FK,J,JG, L,N,P	3-98
LinCMOS, Low Noise, Precision, Chopper Stabilized	4.6	16	0.02	Typ 0.05	1000	2.2	2	TLC2654M	D,FK,J,JG, L,N,P	3-98
Excalibur, High-Speed, Precision	4	40	0.5	25	1000	2	0.9	TLE2021M	D,FK,JG,L,P	3-99
General Purpose	±2	±22	5	500	50	1	0.5	µA741M	FK,J,JG,U	3-106

internally compensated, single

automotive temperature range

(Values specified for $T_A = 25\,°C$)

DESCRIPTION	SUPPLY VOLTAGE (V)		V_{IO} (mV)	I_{IB} (nA)	A_{VD} (V/mV)	B_1 (MHz)	SR (V/µs)	TYPE	PACKAGES	PAGE NO.
	MIN	MAX	MAX	MAX	MIN	TYP	TYP			
High Performance	±5	±22	2	75	50	1	0.5	LM207	D,J,JG, N,P,W	3-10
High Performance	±5	±20	4	250	50	15	70	LM218	D,JG,P	3-14
Chopper-Stabilized	±1.9	±8	0.005	0.03	1000	1.2	4	LTC1052C	J,JG,L,N,P	3-25
Chopper-Stabilized	±1.9	±8	0.005	0.03	1000	1.2	4	LTC7652C	L	3-3
Low Noise, High Speed	±3.5	±22	0.025	40	1000	8	2.8	OP-27E	JG,L,P	3-33
Low Noise, High Speed	±3.5	±22	0.1	80	700	8	2.8	OP-27G	JG,L,P	3-34
Low Noise, High Speed Noncompensated, $A_{VL} \geq 5$	±4	±22	0.025	40	1000	40	17	OP-37E	JG,L,P	3-35
Low Noise, High Speed, Noncompensated, $A_{VL} \geq 5$	±4	±22	0.1	80	700	40	17	OP-37G	JG,P	3-35
BIFET, Low Power, Precision	±3.5	±18	0.8	0.2	5	1.1	2.9	TL031AI	D,JG,L,P	3-40
BIFET, Low Power, Precision	±3.5	±18	1.5	0.2	5	1.1	2.9	TL031I	D,JG,L,P	3-40
BIFET, Precision	±3.5	±18	0.8	0.2	50	3.1	20	TL051AI	D,JG,L,P	3-44
BIFET, Precision	±3.5	±18	1.5	0.2	50	3.1	20	TL051I	D,JG,L,P	3-44
BIFET, Low Power	±3.5	±18	6	0.2	4	1	3.5	TL061I	D,JG,P	3-48
BIFET, Adjustable, Low Power	±1.2	±18	6	0.2	4	1	3.5	TL066I	D,JG,P	3-51
BIFET, Low Noise	±3.5	±18	6	0.2	50	3	13	TL071I	D,JG,P	3-53
BIFET, General Purpose	±3.5	±18	6	0.2	50	3	13	TL081I	D,JG,P	3-58
BIFET, Low Offset Voltage	±3.5	±18	0.5	0.2	50	3	13	TL087I	D,JG,P	3-63
BIFET, Low Offset Voltage	±3.5	±18	1	0.2	50	3	13	TL088I	D,JG,P	3-64
LinCMOS, Programmable, Low Bias	4	18	5	Typ 0.007	50	0.11	0.04	TLC271AI	D,JG,P	3-80
LinCMOS, Programmable, Medium Bias	4	18	5	Typ 0.007	25	0.64	0.56	TLC271AI	D,JG,P	3-81
LinCMOS, Programmable, High Bias	4	18	5	Typ 0.007	10	2.2	4.6	TLC271AI	D,JG,P	3-79
LinCMOS, Programmable, Low Bias	4	18	2	Typ 0.007	50	0.11	0.04	TLC271BI	D,JG,P	3-80
LinCMOS, Programmable, Medium Bias	4	18	2	Typ 0.007	25	0.64	0.56	TLC271BI	D,JG,P	3-81
LinCMOS, Programmable, High Bias	4	18	2	Typ 0.007	10	2.2	4.6	TLC271BI	D,JG,P	3-79
LinCMOS, Programmable, Low Bias	4	18	10	Typ 0.007	50	0.11	0.04	TLC271I	D,JG,P	3-80
LinCMOS, Programmable, Medium Bias	4	18	10	Typ 0.007	25	0.64	0.56	TLC271I	D,JG,P	3-81
LinCMOS, Programmable, High Bias	4	18	10	Typ 0.007	10	2.2	4.6	TLC271I	D,JG,P	3-79
LinCMOS, Low Noise Precision	4.6	16	0.2	Typ 0.001	400	1.9	2.7	TLC2201AI	D,JG,L,P	3-96

internally compensated, single

automotive temperature range (Values specified for $T_A = 25\,°C$)

DESCRIPTION	SUPPLY VOLTAGE (V)		V_{IO} (mV)	I_{IB} (nA)	A_{VD} (V/mV)	B_1 (MHz)	SR (V/μs)	TYPE	PACKAGES	PAGE NO.
	MIN	MAX	MAX	MAX	MIN	TYP	TYP			
LinCMOS, Low Noise Precision, 100% Noise Tested	4.6	16	0.2	Typ 0.001	400	1.9	2.7	TLC2201BI	D,JG,L,P	3-96
LinCMOS, Low Noise Precision	4.6	16	0.5	Typ 0.001	400	1.9	2.7	TLC2201I	D,JG,L,P	3-96
LinCMOS, Precision, Chopper Stabilized	3.8	16	0.001	Typ 0.004	5600	1.9	2.8	TLC2652AI	D,J,JG,L,N,P	3-97
LinCMOS, Precision, Chopper Stabilized	3.8	16	0.003	Typ 0.004	1000	1.9	2.8	TLC2652I	D,J,JG,L,N,P	3-97
LinCMOS, Low Noise, Precision, Chopper Stabilized	4.6	16	0.01	Typ 0.05	5600	1.9	2	TLC2654AI	D,J,JG,L,N,P	3-98
LinCMOS, Low Noise, Precision, Chopper Stabilized	4.6	16	0.02	Typ 0.05	1000	1.9	2	TLC2654I	D,J,JG,L,N,P	3-98
Excalibur, High-Speed, Precision	4	40	0.5	25	1000	2	0.9	TLE2021	D,FK,JG,L,P	3-99

internally compensated, single

industrial temperature range   (Values specified for T_A = 25°C)

DESCRIPTION	SUPPLY VOLTAGE (V) MIN	MAX	V_{IO} (mV) MAX	I_{IB} (nA) MAX	A_{VD} (V/mV) MIN	B_1 (MHz) TYP	SR (V/μs) TYP	TYPE	PACKAGES	PAGE NO.
Chopper-Stabilized	± 1.9	± 8	0.005	0.03	1000	1.2	4	LTC1052C	J,JG,L,N,P	3-25
Chopper-Stabilized	± 1.9	± 8	0.005	0.03	1000	1.2	4	LTC7652C	L	3-3
Low Noise, High Speed	± 3.5	± 22	0.025	40	1000	8	2.8	OP-27E	JG,L,P	3-33
Low Noise, High Speed	± 3.5	± 22	0.1	80	700	8	2.8	OP-27G	JG,L,P	3-34
Low Noise, High Speed, Bipolar, Noncompensated, $A_{VL} \geq 5$	± 4	± 22	0.025	40	1000	40	17	OP-37E	JG,L,P	3-35
Low Noise, High Speed, Bipolar, Noncompensated, $A_{VL} \geq 5$	± 4	± 22	0.1	80	700	40	17	OP-37G	JG,L,P	3-35
BIFET, Low Power, Precision	± 3.5	± 18	0.8	0.2	5	1.1	2.9	TL031AI	D,JG,L,P	3-40
BIFET, Low Power, Precision	± 3.5	± 18	1.5	0.2	5	1.1	2.9	TL031I	D,JG,L,P	3-40
BIFET, Precision	± 3.5	± 18	0.8	0.2	50	3.1	20	TL051AI	D,JG,L,P	3-44
BIFET, Precision	± 3.5	± 18	1.5	0.2	50	3.1	20	TL051I	D,JG,L,P	3-44
BIFET, Low Power	± 3.5	± 18	6	0.2	4	1	3.5	TL061I	D,JG,P	3-48
BIFET, Adjustable, Low-Power	± 1.2	± 18	6	0.2	4	1	3.5	TL066I	D,JG,P	3-51
BIFET, Low Noise	± 3.5	± 18	6	0.2	50	3	13	TL071I	D,JG,P	3-53
BIFET, General Purpose	± 3.5	± 18	6	0.2	50	3	13	TL081I	D,JG,P	3-58
BIFET, Low Offset Voltage	± 3.5	± 18	0.5	0.2	50	3	13	TL087I	D,JG,P	3-63
BIFET, Low Offset Voltage	± 3.5	± 18	1	0.2	50	3	13	TL088I	D,JG,P	3-64
Single LM324, S/S High Performance D/S	3 ± 1.5	30 ± 15	5	− 150	50	0.6	0.3	TL321I	JG,P	3-68
LinCMOS, Programmable, Low Bias	4	18	5	Typ 0.007	50	0.11	0.04	TLC271AI	D,JG,P	3-80
LinCMOS, Programmable, Medium Bias	4	18	5	Typ 0.007	25	0.64	0.56	TLC271AI	D,JG,P	3-81
LinCMOS, Programmable, High Bias	4	18	5	Typ 0.007	10	2.2	4.6	TLC271AI	D,JG,P	3-79
LinCMOS, Programmable, Low Bias	4	18	2	Typ 0.007	50	0.11	0.04	TLC271BI	D,JG,P	3-80
LinCMOS, Programmable, Medium Bias	4	18	2	Typ 0.007	25	0.64	0.56	TLC271BI	D,JG,P	3-81
LinCMOS, Programmable, High Bias	4	18	2	Typ 0.007	10	2.2	4.6	TLC271BI	D,JG,P	3-79
LinCMOS, Programmable, Low Bias	4	18	10	Typ 0.007	50	0.11	0.04	TLC271I	D,JG,P	3-80
LinCMOS, Programmable, Medium Bias	4	18	10	Typ 0.007	25	0.64	0.56	TLC271I	D,JG,P	3-81
LinCMOS, Programmable, High Bias	4	18	10	Typ 0.007	10	2.2	4.6	TLC271I	D,JG,P	3-79
LinCMOS, Precision, Low Noise	4.6	16	0.2	Typ 0.001	400	1.9	2.7	TLC2201AI	D,JG,L,P	3-96
LinCMOS, Precision, Low Noise, 100% Noise Tested	4.6	16	0.2	Typ 0.001	400	1.9	2.7	TLC2201BI	D,JG,L,P	3-96
LinCMOS, Precision, Low Noise	4.6	16	0.5	Typ 0.001	400	1.9	2.7	TLC2201I	D,JG,L,P	3-96
LinCMOS, Precision, Chopper Stabilized	3.8	16	0.001	Typ 0.004	5600	1.9	2.8	TLC2652AI	D,J,JG,L,N,P	3-97
LinCMOS, Precision, Chopper Stabilized	3.8	16	0.003	Typ 0.004	1000	1.9	2.8	TLC2652I	D,J,JG, L,N,P	3-97
LinCMOS, Low-Noise, Precision, Chopper Stabilized	4.6	16	0.01	Typ 0.05	5600	1.9	2	TLC2654AI	D,J,JG, L,N,P	3-98
LinCMOS, Low-Noise, Precision, Chopper Stabilized	4.6	16	0.02	Typ 0.05	1000	1.9	2	TLC2654I	D,J,JG, L,N,P	3-98
Excalibur, High-Speed, Precision	4	40	0.5	25	1000	2	0.9	TLE2021I	D,FK,JG,L,P	3-99

internally compensated, single

commercial temperature range (Values specified for $T_A = 25\,°C$)

DESCRIPTION	SUPPLY VOLTAGE (V)		V_{IO} (mV)	I_{IB} (nA)	A_{VD} (V/mV)	B_1 (MHz)	SR (V/µs)	TYPE	PACKAGES	PAGE NO.
	MIN	MAX	MAX	MAX	MIN	TYP	TYP			
BIFET	±3.5	±18	10	0.2	25	3	13	LF351	D,JG,P	3-5
BIFET	±3.5	±18	2	0.2	25	3	13	LF411C	D,JG,P	3-7
High Performance	±2	±18	7.5	250	25	1	0.5	LM307	D,J,JG, N,P,W	3-12
High Performance	±5	±20	10	250	25	15	70	LM318	D,JG,P	3-14
Precision	±5	±22	0.025	2	450	0.8	0.25	LT1001AC	JG,L,P	3-18
Precision	±5	±22	0.06	4	400	0.8	0.25	LT1001C	JG,L,P	3-18
Low Noise, High Speed, Precision Input	±2.5	±22	0.025	35	7000	8	1.7	LT1007AC	JG,P	3-19
Low Noise, High Speed, Precision Input	±2.5	±22	0.060	55	5000	8	1.7	LT1007C	JG,P	3-19
Ultra Precision	±2.5	±20	0.05	0.15	200	–	0.2	LT1012C	L,P	3-21
Low Noise, High Performance	±4.5	±18	0.08	120	500	75	15	LT1028AC	D,JG,L,P	3-23
Low Noise, High Performance	±4.5	±18	0.13	240	300	75	15	LT1028C	D,JG,L,P	3-23
Low Noise, High Speed, Noncompensated, $A_{VL} \geq 5$	±2.5	±22	0.025	35	7000	60	15	LT1037AC	JG,P	3-24
Low Noise, High Speed, Noncompensated, $A_{VL} \geq 5$	±2.5	±22	0.060	55	5000	60	15	LT1037C	JG,P	3-24
Low Noise, High Performance	±3	±22	4	1500	25	10	13	NE5534	D,JG,P	3-28
Low Noise, High Performance	±3	±22	4	1500	25	10	13	NE5534A	D,JG,P	3-28
Ultra-Low Offset Voltage	±3	±22	0.15	7	120	0.6	0.3	OP-07C	D,JG,P	3-29
Ultra-Low Offset Voltage	±3	±22	0.15	12	120	0.6	0.3	OP-07D	D,JG,P	3-30
Ultra-Low Offset Voltage	±3	±22	0.075	4	200	0.6	0.3	OP-07E	D,JG,P	3-31

internally compensated, single

commercial temperature range

(Values specified for $T_A = 25\,°C$)

DESCRIPTION		SUPPLY VOLTAGE (V)		V_{IO} (mV)	I_{IB} (nA)	A_{VD} (V/mV)	B_1 (MHz)	SR (V/µs)	TYPE	PACKAGES	PAGE NO.
		MIN	MAX	MAX	MAX	MIN	TYP	TYP			
BIFET, Low Power, Precision		±3.5	±18	0.8	0.2	5	1.1	2.9	TL031AC	D,FK,JG,L,P	3-40
BIFET, Low Power, Precision		±3.5	±18	1.5	0.2	5	1.1	2.9	TL031C	D,FK,JG,L,P	3-40
BIFET, Precision		±3.5	±18	0.8	0.2	50	3.1	20	TL051AC	D,FK,JG,L,P	3-44
BIFET, Precision		±3.5	±18	1.5	0.2	50	3.1	20	TL051C	D,FK,JG,L,P	3-44
BIFET, Low Power		±3.5	±18	6	0.2	4	1	3.5	TL061AC	D,JG,P	3-48
BIFET, Low Power		±3.5	±18	3	0.2	4	1	3.5	TL061BC	D,JG,P	3-48
BIFET, Low Power		±3.5	±18	15	0.2	3	1	3.5	TL061C	D,JG,P	3-48
BIFET, Adjustable, Low-Power		±1.2	±18	6	0.2	4	1	3.5	TL066AC	D,JG,P	3-51
BIFET, Adjustable, Low-Power		±1.2	±18	3	0.2	4	1	3.5	TL066BC	D,JG,P	3-51
BIFET, Adjustable, Low-Power		±1.2	±18	15	0.4	3	1	3.5	TL066C	D,JG,P	3-51
BIFET, Low Noise		±3.5	±18	6	0.2	50	3	13	TL071AC	D,JG,P	3-53
BIFET, Low Noise		±3.5	±18	3	0.2	50	3	13	TL071BC	D,JG,P	3-53
BIFET, Low Noise		±3.5	±18	10	0.2	25	3	13	TL071C	D,JG,P	3-53
BIFET, General Purpose		±3.5	±18	6	0.2	50	3	13	TL081AC	D,JG,P	3-58
BIFET, General Purpose		±3.5	±18	3	0.2	50	3	13	TL081BC	D,JG,P	3-58
BIFET, General Purpose		±3.5	±18	15	0.4	25	3	13	TL081C	D,JG,P	3-58
BIFET, Low V_{IO}		±3.5	±18	0.5	0.2	50	3	13	TL087C	D,JG,L,P	3-63
BIFET, Low V_{IO}		±3.5	±18	1	0.2	50	3	13	TL088C	D,JG,L,P	3-64
Single LM324, High	S/S	3	30	7	−250	25	0.6	0.3	TL321C	JG,P	3-68
Performance	D/S	1.5	15								
LinCMOS, Programmable, Low Bias		1.4	18	5	Typ 0.001	30	0.1	0.04	TLC251AC	D,JG,P	3-71
LinCMOS, Programmable, Medium Bias		1.4	18	5	Typ 0.001	20	0.7	0.6	TLC251AC	D,JG,P	3-72
LinCMOS, Programmable, High Bias		1.4	18	5	Typ 0.001	10	2.3	4.5	TLC251AC	D,JG,P	3-70
LinCMOS, Programmable, Low Bias		1.4	18	2	Typ 0.001	30	0.1	0.04	TLC251BC	D,JG,P	3-71
LinCMOS, Programmable, Medium Bias		1.4	18	2	Typ 0.001	20	0.7	0.6	TLC251BC	D,JG,P	3-72
LinCMOS, Programmable, High Bias		1.4	18	2	Typ 0.001	10	2.3	4.5	TLC251BC	D,JG,P	3-70
LinCMOS, Programmable, Low Bias		1.4	18	10	Typ 0.001	30	0.1	0.04	TLC251C	D,JG,P	3-71
LinCMOS, Programmable, Medium Bias		1.4	18	10	Typ 0.001	20	0.7	0.6	TLC251C	D,JG,P	3-72
LinCMOS, Programmable, High Bias		1.4	18	10	Typ 0.001	10	2.3	4.5	TLC251C	D,JG,P	3-70
LinCMOS, Programmable, Low Bias		3	18	5	Typ 0.007	50	0.11	0.04	TLC271AC	D,JG,P	3-80
LinCMOS, Programmable, Medium Bias		3	18	5	Typ 0.007	25	0.64	0.56	TLC271AC	D,JG,P	3-81
LinCMOS, Programmable, High Bias		3	18	5	Typ 0.007	10	2.2	4.6	TLC271AC	D,JG,P	3-79

internally compensated, single

commercial temperature range   (Values specified for T_A = 25°C)

DESCRIPTION	SUPPLY VOLTAGE (V)		V_{IO} (mV)	I_{IB} (nA)	A_{VD} (V/mV)	B_1 (MHz)	SR (V/μs)	TYPE	PACKAGES	PAGE NO.
	MIN	MAX	MAX	MAX	MIN	TYP	TYP			
LinCMOS, Programmable, Low Bias	3	18	2	Typ 0.007	50	0.11	0.04	TLC271BC	D,JG,P	3-80
LinCMOS, Programmable, Medium Bias	3	18	2	Typ 0.007	25	0.64	0.56	TLC271BC	D,JG,P	3-81
LinCMOS, Programmable, High Bias	3	18	2	Typ 0.007	10	2.2	4.6	TLC271BC	D,JG,P	3-79
LinCMOS, Programmable, Low Bias	3	18	10	Typ 0.007	50	0.11	0.04	TLC271C	D,JG,P	3-80
LinCMOS, Programmable, Medium Bias	3	18	10	Typ 0.007	25	0.64	0.56	TLC271C	D,JG,P	3-81
LinCMOS, Programmable, High Bias	3	18	10	Typ 0.007	10	2.2	4.6	TLC271C	D,JG,P	3-79
LinCMOS, Precision, Low Noise	4.6	16	0.2	Typ 0.001	400	1.9	2.7	TLC2201AC	D,JG,L,P	3-96
LinCMOS, Precision, Low Noise, 100% Noise Tested	4.6	16	0.2	Typ 0.001	400	1.9	2.7	TLC2201BC	D,JG,L,P	3-96
LinCMOS, Precision, Low Noise	4.6	16	0.5	Typ 0.001	400	1.9	2.7	TLC2201C	D,JG,L,P	3-96
LinCMOS, Precision, Chopper Stabilized	3.8	16	0.001	Typ 0.004	5600	1.9	2.8	TLC2652AC	D,J,JG,L,N,P	3-97
LinCMOS, Precision, Chopper Stabilized	3.8	16	0.003	Typ 0.004	1000	1.9	2.8	TLC2652C	D,J,JG,L,N,P	3-97
LinCMOS, Low-Noise, Precision, Chopper Stabilized	4.6	16	0.01	Typ 0.05	5600	1.9	2	TLC2654AC	D,J,JG,L,N,P	3-98
LinCMOS, Low-Noise, Precision, Chopper Stabilized	4.6	16	0.02	Typ 0.05	1000	1.9	2	TLC2654C	D,J,JG,L,N,P	3-98
Excalibur, High-Speed, Precision	4	40	0.5	25	1000	2	0.9	TLE2021C	D,FK,JG,L,P	3-99
General Purpose	±2	±18	6	500	20	1	0.5	uA741C	D,JG,P	3-106

OPERATIONAL AMPLIFERS SELECTION GUIDE

internally compensated, dual

military temperature range

(Values specified for T_A = 25 °C)

DESCRIPTION		SUPPLY VOLTAGE (V) MIN	MAX	V_{IO} (mV) MAX	I_{IB} (nA) MAX	A_{VD} (V/mV) MIN	B_1 (MHz) TYP	SR (V/µs) TYP	TYPE	PACKAGES	PAGE NO.
High Gain, Low Power, Bipolar	S/S	3	30	5	-150	50	0.6	0.2	LM158	FK,JG,U	3-17
	D/S	±1.5	±15								
General Purpose		±2	±22	5	500	50	1	0.5	MC1558	FK,JG,U	3-26
Precision		±5	±22	0.15	20	1500	0.7	0.4	LT1013AM	JG,L	3-22
Precision		±5	±22	0.3	30	1200	0.7	0.4	LT1013M	JG,L	3-22
High Performance		±4	±22	5	500	50	3.5	1.7	RM4558	JG	3-37
Low Power		±2	±22	5	100	1	0.5	0.5	TL022M	U	3-39
BIFET, Low Power, Precision		±3.5	±18	0.8	0.2	5	1.1	2.9	TL032AM	FK,JG,L	3-41
BIFET, Low Power, Precision		±3.5	±18	1.5	0.2	5	1.1	2.9	TL032M	FK,JG,L	3-41
BIFET, Precision		±3.5	±18	0.8	0.2	50	3	16	TL052AM	FK,JG,L	3-45
BIFET, Precision		±3.5	±18	1.5	0.2	50	3	16	TL052M	FK,JG,L	3-45
BIFET, Low Power		±3.5	±18	6	0.2	4	1	3.5	TL062M	FK,JG,U	3-49
BIFET, Low Noise		±3.5	±18	6	0.2	35	3	13	TL072M	FK,JG	3-54
BIFET, General Purpose		±3.5	±18	16	0.2	25	3	13	TL082M	FK,JG	3-59
BIFET, General Purpose		±3.5	±18	6	0.2	25	3	13	TL083M	FK,J	3-60
BIFET, General Purpose		±3.5	±18	3	0.4	50	3	13	TL287M	JG,U	3-66
BIFET, General Purpose		±3.5	±18	3	0.4	50	3	13	TL288M	JG,U	3-67
LinCMOS, High Bias		4	18	10	Typ 0.005	10	2.2	5.3	TLC272M	FK,JG	3-82
LinCMOS, High Bias		4	18	0.5	Typ 0.005	10	2.2	5.3	TLC277M	FK,JG	3-84
LinCMOS, Low Bias		4	18	10	Typ 0.005	50	0.1	0.05	TLC27L2M	FK,JG	3-86
LinCMOS, Low Bias		4	18	0.5	Typ 0.005	50	0.1	0.05	TLC27L7M	FK,JG	3-88
LinCMOS, Medium Bias		4	18	10	Typ 0.005	25	0.6	0.6	TLC27M2M	FK,JG	3-90
LinCMOS, Medium Bias		4	18	0.5	Typ 0.005	25	0.6	0.6	TLC27M7M	FK,JG	3-92
LinCMOS, Micro Power, Precision		4	18	0.6	Typ 0.007	500	0.11	0.5	TLC1078M	FK,JG	3-94
Excalibur, High Speed, Precision		4	40	0.5	25	1000	2	0.9	TLE2022M	FK,JG,L	3-100

internally compensated, dual
automotive temperature range

(Values specified for T_A = 25 °C)

DESCRIPTION		SUPPLY VOLTAGE (V) MIN	MAX	VIO (mV) MAX	IIB (nA) MAX	AVD (V/mV) MIN	B₁ (MHz) TYP	SR (V/μs) TYP	TYPE	PACKAGES	PAGE NO.
High Gain, Low Power, Bipolar	S/S	3	30	5	−150	50	0.6	0.2	LM258	D,JG,P,U	3-17
	D/S	±1.5	±1.5								
High Gain, Low Power, Bipolar	S/S	3	30	3	−80	50	0.6	0.2	LM258A	D,JG,P,U	3-17
	D/S	±1.5	±1.5								
High Gain, Low Power, Bipolar	S/S	3	26	7	−250	Typ 100	0.0	0.2	LM2904	D,JG,P,U	3-17
	D/S	±1.5	±13								
High Performance		±4	±18	6	−500	20	3	1.7	RV4558	D,JG,P	3-37
BIFET, Low Power, Precision		±3.5	±18	0.8	0.2	5	1.1	2.9	TL032AI	D,JG,L,P	3-41
BIFET, Low Power, Precision		±3.5	±18	1.5	0.2	5	1.1	2.9	TL032I	D,JG,L,P	3-41
BIFET, Precision		±3.5	±18	0.8	0.2	50	3	16	TL052AI	D,JG,L,P	3-45
BIFET, Precision		±3.5	±18	1.5	0.2	50	3	16	TL052I	D,JG,L,P	3-45
BIFET, Low Power		±3.5	±18	6	0.2	4	1	3.5	TL062I	D,JG,P	3-49
BIFET, Low Noise		±3.5	±18	6	0.2	50	3	13	TL072I	D,JG,P	3-54
BIFET, General Purpose		±3.5	±18	6	0.2	50	3	13	TL082I	D,JG,P	3-59
BIFET, General Purpose		±3.5	±18	6	0.2	50	3	13	TL083I	D,JG,P	3-60
BIFET, General Purpose		±3.5	±18	0.5	0.2	50	3	13	TL287I	D,JG,P	3-66
BIFET, General Purpose		±3.5	+18	1	0.2	50	3	13	TL288I	D,JG,P	3-67
Low Power		±1.5	±18	8	−500	20	1	0.6	TL322I	D,JG,P	3-69
LinCMOS, High Bias		4	18	5	Typ 0.005	10	2.2	5.3	TLC272AI	D,JG,P	3-82
LinCMOS, High Bias		4	18	2	Typ 0.005	10	2.2	5.3	TLC272BI	D,JG,P	3-82
LinCMOS, High Bias		4	18	10	Typ 0.005	10	2.2	5.3	TLC272I	D,JG,P	3-82
LinCMOS, High Bias		4	18	0.5	Typ 0.005	10	2.2	5.3	TLC277I	D,JG,P	3-84
LinCMOS, Low Bias		4	18	5	Typ 0.005	50	0.1	0.05	TLC27L2AI	D,JG,P	3-86
LinCMOS, Low Bias		4	18	2	Typ 0.005	50	0.1	0.05	TLC27L2BI	D,JG,P	3-86
LinCMOS, Low Bias		4	18	10	Typ 0.005	50	0.1	0.05	TLC27L2I	D,JG,P	3-86
LinCMOS, Low Bias		4	18	0.5	Typ 0.005	50	0.1	0.05	TLC27L7I	D,JG,P	3-88
LinCMOS, Medium Bias		4	18	6	Typ 0.005	25	0.6	0.6	TLC27M2AI	D,JG,P	3-90
LinCMOS, Medium Bias		4	18	2	Typ 0.005	25	0.6	0.6	TLC27M2BI	D,JG,P	3-90
LinCMOS, Medium Bias		4	18	10	Typ 0.005	25	0.6	0.6	TLC27M2I	D,JG,P	3-90
LinCMOS, Medium Bias		4	18	0.5	Typ 0.005	25	0.6	0.6	TLC27M7I	D,JG,P	3-92
LinCMOS, Micro Power, Precision		4	18	0.6	Typ 0.007	500	0.11	0.05	TLC1078I	D,JG,P	3-94
Excalibur, High Speed, Precision		4	40	0.5	25	1000	2	0.9	TLE2022I	D,JG,L,P	3-100

473

internally compensated, dual

industrial temperature range (Values specified for T_A = 25 °C

DESCRIPTION	SUPPLY VOLTAGE (V)		V_{IO} (mV)	I_{IB} (nA)	A_{VD} (V/mV)	B_1 (MHz)	SR (V/μs)	TYPE	PACKAGES	PAGE NO.
	MIN	MAX	MAX	MAX	MIN	TYP	TYP			
BIFET, Low Power, Precision	±3.5	±18	0.8	0.2	5	1.1	2.9	TL032AI	D,JG,L,P	3-41
BIFET, Low Power, Precision	±3.5	±18	1.5	0.2	5	1.1	2.9	TL032I	D,JG,L,P	3-41
BIFET, Precision	±3.5	±18	0.8	0.2	50	3	16	TL052AI	D,JG,L,P	3-45
BIFET, Precision	±3.5	±18	1.5	0.2	50	3	16	TL052I	D,JG,L,P	3-45
BIFET, Low Power	±3.5	±18	6	0.2	4	1	3.5	TL062I	D,JG,P	3-49
BIFET, Low Noise	±3.5	±18	6	0.2	50	3	13	TL072I	D,JG,P	3-54
BIFET, General Purpose	±3.5	±18	6	0.2	50	3	13	TL082I	D,JG,P	3-59
BIFET, General Purpose	±3.5	±18	6	0.2	50	3	13	TL083I	D,JG,P	3-60
BIFET, General Purpose	±3.5	±18	0.5	0.2	50	3	13	TL287I	D,JG,P	3-66
BIFET, General Purpose	±3.5	±18	1	0.2	50	3	13	TL288I	D,JG,P	3-67
Low Power	±1.5	±18	8	-500	20	1	0.6	TL322I	D,JG,P	3-69
LinCMOS, High Bias	4	18	5	Typ 0.005	10	2.2	5.3	TLC272AI	D,JG,P	3-82
LinCMOS, High Bias	4	18	2	Typ 0.005	10	2.2	5.3	TLC272BI	D,JG,P	3-82
LinCMOS, High Bias	4	18	10	Typ 0.005	10	2.2	5.3	TLC272I	D,JG,P	3-82
LinCMOS, High Bias	4	18	0.5	Typ 0.005	10	2.2	5.3	TLC277I	D,JG,P	3-84
LinCMOS, Low Bias	4	18	5	Typ 0.005	50	0.1	0.05	TLC27L2AI	D,JG,P	3-86
LinCMOS, Low Bias	4	18	2	Typ 0.005	50	0.1	0.05	TLC27L2BI	D,JG,P	3-86
LinCMOS, Low Bias	4	18	10	Typ 0.005	50	0.1	0.05	TLC27L2I	D,JG,P	3-86
LinCMOS, Low Bias	4	18	0.5	Typ 0.005	50	0.1	0.05	TLC27L7I	D,JG,P	3-87
LinCMOS, Medium Bias	4	18	5	Typ 0.005	25	0.6	0.6	TLC27M2AI	D,JG,P	3-90
LinCMOS, Medium Bias	4	18	2	Typ 0.005	25	0.6	0.6	TLC27M2BI	D,JG,P	3-90
LinCMOS, Medium Bias	4	18	10	Typ 0.005	25	0.6	0.6	TLC27M2I	D,JG,P	3-90
LinCMOS, Medium Bias	4	18	0.5	Typ 0.005	25	0.6	0.6	TLC27M7I	D,JG,P	3-92
LinCMOS, Micro Power, Precision	4	18	0.6	Typ 0.007	500	0.11	0.05	TLC1078I	D,JG.P	3-94
Excalibur, High Speed, Precision	4	40	0.5	25	1000	2	0.9	TLE2022I	D,JG,L,P	3-100

internally compensated, dual

commercial temperature range (Values specified for $T_A = 25\,°C$)

DESCRIPTION		SUPPLY VOLTAGE (V) MIN	MAX	V_{IO} (mV) MAX	I_{IB} (nA) MAX	A_{VD} (V/mV) MIN	B_1 (MHz) TYP	SR (V/μs) TYP	TYPE	PACKAGES	PAGE NO.
BIFET, General Purpose		±3.5	±18	10	0.2	25	3	13	LF353	D,JG,P	3-6
BIFET, Low Offset		±3.5	±18	3	0.2	25	3	13	LF412C	D,JG,P	3-8
High Gain, Low Power, Bipolar	S/S	3	30	7	-250	25	0.6	0.2	LM358	D,JG,P,U	3-17
	D/S	±1.5	±15								
High Gain, Low Power, Bipolar	S/S	3	30	3	-100	25	0.6	0.2	LM358A	D,JG,P,U	3-17
	D/S	±1.5	±15								
Precision		±5	±22	0.15	20	1500	0.7	0.4	LT1013AC	JG,L,P	3-22
Precision		±5	±22	0.3	30	1200	0.7	0.4	LT1013C	JG,L,P	3-22
Precision		±5	±22	0.8	30	1200	0.7	0.4	LT1013D	D,JG,L,P	3-22
General Purpose		±1.5	±18	6	500	20	1	0.5	MC1458	D,JG,P,U	3-26
Low Noise		±3	±20	4	800	25	10	9	NE5532	JG,P	3-28
Low Noise		±3	±20	4	800	25	10	9	NE5532A	JG,P	3-28
High Performance		±4	±18	6	500	20	3	1.7	RC4558	D,JG,P	3-37
High Performance		±4	±18	6	250	20	4	?	RC4559	D,P	3-38
Low Power		±2	±18	5	250	1	0.5	0.5	TL022C	D,JG,P	3-39
BIFET, Low Power, Precision		±3.5	±18	0.2	0.2	5	1.1	2.9	TL032AC	D,JG,L,P	3-41
BIFET, Low Power, Precision		±3.5	±18	1.5	0.2	5	1.1	2.9	TL032C	D,JG,L,P	3-41
BIFET, Precision		±3.5	±18	1.5	0.2	50	3	16	TL052AC	D,JG,L,P	3-45
BIFET, Precision		±3.5	±18	4	0.2	50	3	16	TL052C	D,JG,L,P	3-45
BIFET, Low Power		+3.5	±18	6	0.2	4	1	3.5	TL062AC	D,JG,P	3-49
BIFET, Low Power		±3.5	±18	3	0.2	4	1	3.5	TL062BC	D,JG,P	3-49
BIFET, Low Power		±3.5	±18	15	0.4	3	1	3.5	TL062C	D,JG,P	3-49
BIFET, Low Noise		±3.5	±18	6	0.2	50	3	13	TL072AC	D,JG,P	3-54
BIFET, Low Noise		±3.5	±18	3	0.2	50	3	13	TL072BC	D,JG,P	3-54
BIFET, Low Noise		+3.5	±18	10	0.2	25	3	13	TL072C	D,JG,P	3-54
BIFET, General Purpose		±3.5	±18	6	0.2	50	3	13	TL082AC	D,JG,P	3-59
BIFET, General Purpose		±3.5	±18	3	0.2	50	3	13	TL082BC	D,JG,P	3-59
BIFET, General Purpose		±3.5	±18	15	0.4	25	3	13	TL082C	D,JG,P	3-59
BIFET, General Purpose		±3.5	±18	6	0.2	50	3	13	TL083AC	D,JG,N	3-60

internally compensated, dual

commercial temperature range (Values specified for T_A = 25 °C)

DESCRIPTION	SUPPLY VOLTAGE (V) MIN	MAX	V_{IO} (mV) MAX	I_{IB} (nA) MAX	A_{VD} (V/mV) MIN	B_1 (MHz) TYP	SR (V/μs) TYP	TYPE	PACKAGES	PAGE NO.
BIFET, General Purpose	±3.5	±18	15	0.4	25	3	13	TL083C	D,JG,N	3-60
BIFET, General Purpose	±3.5	±18	0.5	0.2	50	3	13	TL287C	D,JG,P	3-66
BIFET, General Purpose	±3.5	±18	1	0.2	50	3	13	TL288C	D,JG,P	3-67
Low Power	±1.5	±18	10	−500	20	1	0.6	TL322C	D,JG,P	3-69
LinCMOS, High Bias	1.4	18	5	Typ 0.005	10	2.2	5.3	TLC252AC	D,JG,P	3-73
LinCMOS, High Bias	1.4	18	2	Typ 0.005	10	2.2	5.3	TLC252BC	D,JG,P	3-73
LinCMOS, High Bias	1.4	18	10	Typ 0.005	10	2.2	5.3	TLC252C	D,JG,P	3-73
LinCMOS, Low Bias	1.4	18	5	Typ 0.005	30	0.1	0.05	TLC25L2AC	D,JG,P	3-75
LinCMOS, Low Bias	1.4	18	2	Typ 0.005	30	0.1	0.05	TLC25L2BC	D,JG,P	3-75
LinCMOS, Low Bias	1.4	18	10	Typ 0.005	30	0.1	0.05	TLC25L2C	D,JG,P	3-75
LinCMOS, Medium Bias	1.4	18	5	Typ 0.005	20	0.6	0.6	TLC25M2AC	D,JG,P	3-77
LinCMOS, Medium Bias	1.4	18	2	Typ 0.005	20	0.6	0.6	TLC25M2BC	D,JG,P	3-77
LinCMOS, Medium Bias	1.4	18	10	Typ 0.005	20	0.6	0.6	TLC25M2C	D,JG,P	3-77
LinCMOS, High Bias	3	18	5	Typ 0.005	10	2.2	5.3	TLC272AC	D,JG,P	3-82
LinCMOS, High Bias	3	18	2	Typ 0.005	10	2.2	5.3	TLC272BC	D,JG,P	3-82
LinCMOS, High Bias	3	18	10	Typ 0.005	10	2.2	5.3	TLC272C	D,JG,P	3-82
LinCMOS, High Bias	3	18	0.5	Typ 0.005	10	2.2	5.3	TLC277C	D,JG,P	3-84
LinCMOS, Low Bias	3	18	5	Typ 0.005	50	0.1	0.05	TLC27L2AC	D,JG,P	3-86
LinCMOS, Low Bias	3	18	2	Typ 0.005	50	0.1	0.05	TLC27L2BC	D,JG,P	3-86
LinCMOS, Low Bias	3	18	10	Typ 0.005	50	0.1	0.05	TLC27L2C	D,JG,P	3-86
LinCMOS, Low Bias	3	18	0.5	Typ 0.005	50	0.1	0.05	TLC27L7C	D,JG,P	3-88
LinCMOS, Medium Bias	3	18	5	Typ 0.005	25	0.6	0.6	TLC27M2AC	D,JG,P	3-90
LinCMOS, Medium Bias	3	18	2	Typ 0.005	25	0.6	0.6	TLC27M2BC	D,JG,P	3-90
LinCMOS, Medium Bias	3	18	10	Typ 0.005	25	0.6	0.6	TLC27M2C	D,JG,P	3-90
LinCMOS, Medium Bias	3	18	0.5	Typ 0.005	25	0.6	0.6	TLC27M7C	D,JG,P	3-92
LinCMOS, Micro Power, Precision	1.4	18	0.6	Typ 0.007	500	0.11	0.05	TLC1078C	D,JG,P	3-94
Excalibur, High Speed, Precision	4	40	0.5	25	1000	2	0.9	TLE2022C	D,JG,L,P	3-100
General Purpose	±5	±22	6	500	25	1	0.5	uA747C	D,J,N	3-107

internally compensated, quad

military temperature range

(Values specified for $T_A = 25°C$)

DESCRIPTION	SUPPLY VOLTAGE (V)		V_{IO} (mV)	I_{IB} (nA)	A_{VD} (V/mV)	B_1 (MHz)	SR (V/μs)	TYPE	PACKAGES	PAGE NO.
	MIN	MAX	MAX	MAX	MIN	TYP	TYP			
General Purpose	3	30	5	−150	50	0.6	0.13	LM124	FK,J,W	3-15
General Purpose	±4	±22	5	100	50	1	0.5	LM148	FK,J	3-16
QUAD uA741, High Performance	±4	±22	4	400	50	3.5	1.7	RM4136	FK,J,W	3-36
BIFET, Low Power, Precision	±3.5	±18	1.5	0.2	5	1.1	2.9	TL034AM	FK,J	3-42
BIFET, Low Power, Precision	±3.5	±18	4	0.2	5	1.1	2.9	TL034M	FK,J	3-42
Low Power	±2	±22	5	100	72	0.5	0.5	TL044M	FK,J,W	3-43
BIFET, Precision	±3.5	±18	1.5	0.2	50	2.7	16	TL054AM	FK,J	3-46
BIFET, Precision	±3.5	±18	4	0.2	50	2.7	16	TL054M	FK,J	3-46
BIFET, Low Power	±3.5	±18	9	0.2	4	1	3.5	TL064M	FK,J,W	3-50
BIFET, Low Noise	±3.5	±18	9	0.2	35	3	13	TL074M	FK,J,W	3-55
BIFET, General Purpose	±3.5	±18	9	0.2	25	3	13	TL084M	FK,J,W	3-61
LinCMOS, High Bias	4	18	10	Typ 0.005	10	2.2	5.3	TLC274M	FK,J	3-83
LinCMOS, High Bias	4	18	1.2	Typ 0.005	10	2.2	5.3	TLC279M	FK,J	3-85
LinCMOS, Low Bias	4	18	10	Typ 0.005	50	0.1	0.05	TLC27L4M	FK,J	3-87
LinCMOS, Low Bias	4	18	0.9	Typ 0.005	50	0.1	0.05	TLC27L9M	FK,J	3-89
LinCMOS, Medium Bias	4	18	10	Typ 0.005	20	0.6	0.6	TLC27M4M	FK,J	3-91
LinCMOS, Medium Bias	4	18	0.9	Typ 0.005	20	0.6	0.6	TLC27M9M	FK,J	3-93
LinCMOS, Micro Power, Precision	4	18	1.15	Typ 0.007	500	0.11	0.05	TLC1079M	D,JG,P	3-95
Excalibur, High Speed, Precision	4	40	0.5	25	1000	2	0.9	TLE2024M	FK,J	3-101

internally compensated, quad
automotive temperature range

(Values specified for $T_A = 25\,°C$)

DESCRIPTION		SUPPLY VOLTAGE (V) MIN	MAX	V_{IO} (mV) MAX	I_{IB} (nA) MAX	A_{VD} (V/mV) MIN	B_1 (MHz) TYP	SR (V/μs) TYP	TYPE	PACKAGES	PAGE NO.
Norton Amplifier, Bipolar	S/S	4.5	32	—	200	1.2	2.5	0.5	LM2900	J,N	—
	D/S	±2.2	±16								
Extended Temperature Range LM324		3	26	7	−250	Typ 100	0.6	0.3	LM2902	D,J,N,W	3-15
Low Power, Bipolar	S/S	3	36	8	−500	20	1	0.6	MC3303	D,J,N	3-27
	D/S	±1.5	±18								
Quad uA741		±4.5	±18	6	500	20	3	1.7	RV4136	D,J,N,W	3-36
BIFET, Low Power, Precision		±3.5	±18	1.5	0.2	5	1.1	2.9	TL034AI	D,J,N	3-42
BIFET, Low Power, Precision		±3.5	±18	4	0.2	5	1.1	2.9	TL034I	D,J,N	3-42
BIFET, Precision		±3.5	±18	1.5	0.2	50	2.7	16	TL054AI	D,J,N	3-46
BIFET, Precision		±3.5	±18	4	0.2	50	2.7	16	TL054I	D,J,N	3-46
BIFET, Low Power, Precision		±3.5	±18	6	0.2	4	1	3.5	TL064I	D,J,N	3-50
BIFET, Low Noise, Precision		±3.5	±18	6	0.2	50	3	13	TL074I	D,J,N	3-55
BIFET, General Purpose		±3.5	±18	6	0.2	50	3	13	TL084I	D,J,N	3-61
LinCMOS, High Bias		4	18	5	Typ 0.001	10	2.2	5.3	TLC274AI	D,J,N	3-83
LinCMOS, High Bias		4	18	2	Typ 0.001	10	2.2	5.3	TLC274BI	D,J,N	3-83
LinCMOS, High Bias		4	18	10	Typ 0.001	10	2.2	5.3	TLC274I	D,J,N	3-83
LinCMOS, High Bias		4	18	1.2	Typ 0.005	10	2.2	5.3	TLC279I	D,J,N	3-85
LinCMOS, Low Bias		4	18	5	Typ 0.005	50	0.1	0.05	TLC27L4AI	D,J,N	3-87
LinCMOS, Low Bias		4	18	2	Typ 0.005	50	0.1	0.05	TLC27L4BI	D,J,N	3-87
LinCMOS, Low Bias		4	18	10	Typ 0.005	50	0.1	0.05	TLC27L4I	D,J,N	3-87
LinCMOS, Low Bias		4	18	0.9	Typ 0.005	50	0.1	0.05	TLC27L9I	D,J,N	3-89
LinCMOS, Medium Bias		4	18	5	Typ 0.005	25	0.6	0.6	TLC27M4AI	D,J,N	3-91
LinCMOS, Medium Bias		4	18	2	Typ 0.005	25	0.6	0.6	TLC27M4BI	D,J,N	3-91
LinCMOS, Medium Bias		4	18	10	Typ 0.005	25	0.6	0.6	TLC27M4I	D,J,N	3-91
LinCMOS, Medium Bias		4	18	0.9	Typ 0.005	25	0.6	0.6	TLC27M9I	D,J,N	3-93
LinCMOS, Micro Power, Precision		4	18	1.15	Typ 0.007	500	0.11	0.05	TLC1079I	D,JG,P	3-95
Excalibur, High Speed, Precision		4	40	0.5	25	1000	2	0.9	TLE2024I	DW,FK,J,N	3-101

internally compensated, quad

industrial temperature range

(Values specified for T_A = 25 °C)

DESCRIPTION		SUPPLY VOLTAGE (V) MIN	MAX	V_{IO} (mV) MAX	I_{IB} (nA) MAX	A_{VD} (V/mV) MIN	B_1 (MHz) TYP	SR (V/µs) TYP	TYPE	PACKAGES	PAGE NO.
General Purpose, Bipolar		3	30	5	−150	50	0.6	0.3	LM224	D,J,N,W	3-15
General Purpose, Bipolar		3	30	3	−80	50	0.6	0.3	LM224A	D,J,N,W	3-15
General Purpose, Bipolar		±4	±18	6	200	25	1	0.5	LM248	D,J,N	3-16
High Gain, Low Power, Bipolar	S/S	3	32	5	−150	50	0.6	0.2	LM258	D,J,N	3-17
	D/S	±1.5	±22								
High Gain, Low Power, Bipolar	S/S	3	32	3	−80	50	0.6	0.2	LM258A	D,J,N	3-17
	D/S	±1.5	±22								
Single Supply, Norton Amplifier, Bipolar	S/S	4	32	—	200	1.2	2.5	0.5	LM2900	D,J,N	—
	D/S	±2	±16								
BIFET, Low Power		±3.5	±18	1.5	0.2	5	1.1	2.9	TL034AI	D,J,N	3-42
BIFET, Low Power		±3.5	±18	4	0.2	5	1.1	2.9	TL034I	D,J,N	3-42
BIFET, Precision		±3.5	±18	1.5	0.2	50	2.7	16	TL054AI	D,J,N	3-46
BIFET, Precision		±3.5	±18	4	0.2	50	2.7	16	TL054I	D,J,N	3-46
BIFET, Low Power, Precision		±3.5	±18	8	0.2	4	1	3.5	TL064I	D,J,N	3-50
BIFET, Low Noise, Precision		±3.5	±18	6	0.2	50	3	13	TL074I	D,J,N	3-55
BIFET, General Purpose		±3.5	±18	6	0.2	50	3	13	TL084I	D,J,N	3-61
LinCMOS, High Bias		4	18	5	Typ 0.001	10	2.2	5.3	TLC274AI	D,J,N	3-83
LinCMOS, High Bias		4	18	2	Typ 0.001	10	2.2	5.3	TLC274BI	D,J,N	3-83
LinCMOS, High Bias		4	18	10	Typ 0.001	10	2.2	5.3	TLC274I	D,J,N	3-83
LinCMOS, High Bias		4	18	0.9	Typ 0.005	10	2.2	5.3	TLC279I	D,J,N	3-85
LinCMOS, Low Bias		4	18	5	Typ 0.005	50	0.1	0.05	TLC27L4AI	D,J,N	3-87
LinCMOS, Low Bias		4	18	2	Typ 0.005	50	0.1	0.05	TLC27L4BI	D,J,N	3-87
LinCMOS, Low Bias		4	18	10	Typ 0.005	50	0.1	0.05	TLC27L4I	D,J,N	3-87
LinCMOS, Low Bias		4	18	0.9	Typ 0.005	50	0.1	0.05	TLC27L9I	D,J,N	3-89
LinCMOS, Medium Bias		4	18	5	Typ 0.005	25	0.6	0.6	TLC27M4AI	D,J,N	3-91
LinCMOS, Medium Bias		4	18	2	Typ 0.005	25	0.6	0.6	TLC27M4BI	D,J,N	3-91
LinCMOS, Medium Bias		4	18	10	Typ 0.005	25	0.6	0.6	TLC27M4I	D,J,N	3-91
LinCMOS, Medium Bias		4	18	0.9	Typ 0.005	25	0.6	0.6	TLC27M9I	D,J,N	3-93
LinCMOS, Micro Power, Precision		4	18	1.15	Typ 0.007	500	0.11	0.05	TLC1079I	D,JG,P	3-95
Excalibur, High Speed, Precision		4	40	0.5	25	1000	2	0.9	TLE2024I	DW,FK,J,N	3-101

internally compensated, quad

commercial temperature range   (Values specified for T_A = 25 °C)

DESCRIPTION		SUPPLY VOLTAGE (V)		V_{IO} (mV)	I_{IB} (nA)	A_{VD} (V/mV)	B_1 (MHz)	SR (V/µs)	TYPE	PACKAGES	PAGE NO.
		MIN	MAX	MAX	MAX	MIN	TYP	TYP			
BIFET, General Purpose		±3.5	±18	10	0.2	25	3	13	LF347	D,J,N	3-4
General Purpose		3	30	7	−250	25	0.6	0.3	LM324	D,J,N,W	3-15
General Purpose		3	30	3	−100	25	0.6	0.3	LM324A	D,J,N,W	3-15
General Purpose		±4	±18	6	200	25	1	0.5	LM348	D,J,N	3-16
Single Supply, Norton Amplifier,	S/S	4	32	−	200	1.2	2.5	0.5	LM3900	D,J,N	−
Bipolar	D/S	±2	±16								
Low Power, Bipolar	S/S	3	36	10	−500	20	1	0.6	MC3403	D,J,N	3-27
	D/S	±1.5	±18								
Quad uA741, High Performance		±4	±18	6	500	20	3	1.7	RC4136	D,J,N,W	3-36
BIFET, Low Power, Precision		±3.5	±18	1.5	0.2	5	1.1	2.9	TL034AC	D,J,N	3-42
BIFET, Low Power, Precision		±3.5	±18	4	0.2	5	1.1	2.9	TL034C	D,J,N	3-42
General Purpose		±2	±18	5	250	60	0.5	0.5	TL044C	J,N,W	3-43
BIFET, Precision		±3.5	±18	1.5	0.2	50	2.7	16	TL054AC	D,J,N	3-46
BIFET, Precision		±3.5	±18	4	0.2	50	2.7	16	TL054C	D,J,N	3-46
BIFET, Low Power		±3.5	±18	6	0.2	4	1	3.5	TL064AC	D,J,N	3-50
BIFET, Low Power		±3.5	±18	3	0.2	4	1	3.5	TL064BC	D,J,N	3-50
BIFET, Low Power		±3.5	±18	15	0.4	3	1	3.5	TL064C	D,J,N	3-50
BIFET, Low Noise		±3.5	±18	6	0.2	50	3	13	TL074AC	D,J,N	3-55
BIFET, Low Noise		±3.5	±18	3	0.2	50	3	13	TL074BC	D,J,N	3-55
BIFET, Low Noise		±3.5	±18	10	0.2	50	3	13	TL074C	D,J,N	3-55
BIFET, Low Noise		±3.5	±18	10	0.2	25	3	13	TL075C	J,N	3-56
BIFET, General Purpose		±3.5	±18	6	0.2	50	3	13	TL084AC	D,J,N	3-61
BIFET, General Purpose		±3.5	±18	3	0.2	50	3	13	TL084BC	D,J,N	3-61
BIFET, General Purpose		±3.5	±18	15	0.4	25	3	13	TL084C	D,J,N	3-61
BIFET, General Purpose		±3.5	±18	15	0.4	25	3	13	TL085C	J,N	3-62
High Performance, Bipolar		±4	±18	6	500	20	3	2	TL136C	N	3-65
LinCMOS, High Bias		1.4	18	5	Typ 0.005	10	2.2	5.3	TLC254AC	D,J,N	3-74
LinCMOS, High Bias		1.4	18	2	Typ 0.005	10	2.2	5.3	TLC254BC	D,J,N	3-74
LinCMOS, High Bias		1.4	18	10	Typ 0.005	10	2.2	5.3	TLC254C	D,J,N	3-74
LinCMOS, Low Bias		1.4	18	5	Typ 0.005	30	0.1	0.05	TLC25L4AC	D,J,N	3-76
LinCMOS, Low Bias		1.4	18	2	Typ 0.005	30	0.1	0.05	TLC25L4BC	D,J,N	3-76
LinCMOS, Low Bias		1.4	18	10	Typ 0.005	30	0.1	0.05	TLC25L4C	D,J,N	3-76
LinCMOS, Medium Bias		1.4	18	5	Typ 0.005	20	0.6	0.6	TLC25M4AC	D,J,N	3-78
LinCMOS, Medium Bias		1.4	18	2	Typ 0.005	20	0.6	0.6	TLC25M4BC	D,J,N	3-78
LinCMOS, Medium Bias		1.4	18	10	Typ 0.005	20	0.6	0.6	TLC25M4C	D,J,N	3-78

internally compensated, quad

commercial temperature range   (Values specified for T_A = 25°C)

DESCRIPTION	SUPPLY VOLTAGE (V)		V_{IO} (mV)	I_{IB} (nA)	A_{VD} (V/mV)	B_1 (MHz)	SR (V/μs)	TYPE	PACKAGES	PAGE NO.
	MIN	MAX	MAX	MAX	MIN	TYP	TYP			
LinCMOS, High Bias	3	18	5	Typ 0.005	10	2.2	5.3	TLC274AC	D,J,N	3-83
LinCMOS, High Bias	3	18	2	Typ 0.005	10	2.2	5.3	TLC274BC	D,J,N	3-83
LinCMOS, High Bias	3	18	10	Typ 0.005	10	2.2	5.3	TLC274C	D,J,N	3-83
LinCMOS, High Bias	3	18	0.9	Typ 0.005	10	2.2	5.3	TLC279C	D,J,N	3-85
LinCMOS, Low Bias	3	18	5	Typ 0.005	50	0.1	0.05	TLC27L4AC	D,J,N	3-87
LinCMOS, Low Bias	3	18	2	Typ 0.005	50	0.1	0.05	TLC27L4BC	D,J,N	3-87
LinCMOS, Low Bias	3	18	10	Typ 0.005	50	0.1	0.05	TLC27L4C	D,J,N	3-87
LinCMOS, Low Bias	3	18	0.9	Typ 0.005	50	0.1	0.05	TLC27L9C	D,J,N	3-89
LinCMOS, Medium Bias	3	18	5	Typ 0.005	25	0.6	0.6	TLC27M4AC	D,J,N	3-91
LinCMOS, Medium Bias	3	18	2	Typ 0.005	25	0.6	0.6	TLC27M4BC	D,J,N	3-91
LinCMOS, Medium Bias	3	18	10	Typ 0.005	25	0.6	0.6	TLC27M4C	D,J,N	3-91
LinCMOS, Medium Bias	3	18	0.9	Typ 0.005	25	0.7	0.6	TLC27M9C	D,J,N	3-93
LinCMOS, Micro Power, Precision	1.4	18	1.15	Typ 0.007	500	0.11	0.05	TLC1079C	D,J,N	3-95
Excalibur, High Speed, Precision	4	40	0.5	25	1000	2	0.9	TLE2024C	DW,FK,J,N	3-101

APPENDIX A5.2

OPERATIONAL AMPLIFIERS
CROSS-REFERENCE GUIDE

Replacements are based on similarity of electrical and mechanical characteristics as shown in currently published data. Interchangeability in particular applications is not guaranteed. Before using a device as a substitute, the user should compare the specifications of the substitute device with the specifications of the original.

Texas Instruments makes no warranty as to the information furnished and buyer assumes all risk in the use thereof. No liability is assumed for damages resulting from the use of the information contained herein.

Manufacturers are arranged in alphabetical order.

ADVANCED LINEAR DEVICES	SUGGESTED TI REPLACEMENT	PAGE NO.
ALD1701 or ALD1702 or ALD1703	TLC271	3-79,3-80,3-81

ANALOG DEVICES	SUGGESTED TI REPLACEMENT	PAGE NO.
AD510 or AD517	OP-07	3-29,3-30,3-31

FAIRCHILD	DIRECT TI REPLACEMENT	SUGGESTED TI REPLACEMENT	PAGE NO.
uA714		OP-07C	3-29
uA714E		OP-07E	3-31
uA714L		OP-07D	3-30
uA741	uA741		3-106
uA747	uA747		3-107
uA748	uA748		3-108
uA771		TLO71	3-53
uA771A		TLO71B or TLO81B	3-53 or 3-58
uA771B		TLO71A or TLO81A	3-53 or 3-58
uA771L		TLO81	3-58
uA772		TLO72	3-54
uA772A		TLO72B	3-54
uA772B		TLO72A or TLO82A	3-54 or 3-59
uA772L		TLO82	3-59
uA774		TLO74	3-55
uA774B		TLO74A or TLO74B	3-55
uA774L		TLO84	3-61

GENERAL ELECTRIC	SUGGESTED TI REPLACEMENT	PAGE NO.
ICL7611, or ICL7612, or ICL7613	TLC271	3-79,3-80,3-81
ICL7621	TLC272	3-82
ICL7641	TLC274 or TLC27L9	3-83 or 3-89
ICL7642	TLC27M9	3-93

HARRIS	SUGGESTED TI REPLACEMENT	PAGE NO.
HA2515	LM318	3-14
HA5130-5	OP-07E	3-31
HA5135-5	OP-07C	3-29

INTERSIL	SUGGESTED TI REPLACEMENT	PAGE NO.
ICL7611, or ICL7612, or ICL7613	TLC271	3-79,3-80,3-81
ICL7621	TLC272	3-82
ICL7641	TLC274 or TLC27L9	3-83 or 3-89
ICL7642	TLC27M9	3-93

LINEAR TECHNOLOGY	DIRECT TI REPLACEMENT	SUGGESTED TI REPLACEMENT	PAGE NO.
LT1001	LT1001	OP-07C, OP-07D, or OP-07E	3-18,3-29,3-30, or 3-31
LT1007	LT1007		3-19
LT1007A	LT1007A		3-19
LT1037	LT1037		3-24
LT1037A	LT1037A		3-24

MAXIM	SUGGESTED TI REPLACEMENT	PAGE NO.
ICL7611, or ICL7612, or ICL7613	TLC271	3-79,3-80,3-81
ICL7621	TLC272	3-82
ICL7641	TLC274 or TLC27L9	3-83 or 3-89
ICL7642	TLC27M9	3-39

MOTOROLA	DIRECT TI REPLACEMENT	SUGGESTED TI REPLACEMENT	PAGE NO.
MC748	uA748		3-108
MC1458	MC1458		3-26
MC1558	MC1558		3-26
MC1709	uA709		—
MC1741	uA741		3-106
MC1747	uA747		3-107
MC3303	MC3303		3-27
MC3403	MC3403	RC4136	3-27,3-36
MC4558	RC4558		3-37
MC4741	LM348		3-16
MC34001		TL071 or LF351	3-53 or 3-5
MC34002		TL072 or LF353	3-54 or 3-6
MC34004		TL074 or LF347	3-55 or 3-4
MC34004B		TL074A or LF347B	3-55 or 3-4

NATIONAL	DIRECT TI REPLACEMENT	SUGGESTED TI REPLACEMENT	PAGE NO.
LF347	LF347	TL074 or TL084	3-4,3-55 or 3-61
LF347B	LF347B	TL074A, TL074B, or TL084A	3-4,3-55,3-55 or 3-61
LF351	LF351	TL071 or TL081A	3-5,3-53 or 3-58
LF353	LF353	TL072, TL072A, or TL082A	3-6,3-54,3-54, or 3-59
LF411	LF411	TL081A	3-7,3-58
LF411A		TL071A, TL071B, TL081A, or TL081B	3-53,3-53, 3-58, or 3-58
LF412	LF412	TL072A, TL082A, or TL082B	3-8,3-54,3-59, or 3-59
LF441		TL061	3-48
LF441A		TL061A or TL061B	3-48 or 3-48
LF442		TL062	3-49
LF442A		TL062B	3-49
LF444		TL064	3-50
LF444A		TL064A	3-50
LH0044		OP-07C	3-29
LH0044B		OP-07D	3-30
LH0044C		OP-07E	3-31
LM201A	LM201A		3-9
LM207	LM207		3-10
LM218	LM218		3-14
LM224	LM224		3-15
LM248	LM248		3-16
LM258	LM258		3-17

484

(Continued)

NATIONAL	DIRECT TI REPLACEMENT	SUGGESTED TI REPLACEMENT	PAGE NO.
LM301A	LM301A		3-11
LM307	LM307		3-12
LM318	LM318		3-14
LM324	LM324		3-15
LM348	LM348		3-16
LM358	LM358		3-17
LM709	uA709		—
LM741	uA741		3-106
LM883		RC4558	3-37
LM1458	MC1458		3-26
LM2900	LM2900		—
LM2902	LM2902		3-15
LM2904	LM2904		3-17
LM3900	LM3900		—
LMC660		TLC274	3-83

NEC	SUGGESTED TI REPLACEMENT	PAGE NO.
uPC159	LM318	3-14
uPC251	MC1458	3-26
uPC354	OP-07	3-29,3-30,3-31
uPC801	TL071, TL081A, or LF351	3-53,3-58, or 3-5

PMI	DIRECT TI REPLACEMENT	SUGGESTED TI REPLACEMENT	PAGE NO.
OP 02		uA741	3-106
OP-04		uA747	3-107
OP-07C	OP-07C		3-29
OP-07D	OP-07D		3-30
OP-07E	OP-07E		3-31
OP-07F		RC4136	3-36
OP-14C or OP-14E		MC1458	3-26
OP-14J		MC1558	3-26
OP-15F		TL071, TL081A, or LF351	3-53,3-5 or 3-5
OP-27E	OP-27E		3-33
OP-27G	OP-27G		3-34
OP-37E	OP-37E		3-35
OP-37G	OP-37G		3-35
OP-215F		TL072, TL082A, or LF353	3-54,3-58, or 3-6

RAYTHEON	DIRECT TI REPLACEMENT	SUGGESTED TI REPLACEMENT	PAGE NO.
RC4136	RC4136		3-36
RC4156		LM348	3-16
RC4157		LM348	3-16
RC4558	RC4558		3-37
RC4559	RC4559		3-38

RCA	SUGGESTED TI REPLACEMENT	PAGE NO.
CA081A	TL081	3-58
CA081A	TL081A	3-58
CA082	TL082	3-59
CA082A	TL082A	3-59
CA084	TL084	3-61

SIGNETICS	DIRECT TI REPLACEMENT	SUGGESTED TI REPLACEMENT	PAGE NO.
NE532		LM358 or TL022	3-17 or 3-39
NE5532	NE5532		3-28
NE5532A	NE5532A		3-28
NE5534	NE5534		3-28
NE5534A	NE5534A		3-28
SE5534	SE5534		3-28
SE5534A	SE5534A		3-28

SGS-THOMSON	SUGGESTED TI REPLACEMENT	PAGE NO.
TS271	TLC271	3-79,3-80,3-81
TS271A	TLC271A	3-79,3-80,3-81
TS271B	TLC271B	3-79,3-80,3-81
TS272	TLC272	3-82
TS272A	TLC272A	3-82
TS272B	TLC272B	3-82
TS274	TLC274	3-83
TS274A	TLC274A	3-83
TS274B	TLC274B	3-83
TS27L2	TLC27L2	3-86
TS27L2A	TLC27L2A	3-86
TS27L2B	TLC27L2B	3-86
TS27L4	TLC27L4	3-87
TS27L4A	TLC27L4A	3-87
TS27L4B	TLC27L4B	3-87
TS27M2	TLC27M2	3-90
TS27M2A	TLC27M2A	3-90
TS27M2B	TLC27M2B	3-90
TS27M4	TLC27M4	3-91
TS27M4A	TLC27M4A	3-91
TS27M4B	TLC27M4B	3-91

APPENDIX A5.3

Input Offset Voltage (V_{IO})

The d-c voltage that must be applied between the input terminals to force the quiescent d-c output voltage to zero or other level, if specified.

Average Temperature Coefficient of Input Offset Voltage (α_{VIO})

The ratio of the change in input offset voltage to the change in free-air temperature. This is an average value for the specified temperature range.

$$\alpha_{VIO} = \left[\frac{(V_{IO} @ T_{A(1)}) - (V_{IO} @ T_{A(2)})}{T_{A(1)} - T_{A(2)}} \right] \text{ where } T_{A(1)} \text{ and } T_{A(2)} \text{ are the specified temperature extremes.}$$

Input Offset Current (I_{IO})

The difference between the currents into the two input terminals with the output at zero volts.

Average Temperature Coefficient of Input Offset Current (α_{IIO})

The ratio of the change in input offset current to the change in free-air temperature. This is an average value for the specified temperature range.

$$\alpha_{IIO} = \left[\frac{(I_{IO} @ T_{A(1)}) - (I_{IO} @ T_{A(2)})}{T_{A(1)} - T_{A(2)}} \right] \text{ where } T_{A(1)} \text{ and } T_{A(2)} \text{ are the specified temperature extremes.}$$

Input Bias Current (I_{IB})

The average of the currents into the two input terminals with the output at zero volts.

Common-Mode Input Voltage (V_{IC})

The average of the two input voltages.

Common-Mode Input Voltage Range (V_{ICR})

The range of common-mode input voltage that if exceeded will cause the amplifier to cease functioning properly.

Differential Input Voltage (V_{ID})

The voltage at the noninverting input with respect to the inverting input.

Maximum Peak Output Voltage Swing (V_{OM})

The maximum positive or negative peak output voltage that can be obtained without waveform clipping when the quiescent d-c output voltage is zero.

Maximum Peak-to-Peak Output Voltage Swing (V_{OPP})

The maximum peak-to-peak output voltage that can be obtained without waveform clipping when the quiescent d-c output voltage is zero.

Large-Signal Voltage Amplification (A_V)

The ratio of the peak-to-peak output voltage swing to the change in input voltage required to drive the output.

Differential Voltage Amplification (A_{VD})

The ratio of the change in output voltage to the change in differential input voltage producing it.

Maximum-Output-Swing Bandwidth (B_{OM})

The range of frequencies within which the maximum output voltage swing is above a specified value.

Unity-Gain Bandwidth (B_1)

The range of frequencies within which the open-loop voltage amplification is greater than unity.

Phase Margin (ϕ_m)

The absolute value of the open-loop phase shift between the output and the inverting input at the frequency at which the modulus of the open-loop amplification is unity.

Gain Margin (A_m)

The reciprocal of the open-loop voltage amplification at the lowest frequency at which the open-loop phase shift is such that the output is in phase with the inverting input.

Input Resistance (r_i)

The resistance between the input terminals with either input grounded.

Differential Input Resistance (r_{id})

The small-signal resistance between the two ungrounded input terminals.

Output Resistance (r_o)

The resistance between the output terminal and ground.

Input Capacitance (C_i)

The capacitance between the input terminals with either input grounded.

Common-Mode Input Impedance (z_{ic})

The parallel sum of the small-signal impedance between each input terminal and ground.

Output Impedance (z_o)

The small-signal impedance between the output terminal and ground.

Common-Mode Rejection Ratio (k_{CMR}, CMRR)

The ratio of differential voltage amplification to common-mode voltage amplification.
NOTE: This is measured by determining the ratio of a change in input common-mode voltage to the resulting change in input offset voltage.

Supply Voltage Sensitivity (k_{SVS}, $\Delta V_{IO}/\Delta V_{CC}$)

The absolute value of the ratio of the change in input offset voltage to the change in supply voltages producing it.
NOTES: 1. Unless otherwise noted, both supply voltages are varied symmetrically.
2. This is the reciprocal of supply voltage rejection ratio.

Supply Voltage Rejection Ratio (k_{SVR}, $\Delta V_{CC}/\Delta V_{IO}$)

The absolute value of the ratio of the change in supply voltages to the change in input offset voltage.
NOTES: 1. Unless otherwise noted, both supply voltages are varied symmetrically.
2. This is the reciprocal of supply voltage sensitivity.

Equivalent Input Noise Voltage (V_n)

The voltage of an ideal voltage source (having an internal impedance equal to zero) in series with the input terminals of the device that represents the part of the internally generated noise that can properly be represented by a voltage source.

Equivalent Input Noise Current (I_n)

The current of an ideal current source (having an internal impedance equal to infinity) in parallel with the input terminals of the device that represents the part of the internally generated noise that can properly be represented by a current source.

Average Noise Figure (F)

The ratio of (1) the total output noise power within a designated output frequency band when the noise temperature of the input termination(s) is at the reference noise temperature, T_0, at all frequencies to (2) that part of (1) caused by the noise temperature of the designated signal-input termination within a designated signal-input frequency band.

Short-Circuit Output Current (I_{OS})

The maximum output current available from the amplifier with the output shorted to ground, to either supply, or to a specified point.

Supply Current (I_{CC})

The current into the V_{CC} or V_{CC+} terminal of an integrated circuit.

Total Power Dissipation (P_D)

The total d-c power supplied to the device less any power delivered from the device to a load.
NOTE: At no load: $P_D = V_{CC+} \cdot I_{CC+} + V_{CC-} \cdot I_{CC-}$.

Crosstalk Attenuation (V_{o1}/V_{o2})

The ratio of the change in output voltage of a driven channel to the resulting change in output voltage of another channel.

Rise Time (t_r)

The time required for an output voltage step to change from 10% to 90% of its final value.

Total Response Time (Settling Time) (t_{tot})

The time between a step-function change of the input signal level and the instant at which the magnitude of the output signal reaches for the last time a specified level range ($\pm\epsilon$) containing the final output signal level.

Overshoot Factor

The ratio of (1) the largest deviation of the output signal value from its final steady-state value after a step-function change of the input signal, to (2) the absolute value of the difference between the steady-state output signal values before and after the step-function change of the input signal.

Slew Rate (SR)

The average time rate of change of the closed-loop amplifier output voltage for a step-signal input.

MOTOROLA
■ SEMICONDUCTOR ■
TECHNICAL DATA

MC1741
MC1741C

OPERATIONAL AMPLIFIER

SILICON MONOLITHIC
INTEGRATED CIRCUIT

INTERNALLY COMPENSATED, HIGH PERFORMANCE OPERATIONAL AMPLIFIERS

. . . designed for use as a summing amplifier, integrator, or amplifier with operating characteristics as a function of the external feedback components.

- No Frequency Compensation Required
- Short-Circuit Protection
- Offset Voltage Null Capability
- Wide Common-Mode and Differential Voltage Ranges
- Low-Power Consumption
- No Latch Up

NC
Offset Null — V_{CC}
Invt Input — Output
Noninvt Input — Offset Null
V_{EE}
(Top View)

G SUFFIX
METAL PACKAGE
CASE 601

8 1

P1 SUFFIX
PLASTIC PACKAGE
CASE 626

8
1

U SUFFIX
CERAMIC PACKAGE
CASE 693

8
1

D SUFFIX
PLASTIC PACKAGE
CASE 751
(SO-8)

8 1

MAXIMUM RATINGS (T_A = +25°C unless otherwise noted)

Rating	Symbol	MC1741C	MC1741	Unit
Power Supply Voltage	V_{CC} V_{EE}	+18 −18	+22 −22	Vdc Vdc
Input Differential Voltage	V_{ID}	±30		Volts
Input Common Mode Voltage (Note 1)	V_{ICM}	±15		Volts
Output Short Circuit Duration (Note 2)	t_S	Continuous		
Operating Ambient Temperature Range	T_A	0 to +70	−55 to +125	°C
Storage Temperature Range Metal and Ceramic Packages Plastic Packages	T_{stg}	−65 to +150 −55 to +125		°C

NOTES:
1. For supply voltages less than +15 V, the absolute maximum input voltage is equal to the supply voltage.
2. Supply voltage equal to or less than 15 V.

PIN CONNECTIONS

Offset Null — NC
Invt Input — V_{CC}
Noninvt Input — Output
V_{EE} — Offset Null
(Top View)

ORDERING INFORMATION

Device	Alternate	Temperature Range	Package
MC1741CD	—		SO-8
MC1741CG	LM741CH, µA741HC	0°C to +70°C	Metal Can
MC1741CP1	LM741CN, µA741TC		Plastic DIP
MC1741CU	—		Ceramic DIP
MC1741G	—	−55°C to	Metal Can
MC1741U	—	+125°C	Ceramic DIP

EQUIVALENT CIRCUIT SCHEMATIC

ELECTRICAL CHARACTERISTICS (V_{CC} = +15 V, V_{EE} = −15 V, T_A = 25°C unless otherwise noted).

Characteristic	Symbol	MC1741			MC1741C			Unit
		Min	Typ	Max	Min	Typ	Max	
Input Offset Voltage ($R_S \leqslant$ 10 k)	V_{IO}	−	1.0	5.0	−	2.0	6.0	mV
Input Offset Current	I_{IO}	−	20	200	−	20	200	nA
Input Bias Current	I_{IB}	−	80	500	−	80	500	nA
Input Resistance	r_i	0.3	2.0	−	0.3	2.0	−	MΩ
Input Capacitance	C_i	−	1.4	−	−	1.4	−	pF
Offset Voltage Adjustment Range	V_{IOR}	−	±15	−	−	±15	−	mV
Common Mode Input Voltage Range	V_{ICR}	±12	±13	−	±12	±13	−	V
Large Signal Voltage Gain (V_O = ±10 V, $R_L \geqslant$ 2.0 k)	A_v	50	200	−	20	200	−	V/mV
Output Resistance	r_o	−	75	−	−	75	−	Ω
Common Mode Rejection Ratio ($R_S \leqslant$ 10 k)	CMRR	70	90	−	70	90	−	dB
Supply Voltage Rejection Ratio ($R_S \leqslant$ 10 k)	PSRR	−	30	150	−	30	150	μV/V
Output Voltage Swing ($R_L \geqslant$ 10 k) ($R_L \geqslant$ 2 k)	V_O	 ±12 ±10	 ±14 ±13	− − −	 ±12 ±10	 ±14 ±13	V − −	V
Output Short-Circuit Current	I_{os}	−	20	−	−	20	−	mA
Supply Current	I_D	−	1.7	2.8	−	1.7	2.8	mA
Power Consumption	P_C	−	50	85	−	50	85	mW
Transient Response (Unity Gain − Non-Inverting) (V_I = 20 mV, $R_L \geqslant$ 2 k, $C_L <$ 100 pF) Rise Time (V_I = 20 mV, $R_L \geqslant$ 2 k, $C_L <$ 100 pF) Overshoot (V_I = 10 V, $R_L \geqslant$ 2 k, $C_L <$ 100 pF) Slew Rate	t_{TLH} os SR	 − − −	 0.3 15 0.5	 − − −	 − − −	 0.3 15 0.5	 − − −	μs % V/μs

ELECTRICAL CHARACTERISTICS (V_{CC} = +15 V, V_{EE} = −15 V, T_A = T_{low} to T_{high} unless otherwise noted).

Characteristic	Symbol	MC1741			MC1741C			Unit
		Min	Typ	Max	Min	Typ	Max	
Input Offset Voltage ($R_S \leqslant$ 10 kΩ)	V_{IO}	−	1.0	6.0	−	−	7.5	mV
Input Offset Current (T_A = 125°C) (T_A = −55°C) (T_A = 0°C to +70°C)	I_{IO}	 − − −	 7.0 85 −	 200 500 −	 − − −	 − − −	nA − − 300	nA
Input Bias Current (T_A = 125°C) (T_A = −55°C) (T_A = 0°C to +70°C)	I_{IB}	 − − −	 30 300 −	 500 1500 −	 − − −	 − − −	 − − 800	nA
Common Mode Input Voltage Range	V_{ICR}	±12	±13	−	−	−	−	V
Common Mode Rejection Ratio ($R_S \leqslant$ 10 k)	CMRR	70	90	−	−	−	−	dB
Supply Voltage Rejection Ratio ($R_S \leqslant$ 10 k)	PSRR	−	30	150	−	−	−	μV/V
Output Voltage Swing ($R_L \geqslant$ 10 k) ($R_L \geqslant$ 2 k)	V_O	 ±12 ±10	 ±14 ±13	 − −	 ±10	 ±13	 − −	V
Large Signal Voltage Gain ($R_L \geqslant$ 2 k, V_{out} = +10 V)	A_v	25	−	−	15	−	−	V/mV
Supply Currents (T_A = 125°C) (T_A = −55°C)	I_D	 	 1.5 2.0	 2.5 3.3	 	 − −	 − −	mA
Power Consumption (T_A = +125°C) (T_A = −55°C)	P_C	 	 45 60	 75 100	 	 − −	 − −	mW

*T_{high} = 125°C for MC1741 and 70°C for MC1741C
T_{low} = −55°C for MC1741 and 0°C for MC1741C

MC1741, MC1741C

FIGURE 1 – BURST NOISE versus SOURCE RESISTANCE

FIGURE 2 – RMS NOISE versus SOURCE RESISTANCE

FIGURE 3 – OUTPUT NOISE versus SOURCE RESISTANCE

FIGURE 4 – SPECTRAL NOISE DENSITY

FIGURE 5 – BURST NOISE TEST CIRCUIT

Unlike conventional peak reading or RMS meters, this system was especially designed to provide the quick response time essential to burst (popcorn) noise testing.

The test time employed is 10 seconds and the 20 μV peak limit refers to the operational amplifier input thus eliminating errors in the closed-loop gain factor of the operational amplifier under test.

MC1741, MC1741C

(V_{CC} = +15 Vdc, V_{EE} = -15 Vdc, T_A = +25°C unless otherwise noted)

FIGURE 6 – POWER BANDWIDTH
(LARGE SIGNAL SWING versus FREQUENCY)

FIGURE 7 – OPEN LOOP FREQUENCY RESPONSE

FIGURE 8 – POSITIVE OUTPUT VOLTAGE SWING
versus LOAD RESISTANCE

FIGURE 9 – NEGATIVE OUTPUT VOLTAGE SWING
versus LOAD RESISTANCE

FIGURE 10 – OUTPUT VOLTAGE SWING versus
LOAD RESISTANCE (Single Supply Operation)

FIGURE 11 – SINGLE SUPPLY INVERTING AMPLIFIER

493

FIGURE 12 — NONINVERTING PULSE RESPONSE

FIGURE 13 — TRANSIENT REPONSE TEST CIRCUIT

FIGURE 14 — OPEN LOOP VOLTAGE GAIN
versus SUPPLY VOLTAGE

National Semiconductor

LF412A/LF412 Low Offset, Low Drift Dual JFET Input Operational Amplifier

General Description

These devices are low cost, high speed, JFET input operational amplifiers with very low input offset voltage and guaranteed input offset voltage drift. They require low supply current yet maintain a large gain bandwidth product and fast slew rate. In addition, well matched high voltage JFET input devices provide very low input bias and offset currents. LF412 dual is pin compatible with the LM1558, allowing designers to immediately upgrade the overall performance of existing designs.

These amplifiers may be used in applications such as high speed integrators, fast D/A converters, sample and hold circuits and many other circuits requiring low input offset voltage and drift, low input bias current, high input impedance, high slew rate and wide bandwidth.

Features

■ Internally trimmed offset voltage	1 mV (max)
■ Input offset voltage drift	10 µV/°C (max)
■ Low input bias current	50 pA
■ Low input noise current	0.01 pA/√Hz
■ Wide gain bandwidth	3 MHz (min)
■ High slew rate	10V/µs (min)
■ Low supply current	1.8 mA/Amplifier
■ High input impedance	$10^{12}\Omega$
■ Low total harmonic distortion $A_V = 10$, $R_L = 10k$, $V_O = 20$ Vp-p, BW = 20 Hz-20 kHz	≤0.02%
■ Low 1/f noise corner	50 Hz
■ Fast settling time to 0.01%	2 µs

Typical Connection

Simplified Schematic

Ordering Information

LF412XYZ

- **X** indicates electrical grade
- **Y** indicates temperature range
  - "M" for military
  - "C" for commercial
- **Z** indicates package type
  - "H" or "N"

Connection Diagrams

Metal Can Package

Note. Pin 4 connected to case.
TOP VIEW

Order Number LF412AMH, LF412MH, LF412ACH or LF412CH
See NS Package Number H08B

Dual-In-Line Package

TOP VIEW

TL/H/5656-1

Order Number LF412ACJ, LF412CJ, LF412ACN or LF412CN
See NS Package Number J08A or N08E

3-83

Absolute Maximum Ratings

If Military/Aerospace specified devices are required, please contact the National Semiconductor Sales Office/Distributors for availability and specifications.
(Note 9)

	LF412A	LF412		H Package	N Package
Supply Voltage	±22V	±18V	Power Dissipation (Note 10)	(Note 3)	670 mW
Differential Input Voltage	±38V	±30V	T_j max	150°C	115°C
Input voltage Range			θ_{jA} (Typical)	152°C/W	115°C/W
(Note 1)	±19V	±15V	Operating Temp. Range	(Note 4)	(Note 4)
Output Short Circuit			Storage Temp.	$-65°C \le T_A \le 150°C$	$-65°C \le T_A \le 150°C$
Duration (Note 2)	Continuous	Continuous	Range		
			Lead Temp.		
			(Soldering, 10 sec.)	260°C	260°C
			ESD rating to be determined.		

DC Electrical Characteristics (Note 5)

Symbol	Parameter	Conditions		LF412A			LF412			Units
				Min	Typ	Max	Min	Typ	Max	
V_{OS}	Input Offset Voltage	$R_S = 10 \text{ k}\Omega$, $T_A = 25°C$			0.5	1.0		1.0	3.0	mV
$\Delta V_{OS}/\Delta T$	Average TC of Input Offset Voltage	$R_S = 10 \text{ k}\Omega$ (Note 6)			7	10		7	20 (Note 6)	μV/°C
I_{OS}	Input Offset Current	$V_S = \pm 15V$ (Notes 5 and 7)	$T_j = 25°C$		25	100		25	100	pA
			$T_j = 70°C$			2			2	nA
			$T_j = 125°C$			25			25	nA
I_B	Input Bias Current	$V_S = \pm 15V$ (Notes 5 and 7)	$T_j = 25°C$		50	200		50	200	pA
			$T_j = 70°C$			4			4	nA
			$T_j = 125°C$			50			50	nA
R_{IN}	Input Resistance	$T_j = 25°C$			10^{12}			10^{12}		Ω
A_{VOL}	Large Signal Voltage Gain	$V_S = \pm 15V$, $V_O = \pm 10V$, $R_L = 2k$, $T_A = 25°C$		50	200		25	200		V/mV
		Over Temperature		25	200		15	200		V/mV
V_O	Output Voltage Swing	$V_S = \pm 15V$, $R_L = 10k$		±12	±13.5		±12	±13.5		V
V_{CM}	Input Common-Mode Voltage Range			±16	+19.5		±11	+14.5		V
					−16.5			−11.5		V
CMRR	Common-Mode Rejection Ratio	$R_S \le 10k$		80	100		70	100		dB
PSRR	Supply Voltage Rejection Ratio	(Note 8)		80	100		70	100		dB
I_S	Supply Current	$V_O = 0V$, $R_L = \infty$			3.6	5.6		3.6	6.5	mA

AC Electrical Characteristics (Note 5)

Symbol	Parameter	Conditions	LF412A			LF412			Units
			Min	Typ	Max	Min	Typ	Max	
	Amplifier to Amplifier Coupling	$T_A = 25°C$, f = 1 Hz-20 kHz (Input Referred)		−120			−120		dB
SR	Slew Rate	$V_S = \pm 15V$, $T_A = 25°C$	10	15		8	15		V/μs
GBW	Gain-Bandwidth Product	$V_S = \pm 15V$, $T_A = 25°C$	3	4		2.7	4		MHz
e_n	Equivalent Input Noise Voltage	$T_A = 25°C$, $R_S = 100\Omega$, f = 1 kHz		25			25		nV/$\sqrt{Hz}$
i_n	Equivalent Input Noise Current	$T_A = 25°C$, f = 1 kHz		0.01			0.01		pA/$\sqrt{Hz}$

Note 1: Unless otherwise specified the absolute maximum negative input voltage is equal to the negative power supply voltage.

Note 2: Any of the amplifier outputs can be shorted to ground indefintely, however, more than one should not be simultaneously shorted as the maximum junction temperature will be exceeded.

Note 3: For operating at elevated temperature, these devices must be derated based on a thermal resistance of θ_{jA}.

Note 4: These devices are available in both the commercial temperature range $0°C \leq T_A \leq 70°C$ and the military temperature range $-55°C \leq T_A \leq 125°C$. The temperature range is designated by the position just before the package type in the device number. A "C" indicates the commercial temperature range and an "M" indicates the military temperature range. The military temperature range is available in "H" package only. In all cases the maximum operating temperature is limited by internal junction temperature T_j max.

Note 5: Unless otherwise specified, the specifications apply over the full temperature range and for $V_S = \pm 20V$ for the LF412A and for $V_S = \pm 15V$ for the LF412. V_{OS}, I_B, and I_{OS} are measured at $V_{CM} = 0$.

Note 6: The LF412A is 100% tested to this specification. The LF412 is sample tested on a per amplifier basis to insure at least 85% of the amplifiers meet this specification.

Note 7: The input bias currents are junction leakage currents which approximately double for every 10°C increase in the junction temperature, T_j. Due to limited production test time, the input bias currents measured are correlated to junction temperature. In normal operation the junction temperature rises above the ambient temperature as a result of internal power dissipation, P_D. $T_j = T_A + \theta_{jA} P_D$ where θ_{jA} is the thermal resistance from junction to ambient. Use of a heat sink is recommended if input bias current is to be kept to a minimum.

Note 8: Supply voltage rejection ratio is measured for both supply magnitudes increasing or decreasing simultaneously in accordance with common practice. $V_S = \pm 6V$ to $\pm 15V$.

Note 9: Refer to RETS412AX for LF412AMH military specifications and to RETS412X for LF412MH military specifications.

Note 10: Max. Power Dissipation is defined by the package characteristics. Operating the part near the Max. Power Dissipation may cause the part to operate outside guaranteed limits.

Typical Performance Characteristics

TL/H/5656–2

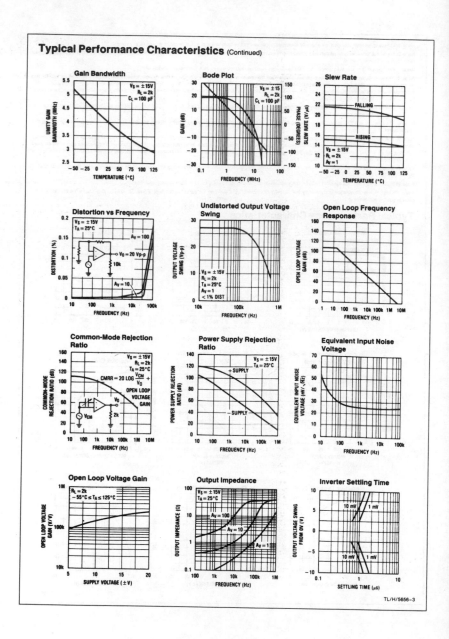

TL/H/5656-3

Pulse Response $R_L = 2$ kΩ, $C_L = 10$ pF

Small Signal Inverting

Small Signal Non-Inverting

Large Signal Inverting

Large Signal Non-Inverting

Current Limit ($R_L = 100\Omega$)

TL/H/5656-4

Application Hints

The LF412 series of JFET input dual op amps are internally trimmed (BI-FET II™) providing very low input offset voltages and guaranteed input offset voltage drift. These JFETs have large reverse breakdown voltages from gate to source and drain eliminating the need for clamps across the inputs. Therefore, large differential input voltages can easily be accommodated without a large increase in input current. The maximum differential input voltage is independent of the supply voltages. However, neither of the input voltages should be allowed to exceed the negative supply as this will cause large currents to flow which can result in a destroyed unit.

Exceeding the negative common-mode limit on either input will cause a reversal of the phase to the output and force the amplifier output to the corresponding high or low state.

Exceeding the negative common-mode limit on both inputs will force the amplifier output to a high state. In neither case does a latch occur since raising the input back within the common-mode range again puts the input stage and thus the amplifier in a normal operating mode.

Exceeding the positive common-mode limit on a single input will not change the phase of the output, however, if both inputs exceed the limit, the output of the amplifier may be forced to a high state.

The amplifiers will operate with a common-mode input voltage equal to the positive supply; however, the gain bandwidth and slew rate may be decreased in this condition. When the negative common-mode voltage swings to within 3V of the negative supply, an increase in input offset voltage may occur.

Application Hints (Continued)

Each amplifier is individually biased by a zener reference which allows normal circuit operation on ±6.0V power supplies. Supply voltages less than these may result in lower gain bandwidth and slew rate.

The amplifiers will drive a 2 kΩ load resistance to ±10V over the full temperature range. If the amplifier is forced to drive heavier load currents, however, an increase in input offset voltage may occur on the negative voltage swing and finally reach an active current limit on both positive and negative swings.

Precautions should be taken to ensure that the power supply for the integrated circuit never becomes reversed in polarity or that the unit is not inadvertently installed backwards in a socket as an unlimited current surge through the resulting forward diode within the IC could cause fusing of the internal conductors and result in a destroyed unit.

Because these amplifiers are JFET rather than MOSFET input op amps they do not require special handling.

As with most amplifiers, care should be taken with lead dress, component placement and supply decoupling in order to ensure stability. For example, resistors from the output to an input should be placed with the body close to the input to minimize "pick-up" and maximize the frequency of the feedback pole by minimizing the capacitance from the input to ground.

A feedback pole is created when the feedback around any amplifier is resistive. The parallel resistance and capacitance from the input of the device (usually the inverting input) to AC ground set the frequency of the pole. In many instances the frequency of this pole is much greater than the expected 3 dB frequency of the closed loop gain and consequently there is negligible effect on stability margin. However, if the feedback pole is less than approximately 6 times the expected 3 dB frequency a lead capacitor should be placed from the output to the input of the op amp. The value of the added capacitor should be such that the RC time constant of this capacitor and the resistance it parallels is greater than or equal to the original feedback pole time constant.

Typical Application

Single Supply Sample and Hold

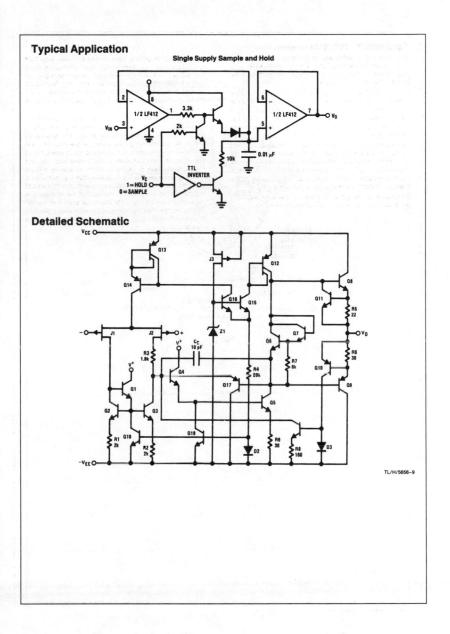

Detailed Schematic

TL/H/5656-9

501

Physical Dimensions inches (millimeters)

Metal Can Package (H)
Order Number LF412AMH, LF412MH, LF412ACH or LF412CH
NS Package Number H08B

Physical Dimensions inches (millimeters) (Continued)

Dual-In-Line Package (J)
Order Number LF412ACJ or LF412CJ
NS Package Number J08A

J08A (REV J)

© 1989 National Semiconductor Corporation

503

Physical Dimensions inches (millimeters) (Continued)

Dual-In-Line Package (N)
Order Number LF412ACN or LF412CN
NS Package Number N08E

LIFE SUPPORT POLICY

NATIONAL'S PRODUCTS ARE NOT AUTHORIZED FOR USE AS CRITICAL COMPONENTS IN LIFE SUPPORT DEVICES OR SYSTEMS WITHOUT THE EXPRESS WRITTEN APPROVAL OF THE PRESIDENT OF NATIONAL SEMICONDUCTOR CORPORATION. As used herein:

1. Life support devices or systems are devices or systems which, (a) are intended for surgical implant into the body, or (b) support or sustain life, and whose failure to perform, when properly used in accordance with instructions for use provided in the labeling, can be reasonably expected to result in a significant injury to the user.

2. A critical component is any component of a life support device or system whose failure to perform can be reasonably expected to cause the failure of the life support device or system, or to affect its safety or effectiveness.

National Semiconductor Corporation
2900 Semiconductor Drive
P.O. Box 58090
Santa Clara, CA 95052-8090
Tel: (408) 721-5000
TWX: (910) 339-9240

National Semiconductor GmbH
Westendstrasse 193-195
D-8000 Munchen 21
West Germany
Tel: (089) 5 70 95 01
Telex: 522772

NS Japan Ltd.
Sanseda Bldg. 5F
4-15 Nishi Shinjuku
Shinjuku-Ku,
Tokyo 160, Japan
Tel: 3-299-7001
FAX: 3-299-7000

National Semiconductor
Hong Kong Ltd.
Southeast Asia Marketing
Austin Tower, 4th Floor
22-26A Austin Avenue
Tsimshatsui, Kowloon, H.K.
Tel: 3-7231290, 3-7243645
Cable: NSSEAMKTG
Telex: 52996 NSSEA HX

National Semiconductores
Do Brasil Ltda.
Av. Brg. Faria Lima, 830
8 Andar
01452 Sao Paulo, SP. Brasil
Tel: (55/11) 212-5066
Telex: 391-1131931 NSBR BR

National Semiconductor
(Australia) PTY, Ltd.
21/3 High Street
Bayswater, Victoria 3153
Australia
Tel: (03) 729-6333
Telex: AA32096

504

APPENDIX A5.6

- Extremely Low Offset Voltage ... 1 µV Max

- Extremely Low Change in Offset Voltage with
  Temperature ... 0.003 µV/°C Typ

- Low Input Offset Current ... 500 pA Max
  at T_A = −55°C to 125°C

- AVD ... 135 dB Min

- CMRR and kSVR ... 120 dB Min

- Single-Supply Operation

- Common-Mode Input Voltage Range Includes
  the Negative Rail

- No Noise Degradation with External
  Capacitors Connected to V_{DD-}

description

The TLC2652 and TLC2652A are high-precision chopper-stabilized operational amplifiers using Texas Instruments **Advanced LinCMOS™** process. This process in conjunction with unique chopper-stabilization circuitry produces operational amplifiers whose performance matches or exceeds that of similar devices available today.

D008, JG, or P PACKAGE
(TOP VIEW)

```
       ___ ___
CXA [1  U  8] CXB
IN − [2     7] VDD +
IN + [3     6] OUT
VDD − [4    5] CLAMP
```

D014, J, or N PACKAGE
(TOP VIEW)

```
       ___ ___
CXB  [1  U  14] INT/EXT
CXA  [2     13] CLK IN
NC   [3     12] CLK OUT
IN − [4     11] VDD +
IN + [5     10] OUT
NC   [6      9] CLAMP
VDD − [7     8] C RETURN
```

FK PACKAGE
(TOP VIEW)

```
         CXA CXB NC INT/EXT CLK IN
          3   2  1   20    19
NC   [4                       18] CLK OUT
NC   [5                       17] NC
IN − [6                       16] VDD +
NC   [7                       15] NC
IN + [8                       14] OUT
          9  10  11  12  13
          NC VDD− NC C RETURN CLAMP
```

L PACKAGE
(TOP VIEW)

```
           CXB
      CXA  (1)(8) VDD +
IN − (2)      (6) OUT
   IN + (3)(4)(5) CLAMP
           VDD −
```

Pin 4 of the L package is in electrical contact with the case.

NC − No internal connection

AVAILABLE OPTIONS

T_A	V_{IO} max at 25°C	PACKAGE							
		8-PIN				14-PIN			20-PIN
		SMALL-OUTLINE (D008)	PLASTIC DIP (P)	CERAMIC DIP (JG)	METAL CAN (L)	SMALL-OUTLINE (D014)	PLASTIC DIP (N)	CERAMIC DIP (J)	CHIP CARRIER (FK)
0°C to 70°C	1 µV	TLC2652AC-8D	TLC2652ACP	TLC2652ACJG	TLC2652ACL	TLC2652AC-14D	TLC2652ACN	TLC2652ACJ	—
	3 µV	TLC2652C-8D	TLC2652CP	TLC2652CJG	TLC2652CL	TLC2652C-14D	TLC2652CN	TLC2652CJ	—
−40°C to 85°C	1 µV	TLC2652AI-8D	TLC2652AIP	TLC2652AIJG	TLC2652AIL	TLC2652AI-14D	TLC2652AIN	TLC2652AIJ	—
	3 µV	TLC2652I-8D	TLC2652IP	TLC2652IJG	TLC2652IL	TLC2652I-14D	TLC2652IN	TLC2652IJ	—
−55°C to 125°C	1 µV	TLC2652AM-8D	TLC2652AMP	TLC2652AMJG	TLC2652AML	TLC2652AM-14D	TLC2652AMN	TLC2652AMJ	TLC2652AMFK
	3 µV	TLC2652M-8D	TLC2652MP	TLC2652MJG	TLC2652ML	TLC2652M-14D	TLC2652MN	TLC2652MJ	TLC2652MFK

D008 and D014 packages are available taped and reeled. Add "R" suffix to device type when ordering (e.g., TLC2652AC-8DR).
Advanced LinCMOS is a trademark of Texas Instruments Incorporated.

TLC2652, TLC2652A
Advanced LinCMOS™ PRECISION CHOPPER-STABILIZED
OPERATIONAL AMPLIFIERS

description (continued)

Chopper stabilization techniques make possible extremely high dc precision by continuously nulling input offset voltage even during variations in temperature, time, common-mode voltage, and power supply voltage. In addition, low-frequency noise voltage is significantly reduced. This high precision, coupled with the extremely high input impedance of the CMOS input stage, makes the TLC2652 and TLC2652A an ideal choice for low-level signal processing applications such as strain gauges, thermocouples, and other transducer amplifiers. (For applications that require extremely low noise and higher usable bandwidth, use the TLC2654 or TLC2654A device, which has a chopping frequency of 10 kHz.)

The TLC2652 and TLC2652A input common-mode range includes the negative rail, thereby providing superior performance in either single-supply or split-supply applications, even at power supply voltage levels as low as ±1.9 V.

Two external capacitors are required for operation of the device; however, the on-chip chopper control circuitry is transparent to the user. On devices in the 14-pin and 20-pin packages, the control circuitry is made accessible to allow the user the option of controlling the clock frequency with an external frequency source. In addition, the clock threshold level of the TLC2652 and TLC2652A require no level shifting when used in the single-supply configuration with a normal CMOS or TTL clock input.

Innovative circuit techniques are used on the TLC2652 and TLC2652A to allow exceptionally fast overload recovery time. If desired, an output clamp pin is available to reduce the recovery time even further.

The device inputs and output are designed to withstand −100-mA surge currents without sustaining latchup. Additionally, the TLC2652 and TLC2652A incorporate internal ESD protection circuits that prevent functional failures at voltages up to 2000 V as tested under MIL-STD-883C, Method 3015.2; however, care should be exercised in handling these devices as exposure to ESD may result in degradation of the device parametric performance.

The M-suffix devices are characterized for operation over the full military temperature range of −55°C to 125°C. The I-suffix devices are characterized for operation from −40°C to 85°C. The C-suffix devices are characterized for operation from 0°C to 70°C.

functional block diagram

DISTRIBUTION OF TLC2652
INPUT OFFSET VOLTAGE

absolute maximum ratings over operating free-air temperature (unless otherwise noted)

Supply voltage, V_{DD+} (see Note 1) . . . . 8 V
Supply voltage, V_{DD-} (see Note 1) . . . . −8 V
Differential input voltage (see Note 2) . . . . ±16 V
Input voltage range, V_I (any input, see Note 1) . . . . ±8 V
Voltage on CLK IN and INT/$\overline{EXT}$ pins . . . . V_{DD-} to V_{DD-} +5.2 V
Input current, I_I (each input) . . . . ±5 mA
Output current, I_O . . . . ±50 mA
Duration of short-circuit current at (or below) 25°C (see Note 3) . . . . Unlimited
Current into CLK IN and INT/$\overline{EXT}$ pins . . . . ±5 mA
Continuous total dissipation . . . . See Dissipation Rating Table
Operating free-air temperature. T_A: M-suffix . . . . −55°C to 125°C
I-suffix . . . . −40°C to 85°C
C-suffix . . . . 0°C to 70°C
Storage temperature range . . . . −65°C to 150°C
Case temperature for 60 seconds: FK package . . . . 260°C
Lead temperature 1,6 mm (1/16 inch) from case for 10 seconds: D, N, or P package . . . . 260°C
Lead temperature 1,6 mm (1/16 inch) from case for 60 seconds. J, JG, or L package . . . . 300°C

NOTES: 1. All voltage values, except differential voltages, are with respect to the midpoint between V_{DD+} and V_{DD-}.
2. Differential voltages are at the noninverting input with respect to the inverting input.
3. The output may be shorted to either supply. Temperature and/or supply voltages must be limited to ensure that the maximum dissipation rating is not exceeded.

DISSIPATION RATING TABLE

PACKAGE	$T_A \leq 25°C$ POWER RATING	DERATING FACTOR ABOVE $T_A = 25°C$	$T_A = 70°C$ POWER RATING	$T_A = 85°C$ POWER RATING	$T_A = 125°C$ POWER RATING
D008	725 mW	5.8 mW/°C	464 mW	377 mW	145 mW
D014	950 mW	7.6 mW/°C	608 mW	494 mW	190 mW
FK	1375 mW	11 mW/°C	880 mW	715 mW	275 mW
J	1375 mW	11 mW/°C	880 mW	715 mW	275 mW
JG	1050 mW	8.4 mW/°C	672 mW	546 mW	210 mW
L	650 mW	5.2 mW/°C	416 mW	338 mW	130 mW
N	1575 mW	12.6 mW/°C	1008 mW	819 mW	315 mW
P	1000 mW	8.0 mW/°C	640 mW	520 mW	200 mW

recommended operating conditions

	M-SUFFIX			I-SUFFIX			C-SUFFIX			UNIT
	MIN	NOM	MAX	MIN	NOM	MAX	MIN	NOM	MAX	
Supply voltage. V_{DD}	±1.9		±8	±1.9		±8	±1.9		±8	V
Common-mode input voltage. V_{IC}	V_{DD-}		V_{DD+} −1.9	V_{DD-}		V_{DD+} −1.9	V_{DD-}		V_{DD+} −1.9	V
Clock Input voltage	V_{DD-}		V_{DD-} +5	V_{DD-}		V_{DD-} +5	V_{DD-}		V_{DD-} +5	V
Operating free-air temperature. T_A	−55		125	−40		85	0		70	C

electrical characteristics at specified free-air temperature, $V_{DD\pm}$ = ± 5 V (unless otherwise noted)

PARAMETER		TEST CONDITIONS[†]		TLC2652AM			TLC2652M			UNIT
				MIN	TYP	MAX	MIN	TYP	MAX	
V_{IO}	Input offset voltage		25°C		0.5	1		0.6	3	µV
			Full range			4			6	
α_{VIO}	Temperature coefficient of input offset voltage		−55°C to 125°C		0.003	0.03		0.003	0.03	µV/°C
	Input offset voltage long-term drift (see Note 4)	V_{IC} = 0, R_S = 50 Ω	25°C		0.003	0.02		0.003	0.06	µV/mo
I_{IO}	Input offset current		25°C		2			2		pA
			Full range			500			500	
I_{IB}	Input bias current		25°C		4			4		pA
			Full range			500			500	
V_{ICR}	Common-mode input voltage range	R_S = 50 Ω	Full range	−5 to 3.1			−5 to 3.1			V
V_{OM+}	Maximum positive peak output voltage swing	R_L = 10 kΩ, See Note 5	25°C	4.7	4.8		4.7	4.8		V
			Full range	4.7			4.7			
V_{OM-}	Maximum negative peak output voltage swing	R_L = 10 kΩ, See Note 5	25°C	−4.7	−4.9		−4.7	−4.9		V
			Full range	−4.7			−4.7			
A_{VD}	Large-signal differential voltage amplification	V_O = ± 4 V, R_L = 10 kΩ	25°C	135	150		120	150		dB
			Full range	120			120			
f_{ch}	Internal chopping frequency		25°C		450			450		Hz
	Clamp on-state current	R_L = 100 kΩ	25°C	25			25			µA
			Full range	25			25			
	Clamp off-state current	V_O = −4 V to 4 V	25°C			100			100	pA
			Full range			100			100	
CMRR	Common-mode rejection ratio	V_O = 0, V_{IC} = V_{ICR} min, R_S = 50 Ω	25°C	120	140		120	140		dB
			Full range	120			120			
k_{SVR}	Supply-voltage rejection ratio (Δ$V_{DD\pm}$ / ΔV_{IO})	$V_{DD\pm}$ = ± 1.9 V to ± 8 V, V_O = 0, R_S = 50 Ω	25°C	120	135		120	135		dB
			Full range	120			120			
I_{DD}	Supply current	V_O = 0, No load	25°C		1.5	2.4		1.5	2.4	mA
			Full range			2.5			2.5	

[†]Full range is −55°C to 125°C.

NOTES: 4. Typical values are based on the input offset voltage shift observed through 168 hours of operating life test at T_A = 150°C extrapolated to T_A = 25°C using the Arrhenius equation and assuming an activation energy of 0.96 eV.
5. Output clamp is not connected.

operating characteristics at specified free-air temperature, $V_{DD\pm} = \pm 5$ V (unless otherwise noted)

PARAMETER		TEST CONDITIONS[†]		TLC2652AM			TLC2652M			UNIT
				MIN	TYP	MAX	MIN	TYP	MAX	
SR +	Positive slew rate at unity gain	$V_O = \pm 2.3$ V, $R_L = 10$ kΩ, $C_L = 100$ pF	25°C	2	2.8		2	2.8		V/μs
			Full range	1.3			1.3			
SR −	Negative slew rate at unity gain		25°C	2.3	3.1		2.3	3.1		V/μs
			Full range	1.6			1.6			
V_n	Equivalent input noise voltage	f = 10 Hz	25°C		94			94		nV/√Hz
		f = 1 kHz	25°C		23			23		
V_{NPP}	Peak-to-peak equivalent input noise voltage	f = 0 to 1 Hz	25°C		0.8			0.8		μV
		f = 0 to 10 Hz	25°C		2.8			2.8		
I_n	Equivalent input noise current	f = 1 kHz	26°C		0.004			0.004		pA/√Hz
	Gain-bandwidth product	f = 10 kHz, $R_L = 10$ kΩ, $C_L = 100$ pF	25°C		1.9			1.9		MHz
ϕ_m	Phase margin at unity gain	$R_L = 10$ kΩ, $C_L = 100$ pF	25°C		48°			48°		

[†]Full range is − 55°C to 125°C.

electrical characteristics at specified free-air temperature, $V_{DD\pm} = \pm 5$ V (unless otherwise noted)

PARAMETER		TEST CONDITIONS[†]	TLC2652AI			TLC2652I			UNIT
			MIN	TYP	MAX	MIN	TYP	MAX	
V_{IO}	Input offset voltage	25°C		0.5	1		0.6	3	μV
		Full range			2.95			4.95	
α_{VIO}	Temperature coefficient of input offset voltage	$V_{IC} = 0$, $R_S = 50\ \Omega$ −40°C to 85°C		0.003	0.03		0.003	0.03	μV/°C
	Input offset voltage long-term drift (see Note 4)	25°C		0.003	0.02		0.003	0.06	μV/mo
I_{IO}	Input offset current	25°C		2			2		pA
		Full range			150			150	
I_{IB}	Input bias current	25°C		4			4		pA
		Full range			150			150	
V_{ICR}	Common-mode input voltage range	$R_S = 50\ \Omega$ Full range	−5 to 3.1			−5 to 3.1			V
V_{OM+}	Maximum positive peak output voltage swing	$R_L = 10\ k\Omega$, See Note 5 25°C	4.7	4.8		4.7	4.8		V
		Full range	4.7			4.7			
V_{OM-}	Maximum negative peak output voltage swing	$R_L = 10\ k\Omega$, See Note 5 25°C	−4.7	−4.9		−4.7	−4.9		V
		Full range	−4.7			−4.7			
A_{VD}	Large-signal differential voltage amplification	$V_O = \pm 4$ V, $R_L = 10\ k\Omega$ 25°C	135	150		120	150		dB
		Full range	125			120			
f_{ch}	Internal chopping frequency	25°C		450			450		Hz
	Clamp on-state current	$R_L = 100\ k\Omega$ 25°C	25			25			μA
		Full range	25			25			
	Clamp off-state current	$V_O = -4$ V to 4 V 25°C			100			100	pA
		Full range			100			100	
CMRR	Common-mode rejection ratio	$V_O = 0$, $V_{IC} = V_{ICR}$ min, $R_S = 50\ \Omega$ 25°C	120	140		120	140		dB
		Full range	120			120			
k_{SVR}	Supply-voltage rejection ratio ($\Delta V_{DD\pm}/\Delta V_{IO}$)	$V_{DD\pm} = \pm 1.9$ V to ± 8 V, $V_O = 0$, $R_S = 50\ \Omega$ 25°C	120	135		120	135		dB
		Full range	120			120			
I_{DD}	Supply current	$V_O = 0$, No load 25°C		1.5	2.4		1.5	2.4	mA
		Full range			2.5			2.5	

[†]Full range is −40°C to 85°C.

NOTES: 4. Typical values are based on the input offset voltage shift observed through 168 hours of operating life test at $T_A = 150°C$ extrapolated to $T_A = 25°C$ using the Arrhenius equation and assuming an activation energy of 0.96 eV.

5. Output clamp is not connected.

operating characteristics at specified free-air temperature, $V_{DD\pm} = \pm 5$ V (unless otherwise noted)

PARAMETER		TEST CONDITIONS[†]		TLC2652AI			TLC2652I			UNIT
				MIN	TYP	MAX	MIN	TYP	MAX	
SR +	Positive slew rate at unity gain	$V_O = \pm 2.3$ V, $R_L = 10$ kΩ, $C_L = 100$ pF	25°C	2	2.8		2	2.8		V/µs
			Full range	1.4			1.4			
SR −	Negative slew rate at unity gain		25°C	2.3	3.1		2.3	3.1		V/µs
			Full range	1.7			1.7			
V_n	Equivalent input noise voltage (see Note 6)	f = 10 Hz	25°C		94	140		94		nV/√Hz
		f = 1 kHz	25°C		23	35		23		
V_{NPP}	Peak-to-peak equivalent input noise voltage	f = 0 to 1 Hz	25°C		0.8			0.8		µV
		f = 0 to 10 Hz	25°C		2.8			2.8		
I_n	Equivalent input noise current	f = 1 kHz	25°C		0.004			0.004		pA/√Hz
	Gain-bandwidth product	f = 10 kHz, $R_L = 10$ kΩ, $C_L = 100$ pF	25°C		1.9			1.9		MHz
ϕ_m	Phase margin at unity gain	$R_L = 10$ kΩ, $C_L = 100$ pF	25°C		48°			48°		

[†]Full range is − 40°C to 85°C.
NOTE 6: This parameter is tested on a sample basis for the TLC2652A. For other test requirements, please contact the factory. This statement has no bearing on testing or nontesting of other parameters.

TLC2652C, TLC2652AC
Advanced LinCMOS™ PRECISION CHOPPER-STABILIZED
OPERATIONAL AMPLIFIERS

electrical characteristics at specified free-air temperature, $V_{DD\pm} = \pm 5$ V (unless otherwise noted)

PARAMETER		TEST CONDITIONS[†]		TLC2652AC			TLC2652C			UNIT
				MIN	TYP	MAX	MIN	TYP	MAX	
V_{IO}	Input offset voltage		25°C		0.5	1		0.6	3	μV
			Full range			2.35			4.35	
α_{VIO}	Temperature coefficient of input offset voltage		0°C to 70°C		0.003	0.03		0.003	0.03	μV/°C
	Input offset voltage long-term drift (see Note 4)	$V_{IC} = 0$, $R_S = 50\ \Omega$	25°C		0.003	0.02		0.003	0.06	μV/mo
I_{IO}	Input offset current		25°C		2			2		pA
			Full range			100			100	
I_{IB}	Input bias current		25°C		4			4		pA
			Full range			100			100	
V_{ICR}	Common-mode input voltage range	$R_S = 50\ \Omega$	Full range	−5 to 3.1			−5 to 3.1			V
V_{OM+}	Maximum positive peak output voltage swing	$R_L = 10$ kΩ, See Note 5	25°C	4.7	4.8		4.7	4.8		V
			Full range	4.7			4.7			
V_{OM-}	Maximum negative peak output voltage swing	$R_L = 10$ kΩ, See Note 5	25°C	−4.7	−4.9		−4.7	−4.9		V
			Full range	−4.7			−4.7			
A_{VD}	Large-signal differential voltage amplification	$V_O = \pm 4$ V, $R_L = 10$ kΩ	25°C	135	150		120	150		dB
			Full range	130			120			
f_{ch}	Internal chopping frequency		25°C		450			450		Hz
	Clamp on-state current	$R_L = 100$ kΩ	25°C	25			25			μA
			Full range	25			25			
	Clamp off-state current	$V_O = -4$ V to 4 V	25°C			100			100	pA
			Full range			100			100	
CMRR	Common-mode rejection ratio	$V_O = 0$, $V_{IC} = V_{ICR}$ min, $R_S = 50\ \Omega$	25°C	120	140		120	140		dB
			Full range	120			120			
k_{SVR}	Supply-voltage rejection ratio ($\Delta V_{DD\pm} / \Delta V_{IO}$)	$V_{DD\pm} = \pm 1.9$ V to ± 8 V, $V_O = 0$, $R_S = 50\ \Omega$	25°C	120	135		120	135		dB
			Full range	120			120			
I_{DD}	Supply current	$V_O = 0$, No load	25°C		1.5	2.4		1.5	2.4	mA
			Full range			2.5			2.5	

[†]Full range is 0°C to 70°C.

NOTES: 4. Typical values are based on the input offset voltage shift observed through 168 hours of operating life test at $T_A = 150$°C extrapolated to $T_A = 25$°C using the Arrhenius equation and assuming an activation energy of 0.96 eV.

5. Output clamp is not connected.

operating characteristics at specified free-air temperature, $V_{DD\pm} = \pm 5$ V (unless otherwise noted)

PARAMETER		TEST CONDITIONS[†]		TLC2652AC			TLC2652C			UNIT
				MIN	TYP	MAX	MIN	TYP	MAX	
SR +	Positive slew rate at unity gain	$V_O = \pm 2.3$ V, $R_L = 10$ kΩ, $C_L = 100$ pF	25°C	2	2.8		2	2.8		V/µs
			Full range	1.5			1.5			
SR –	Negative slew rate at unity gain		25°C	2.3	3.1		2.3	3.1		V/µs
			Full range	1.8			1.8			
V_n	Equivalent input noise voltage (see Note 6)	f = 10 Hz	25°C		94	140		94		nV/√Hz
		f = 1 kHz	25°C		23	35		23		
V_{NPP}	Peak-to-peak equivalent input noise voltage	f = 0 to 1 Hz	25°C		0.8			0.8		µV
		f = 0 to 10 Hz	25°C		2.8			2.8		
I_n	Equivalent input noise current	f = 1 kHz	25°C		0.004			0.004		pA/√Hz
	Gain-bandwidth product	f = 10 kHz, $R_L = 10$ kΩ, $C_L = 100$ pF	25°C		1.9			1.9		MHz
ϕ_m	Phase margin at unity gain	$R_L = 10$ kΩ, $C_L = 100$ pF	25°C		48°			48°		

[†]Full range is 0°C to 70°C.
NOTE 6: This parameter is tested on a sample basis for the TLC2652A. For other test requirements, please contact the factory. This statement has no bearing on testing or nontesting of other parameters.

TYPICAL CHARACTERISTICS

table of graphs

				FIGURE
V_{IO}	Normalized input offset voltage	vs	Chopping frequency	1
I_{IB}	Input bias current	vs	Common-mode input voltage	2
		vs	Chopping frequency	3
		vs	Temperature	4
I_{IO}	Input offset current	vs	Chopping frequency	5
		vs	Temperature	6
	Clamp current	vs	Output voltage	7
V_{OPP}	Maximum peak-to-peak output voltage swing	vs	Frequency	8
V_{OM}	Maximum peak output voltage swing	vs	Output current	9, 10
		vs	Temperature	11, 12
A_{VD}	Differential voltage amplification	vs	Frequency	13
		vs	Temperature	14
f_{ch}	Chopping frequency	vs	Supply voltage	15
		vs	Temperature	16
I_{DD}	Supply current	vs	Supply voltage	17
		vs	Temperature	18
I_{OS}	Short-circuit output current	vs	Supply voltage	19
		vs	Temperature	20
SR	Slew rate	vs	Supply voltage	21
		vs	Temperature	22
	Pulse response		Small-signal	23
			Large-signal	24
V_{NPP}	Peak-to-peak equivalent input noise voltage	vs	Chopping frequency	25, 26
V_n	Equivalent input noise voltage	vs	Frequency	27
	Gain-bandwidth product	vs	Supply voltage	28
		vs	Temperature	29
ϕ_m	Phase margin	vs	Supply voltage	30
		vs	Temperature	31
		vs	Load capacitance	32
	Phase shift	vs	Frequency	13

TYPICAL CHARACTERISTICS†

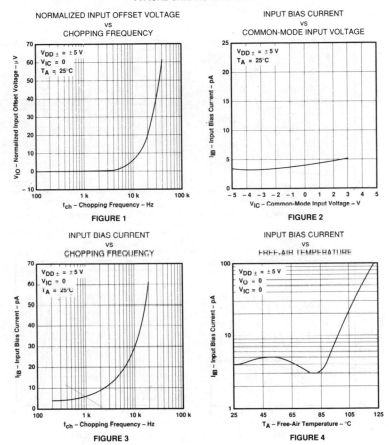

NORMALIZED INPUT OFFSET VOLTAGE
vs
CHOPPING FREQUENCY

FIGURE 1

INPUT BIAS CURRENT
vs
COMMON-MODE INPUT VOLTAGE

FIGURE 2

INPUT BIAS CURRENT
vs
CHOPPING FREQUENCY

FIGURE 3

INPUT BIAS CURRENT
vs
FREE-AIR TEMPERATURE

FIGURE 4

†Data at high and low temperatures are applicable only within the rated operating free-air temperature ranges of the various devices.

TYPICAL CHARACTERISTICS†

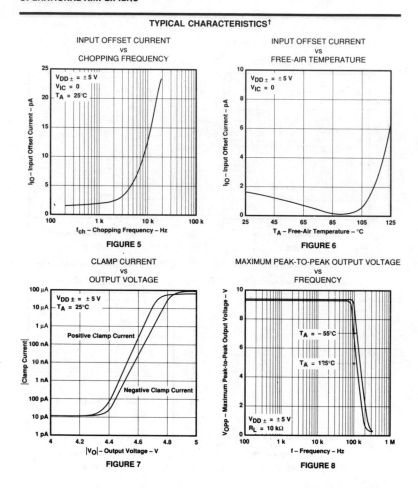

INPUT OFFSET CURRENT
vs
CHOPPING FREQUENCY

FIGURE 5

INPUT OFFSET CURRENT
vs
FREE-AIR TEMPERATURE

FIGURE 6

CLAMP CURRENT
vs
OUTPUT VOLTAGE

FIGURE 7

MAXIMUM PEAK-TO-PEAK OUTPUT VOLTAGE
vs
FREQUENCY

FIGURE 8

†Data at high and low temperatures are applicable only within the rated operating free-air temperature ranges of the various devices.

516

TYPICAL CHARACTERISTICS[†]

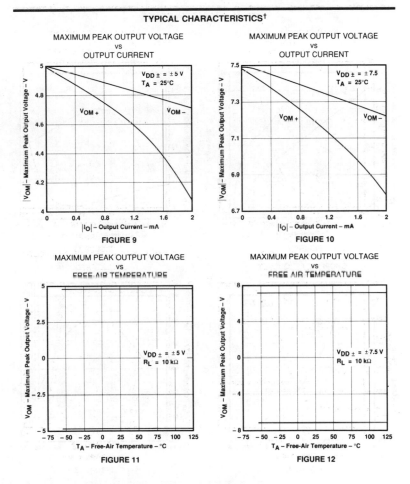

MAXIMUM PEAK OUTPUT VOLTAGE
vs
OUTPUT CURRENT

FIGURE 9

MAXIMUM PEAK OUTPUT VOLTAGE
vs
OUTPUT CURRENT

FIGURE 10

MAXIMUM PEAK OUTPUT VOLTAGE
vs
FREE-AIR TEMPERATURE

FIGURE 11

MAXIMUM PEAK OUTPUT VOLTAGE
vs
FREE-AIR TEMPERATURE

FIGURE 12

[†]Data at high and low temperatures are applicable only within the rated operating free-air temperature ranges of the various devices.

TLC2652, TLC2652A
Advanced LinCMOS™ PRECISION CHOPPER-STABILIZED
OPERATIONAL AMPLIFIERS

TYPICAL CHARACTERISTICS†

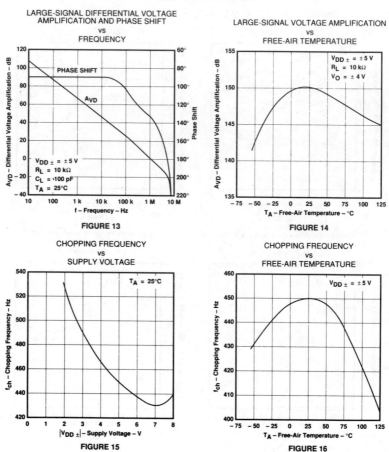

LARGE-SIGNAL DIFFERENTIAL VOLTAGE
AMPLIFICATION AND PHASE SHIFT
vs
FREQUENCY

FIGURE 13

LARGE-SIGNAL VOLTAGE AMPLIFICATION
vs
FREE-AIR TEMPERATURE

FIGURE 14

CHOPPING FREQUENCY
vs
SUPPLY VOLTAGE

FIGURE 15

CHOPPING FREQUENCY
vs
FREE-AIR TEMPERATURE

FIGURE 16

†Data at high and low temperatures are applicable only within the rated operating free-air temperature ranges of the various devices.

TYPICAL CHARACTERISTICS[†]

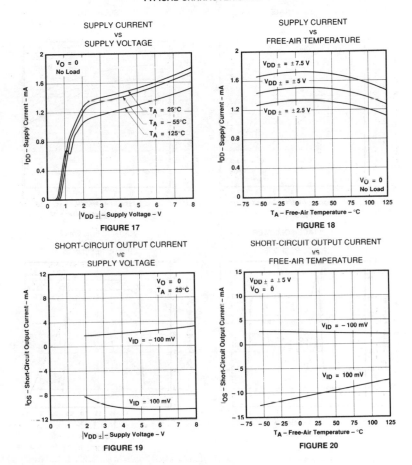

SUPPLY CURRENT
vs
SUPPLY VOLTAGE

FIGURE 17

SUPPLY CURRENT
vs
FREE-AIR TEMPERATURE

FIGURE 18

SHORT-CIRCUIT OUTPUT CURRENT
vs
SUPPLY VOLTAGE

FIGURE 19

SHORT-CIRCUIT OUTPUT CURRENT
vs
FREE-AIR TEMPERATURE

FIGURE 20

[†]Data at high and low temperatures are applicable only within the rated operating free-air temperature ranges of the various devices.

TLC2652, TLC2652A
Advanced LinCMOS™ PRECISION CHOPPER-STABILIZED OPERATIONAL AMPLIFIERS

TYPICAL CHARACTERISTICS†

SLEW RATE
vs
SUPPLY VOLTAGE

FIGURE 21

SLEW RATE
vs
FREE-AIR TEMPERATURE

FIGURE 22

VOLTAGE-FOLLOWER
SMALL-SIGNAL
PULSE RESPONSE

FIGURE 23

VOLTAGE-FOLLOWER
LARGE-SIGNAL
PULSE RESPONSE

FIGURE 24

†Data at high and low temperatures are applicable only within the rated operating free-air temperature ranges of the various devices.

520

TYPICAL CHARACTERISTICS

PEAK-TO-PEAK INPUT NOISE VOLTAGE
vs
CHOPPING FREQUENCY

FIGURE 25

PEAK-TO-PEAK INPUT NOISE VOLTAGE
vs
CHOPPING FREQUENCY

FIGURE 26

EQUIVALENT INPUT NOISE VOLTAGE
vs
FREQUENCY

FIGURE 27

GAIN-BANDWIDTH PRODUCT
vs
SUPPLY VOLTAGE

FIGURE 28

521

TYPICAL CHARACTERISTICS†

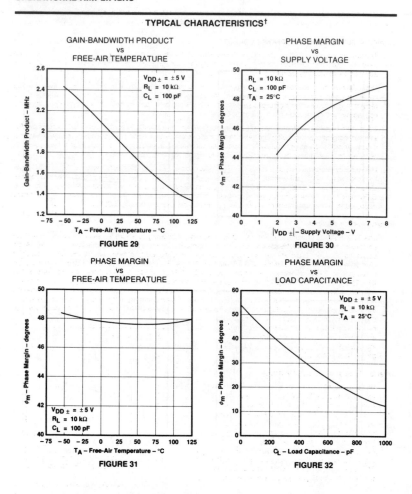

GAIN-BANDWIDTH PRODUCT
vs
FREE-AIR TEMPERATURE

FIGURE 29

PHASE MARGIN
vs
SUPPLY VOLTAGE

FIGURE 30

PHASE MARGIN
vs
FREE-AIR TEMPERATURE

FIGURE 31

PHASE MARGIN
vs
LOAD CAPACITANCE

FIGURE 32

†Data at high and low temperatures are applicable only within the rated operating free-air temperature ranges of the various devices.

522

TYPICAL APPLICATION DATA

capacitor selection and placement

The two important factors to consider when selecting external capacitors C_{XA} and C_{XB} are leakage and dielectric absorption. Both factors can cause system degradation that can negate the performance advantages realized by using the TLC2652.

Degradation from capacitor leakage becomes more apparent with increasing temperatures. Low-leakage capacitors and standoffs are recommended for operation at $T_A = 125°C$. In addition, guardbands are recommended around the capacitor connections on both sides of the printed circuit board to alleviate problems caused by surface leakage on circuit boards.

Capacitors with high dielectric absorption tend to take several seconds to settle upon application of power, which directly affects input offset voltage. In applications where fast settling of input offset voltage is needed, it is recommended that high-quality film capacitors, such as mylar, polystyrene, or polypropylene, be used. In other applications, however, a ceramic or other low-grade capacitor may suffice.

Unlike many choppers available today, the TLC2652 is designed to function with values of C_{XA} and C_{XB} in the range of 0.1 µF to 1 µF without degradation to input offset voltage or input noise voltage. These capacitors should be located as close as possible to the C_{XA} and C_{XB} pins and returned to either the V_{DD-} pin or the C RETURN pin. Note that on many choppers, connecting these capacitors to the V_{DD-} pin will cause degradation in the noise performance. This problem is eliminated on the TLC2652.

internal/external clock

The TLC2652 has an internal clock that sets the chopping frequency to a nominal value of 450 Hz. On 8-pin packages, the chopping frequency can only be controlled by the internal clock; however, on all 14-pin packages and the 20-pin FK package, the device chopping frequency may be set by the internal clock or controlled externally by use of the INT/$\overline{\text{EXT}}$ and CLK IN pins. To use the internal 450-Hz clock, no connection is necessary. If external clocking is desired, connect the INT/$\overline{\text{EXT}}$ pin to V_{DD-} and the external clock to CLK IN. The external clock trip point is 2.5 V above the negative rail; however, the CLK IN pin may be driven from the negative rail to 5 V above the negative rail. If this level is exceeded, damage could occur to the device unless the current into the CLK IN pin is limited to ±5 mA. When operating in the single-supply configuration, this feature allows the TLC2652 to be driven directly by 5-V TTL and CMOS logic. A divide-by-two frequency divider interfaces with the CLK IN pin and sets the chopping frequency. The chopping frequency appears on the CLK OUT pin. The duty cycle of the external clock is not critical but should be kept between 30% and 60%.

overload recovery/output clamp

When large differential input voltage conditions are applied to the TLC2652, the nulling loop will attempt to prevent the output from saturating by driving C_{XA} and C_{XB} to internally-clamped voltage levels. Once the overdrive condition is removed, a period of time is required to allow the built-up charge to dissipate. This time period is defined as overload recovery time (see Figure 33). Typical overload recovery time for the TLC2652 is significantly faster than competitive products; however, if required, this time can be

FIGURE 33. OVERLOAD RECOVERY

523

TYPICAL APPLICATION DATA

reduced further by using internal clamp circuitry accessible through the CLAMP pin.
The clamp is simply a switch that is automatically activated when the output is approximately 1 V from either supply rail. When connected to the inverting input (in parallel with the closed-loop feedback resistor), the closed-loop gain is reduced and the TLC2652 output is prevented from going into saturation. Since the output must source or sink current through the switch (see Figure 7), the maximum output voltage swing is slightly reduced.

thermoelectric effects

To take advantage of the extremely low offset voltage drift of the TLC2652, care must be taken to compensate for the thermoelectric effects present when two dissimilar metals are brought into contact with each other (such as device leads being soldered to a printed circuit board). Dissimilar metal junctions can produce thermoelectric voltages in the range of several microvolts per degree Celsius (orders of magnitude greater than the 0.003 μV/°C typical of the TLC2652).

To help minimize thermoelectric effects, careful attention should be paid to component selection and circuit board layout. Avoid the use of nonsoldered connections (such as sockets, relays, switches, etc.) in the input signal path. Cancel thermoelectric effects by duplicating the number of components and junctions in each device input. The use of low-thermoelectric-coefficient components, such as wire-wound resistors, is also beneficial.

latchup avoidance

Because CMOS devices are susceptable to latchup due to their inherent parasitic thyristors, the TLC2652 inputs and output were designed to withstand −100-mA surge currents without sustaining latchup; however, techniques to reduce the chance of latchup should be used whenever possible. Internal protection diodes should not, by design, be forward biased. Applied input and output voltages should not exceed the supply voltage by more than 300 mV. Care should be exercised when using capacitive coupling on pulse generators. Supply transients should be shunted by the use of decoupling capacitors (0.1 μF typical) located across the supply rails as close to the device as is possible.

The current path established if latchup occurs is usually between the supply rails and is limited only by the impedance of the power supply and the forward resistance of the parasitic thyristor. The chance of latchup occurring increases with increasing temperature and supply voltage.

electrostatic discharge protection

The TLC2652 incorporates internal ESD protection circuits that prevent functional failures at voltages at or below 2000 V. Care should be exercised in handling these devices as exposure to ESD may result in degradation of the device parametric performance.

theory of operation

Chopper-stabilized operational amplifiers offer the best dc performance of any monolithic operational amplifier. This superior performance is the result of using two operational amplifiers – a main amplifier and a nulling amplifier – plus oscillator-controlled logic and two external capacitors to create a system that behaves as a single amplifier. With this approach, the TLC2652 achieves submicrovolt input offset voltage, submicrovolt noise voltage, and offset voltage variations with temperature in the nV/°C range.

The TLC2652 on-chip control logic produces two dominant clock phases; a nulling phase and an amplifying phase. The term *chopper-stabilized* derives from the process of switching between these two clock phases. Figure 34 shows a simplified block diagram of the TLC2652. Switches A and B are make-before-break types. During the nulling phase, switch A is closed, shorting the nulling amplifier inputs together and allowing the

TYPICAL APPLICATION DATA

FIGURE 34. TLC2652 SIMPLIFIED BLOCK DIAGRAM

nulling amplifier to reduce its own input offset voltage by feeding its output signal back to an inverting input node. Simultaneously, external capacitor C_{XA} stores the nulling potential to allow the offset voltage of the amplifier to remain nulled during the amplifying phase.

During the amplifying phase, switch B is closed, connecting the output of the nulling amplifier to a noninverting input of the main amplifier. In this configuration, the input offset voltage of the main amplifier is nulled. Also, external capacitor C_{XB} stores the nulling potential to allow the offset voltage of the main amplifier to remain nulled during the next nulling phase.

This continuous chopping process allows offset voltage nulling during variations in time and temperature and over the common-mode input voltage range and power-supply range. In addition, because the low-frequency signal path is through both the null and main amplifiers, extremely high gain is achieved.

The low-frequency noise of a chopper amplifier depends on the magnitude of the component noise prior to chopping and the capability of the circuit to reduce this noise while chopping. The use of the Advanced LinCMOS process with its low-noise analog MOS transistors and patent pending input stage design significantly reduces the input noise voltage.

The primary source of nonideal operation in chopper-stabilized amplifiers is error charge from the switches.

As charge imbalance accumulates on critical nodes, input offset voltage can increase, especially with increasing chopping frequency. This problem has been significantly reduced in the TLC2652 by use of a patent-pending compensation circuit and the Advanced LinCMOS process.

The TLC2652 incorporates a feed-forward design that ensures continuous frequency response. Essentially, the gain magnitude of the nulling amplifier and compensation network crosses unity at the break frequency of the main amplifier. As a result, the high-frequency response of the system is the same as the frequency response of the main amplifier. This approach also ensures that the slewing characteristics remain the same during both the nulling and amplifying phases.

APPENDIX A5.7

TLE2024, TLE2024A, TLE2024B
EXCALIBUR HIGH-SPEED LOW-POWER PRECISION
QUAD OPERATIONAL AMPLIFIERS

D3265, MAY 1989

available features

- Supply Current ... 1 mA Max
- High Unity-Gain Bandwidth ... 2.8 MHz Typ
- High Slew Rate ... 0.7 V/μs Min
- Supply Current Change Over Military Temp Range ... 50 μA Typ
- Specified for Both 5 V/Gnd and ± 15 V Operation
- Phase-Reversal Protection

- High Open-Loop Gain ... 7 V/μV (137 dB) Typ
- Low Offset Voltage ... 500 μV Max
- Offset Voltage Drift with Time ... 0.005 μV/mo Typ
- Low Input Bias Current ... 50 nA Max
- Low Noise Voltage ... 19 nV/√Hz Typ at f = 10 Hz

description

The TLE2024, TLE2024A, and TLE2024B devices are precision, high-speed, low-power operational amplifiers using Texas Instruments patent-pending Excalibur process. These devices combine the best features of the OP421 with highly improved slew rate, unity-gain bandwidth, and input offset voltage.

The complementary bipolar Excalibur process utilizes isolated vertical P-N-P transistors that yield dramatic improvement in unity-gain bandwidth and slew rate over similar devices.

The addition of a patent-pending bias circuit in conjunction with this process results in extremely stable parameters with both time and temperature. This means that a "precision" device remains a precision device even with changes in temperature and over years of use.

This combination of excellent dc performance with a common-mode input voltage range that

LARGE-SIGNAL DIFFERENTIAL VOLTAGE
AMPLIFICATION and PHASE SHIFT
vs
FREQUENCY

AVAILABLE OPTIONS

T_A	V_{IO} max AT 25°C	PACKAGE			
		SMALL-OUTLINE (DW)	CHIP CARRIER (FK)	CERAMIC DIP (J)	PLASTIC DIP (N)
0°C to 70°C	500 μV	TLE2024BCDW	——	TLE2024BCJ	TLE2024BCN
	750 μV	TLE2024ACDW	——	TLE2024ACJ	TLE2024ACN
	1000 μV	TLE2024CDW	——	TLE2024CJ	TLE2024CN
– 40°C to 85°C	500 μV	TLE2024BIDW	——	TLE2024BIJ	TLE2024BIN
	750 μV	TLE2024AIDW	——	TLE2024AIJ	TLE2024AIN
	1000 μV	TLE2024IDW	——	TLE2024IJ	TLE2024IN
– 55°C to 125°C	500 μV	TLE2024BMDW	——	TLE2024BMJ	TLE2024BMN
	750 μV	TLE2024AMDW	TLE2024AMFK	TLE2024AMJ	TLE2024AMN
	1000 μV	TLE2024MDW	TLE2024MFK	TLE2024MJ	TLE2024MN

D packages are available taped and reeled. Add "R" suffix to device type (e.g.,TLE2024BCDWR).

description (continued)

includes the negative rail makes these devices the ideal choice for low-level signal conditioning applications in either single-supply or split-supply configurations. In addition, these devices offer phase-reversal protection circuitry that eliminates an unexpected change in output states when one of the inputs goes below the negative supply rail.

A variety of available packaging options includes small-outline and chip carrier versions for high-density systems applications.

The M-suffix devices are characterized for operation over the full military temperature range of −55°C to 125°C. The I-suffix devices are characterized for operation from −40°C to 85°C. The C-suffix devices are characterized for operation from 0°C to 70°C.

NC – No internal connection

equivalent schematic (each amplifier)

absolute maximum ratings over operating free-air temperature (unless otherwise noted)

Supply voltage, V_{CC+} (see Note 1) . . . . 20 V
Supply voltage, V_{CC-} (see Note 1) . . . . −20 V
Differential input voltage (see Note 2) . . . . ±0.6 V
Input voltage range, V_I (any input, see Note 1) . . . . $V_{CC±}$
Input current, I_I (each input) . . . . +1 mA
Output current, I_O (each output) . . . . ±40 mA
Total current into V_{CC+} terminal . . . . 80 mA
Total current out of V_{CC-} terminal . . . . 80 mA
Duration of short-circuit current at (or below) 25°C (see Note 3) . . . . unlimited
Continuous total dissipation . . . . See Dissipation Rating Table
Operating free-air temperature, T_A: M suffix . . . . −55°C to 125°C
                                       I-suffix . . . . −40°C to 85°C
                                       C-suffix . . . . 0°C to 70°C
Storage temperature range . . . . −65°C to 150°C
Case temperature for 60 seconds: FK package . . . . 260°C
Lead temperature 1,6 mm (1/16 inch) from case for 10 seconds: DW, or N package . . . . 260°C
Lead temperature 1,6 mm (1/16 inch) from case for 60 seconds: J package . . . . 300°C

NOTES: 1. All voltage values, except differential voltages, are with respect to the midpoint between V_{CC+} and V_{CC-}.
        2. Differential voltages are at the noninverting input with respect to the inverting input. Excessive current will flow if a differential input
           voltage in excess of approximately ± 600 mV is applied between the inputs unless some limiting resistance is used.
        3. The output may be shorted to either supply. Temperature and/or supply voltages must be limited to ensure that the maximum
           dissipation rating is not exceeded.

DISSIPATION RATING TABLE

PACKAGE	T_A ≤ 25°C POWER RATING	DERATING FACTOR ABOVE T_A = 25°C	T_A = 70°C POWER RATING	T_A = 85°C POWER RATING	T_A = 125°C POWER RATING
DW	950 mW	8.1 mW/°C	586 mW	464 mW	140 mW
FK	1375 mW	11.0 mW/°C	880 mW	715 mW	275 mW
J	1375 mW	11.0 mW/°C	880 mW	715 mW	275 mW
N	1575 mW	12.6 mW/°C	1008 mW	819 mW	315 mW

recommended operating conditions

		M-SUFFIX			I-SUFFIX			C-SUFFIX			UNIT
		MIN	NOM	MAX	MIN	NOM	MAX	MIN	NOM	MAX	
Supply voltage, V_{CC}		± 2		± 20	± 2		± 20	± 2		± 20	V
Common-mode input voltage, V_{IC}	V_{CC} = 5 V	0		3.2	0		3.2	0		3.5	V
	$V_{CC±}$ = ± 15 V	− 15		13.2	− 15		13.2	− 15		13.5	
Operating free-air temperature, T_A		− 55		125	− 40		85	0		70	°C

TLE2024M, TLE2024AM, TLE2024BM
EXCALIBUR HIGH-SPEED LOW-POWER PRECISION
QUAD OPERATIONAL AMPLIFIERS

electrical characteristics at specified free-air temperature, $V_{CC}\pm = \pm 15$ V (unless otherwise noted)

PARAMETER	TEST CONDITIONS	T_A†	TLE2024 MIN	TLE2024 TYP	TLE2024 MAX	TLE2024A MIN	TLE2024A TYP	TLE2024A MAX	TLE2024B MIN	TLE2024B TYP	TLE2024B MAX	UNIT
V_{IO} Input offset voltage	$V_{IC} = 0$, $R_S = 50\,\Omega$	25°C			1000			750			500	µV
		Full range			1200			950			700	
α_{VIO} Temperature coefficient of input offset voltage		Full range		2			2			2		µV/°C
Input offset voltage long-term drift (see Note 4)		25°C		0.006			0.006			0.006		µV/mo
I_{IO} Input offset current		25°C		0.6	6		0.5	5		0.4	4	nA
		Full range			10			9			8	
I_{IB} Input bias current		25°C		50	60		45	55		40	50	nA
		Full range			80			75			70	
V_{ICR} Common-mode input voltage range	$R_S = 50\,\Omega$	25°C	-15 to 13.5	14		-15 to 13.5	14		-15 to 13.5	14		V
		Full range	-15 to 13.2			-15 to 13.2			-15 to 13.2			
V_{OM+} Maximum positive peak output voltage swing	$R_L = 10\,k\Omega$	25°C	13.8	14.1		13.9	14.2		14	14.3		V
		Full range	13.7			13.7			13.8			
V_{OM-} Maximum negative peak output voltage swing	$R_L = 10\,k\Omega$	25°C	-13.7	-14.1		-13.7	-14.1		-13.7	-14.1		V
		Full range	-13.6			-13.6			-13.6			
A_{VD} Large-signal differential voltage amplification	$V_O = \pm 10$ V, $R_L = 10\,k\Omega$	25°C	0.4	2		0.8	4		1	7		V/µV
		Full range	0.4			0.8			1			
CMRR Common-mode rejection ratio	$V_{IC} = V_{ICR}$ min, $R_S = 50\,\Omega$	25°C	92	102		94	105		97	108		dB
		Full range	88			90			93			
k_{SVR} Supply-voltage rejection ratio ($\Delta V_{CC\pm} / \Delta V_{IO}$)	$V_{CC\pm} = \pm 2.5$ V to ± 15 V	25°C	98	112		100	115		103	117		dB
		Full range	93			95			98			
I_{CC} Supply current	$V_O = 0$, No load	25°C		940	1200		940	1200		940	1200	µA
		Full range			1200			1200			1200	
ΔI_{CC} Supply current change over operating temperature range		Full range		85			85			85		µA

† Full range is −55°C to 125°C

NOTE 4: Typical values are based on the input offset voltage shift observed through 168 hours of operating life test at $T_A = 150$°C extrapolated to $T_A = 25$°C using the Arrhenius equation and assuming an activation energy of 0.96 eV.

electrical characteristics at specified free-air temperature, V_{CC} = 5 V (unless otherwise noted)

PARAMETER	TEST CONDITIONS	T_A†	TLE2024 MIN	TLE2024 TYP	TLE2024 MAX	TLE2024A MIN	TLE2024A TYP	TLE2024A MAX	TLE2024B MIN	TLE2024B TYP	TLE2024B MAX	UNIT
V_{IO} Input offset voltage	V_{IC} = 0, R_S = 50 Ω	25°C		2	1100		2	850		2	600	µV
		Full range			1300			1050			800	
α_{VIO} Temperature coefficient of input offset voltage		Full range		2			2			2		µV/°C
Input offset voltage long-term drift (see Note 4)		25°C		0.005			0.005			0.005		µV/mo
I_{IO} Input offset current		25°C		0.6	6		0.5	5		0.4	4	nA
		Full range			10			9			8	
I_{IB} Input bias current		25°C		45	60		40	55		35	50	nA
		Full range			80			75			70	
V_{ICR} Common-mode input voltage range	R_S = 50 Ω	25°C	0 to 3.5	−0.3 to 4		0 to 3.5	−0.3 to 4		0 to 3.5	−0.3 to 4		V
		Full range	0 to 3.2			0 to 3.2			0 to 3.2			
V_{OH} High-level output voltage	R_L = 10 kΩ	25°C	3.9	4.2		3.9	4.2		4	4.3		V
		Full range	3.7			3.7			3.8			
V_{OL} Low-level output voltage		25°C		0.7	0.8		0.7	0.8		0.7	0.8	V
		Full range			0.95			0.95			0.95	
A_{VD} Large-signal differential voltage amplification	V_O = 1.4 V to 4 V, R_L = 10 kΩ	25°C	0.2	1.5		0.3	1.5		0.4	1.5		V/µV
		Full range	0.1			0.1			0.1			
$CMRR$ Common-mode rejection ratio	V_{IC} = V_{ICR} min, R_S = 50 Ω	25°C	80	90		82	92		85	95		dB
		Full range	80			82			85			
k_{SVR} Supply-voltage rejection ratio ($\Delta V_{CC}/\Delta V_{IO}$)	V_{CC} = 5 V to 30 V	25°C	98	112		100	115		103	117		dB
		Full range	93			95			98			
I_{CC} Supply current	V_O = 2.5 V, No load	25°C		700	1000		700	1300		700	1000	µA
		Full range			1000			1300			1000	
ΔI_{CC} Supply current change over operating temperature		Full range		50			50			50		µA

†Full range is −55°C to 125°C
NOTE 4: Typical values are based on the input offset voltage shift observed through 168 hours of operating life test at T_A = 150°C extrapolated to T_A = 25°C using the Arrhenius equation and assuming an activation energy of 0.96 eV.

TLE2024M, TLE2024AM, TLE2024BM
EXCALIBUR HIGH-SPEED LOW-POWER PRECISION
QUAD OPERATIONAL AMPLIFIERS

operating characteristics at specified free-air temperature, $V_{CC} \pm = \pm 15$ V (unless otherwise noted)

PARAMETER		TEST CONDITIONS	T_A†	TLE2024 MIN	TYP	MAX	TLE2024A MIN	TYP	MAX	TLE2024B MIN	TYP	MAX	UNIT
SR	Slew rate at unity gain	$V_O = \pm 10$ V, See Figure 1	25°C	0.7	0.9		0.7	0.9		0.7	0.9		V/µs
			Full range	0.6			0.6			0.6			
V_n	Equivalent input noise voltage (see Figure 2)	f = 10 Hz	25°C		19			19			19		nV/√Hz
		f = 1 kHz	25°C		15			15			15		
$V_{N(PP)}$	Peak-to-peak equivalent input noise voltage	f = 0.1 to 1 Hz	25°C		0.16			0.16			0.16		µV
		f = 0.1 to 10 Hz			0.47			0.47			0.47		
I_n	Equivalent input noise current		25°C		0.1			0.1			0.1		pA/√Hz
B_1	Unity-gain bandwidth	See Figure 3	25°C		2.8			2.8			2.8		MHz
ϕ_m	Phase margin at unity gain	See Figure 3	25°C		52°			52°			52°		

†Full range is −55°C to 125°C

operating characteristics at specified free-air temperature, $V_{CC} = 5$ V (unless otherwise noted)

PARAMETER		TEST CONDITIONS	T_A	TLE2024 MIN	TYP	MAX	TLE2024A MIN	TYP	MAX	TLE2024B MIN	TYP	MAX	UNIT
SR	Slew rate at unity gain	$V_O = 1$ V to 3 V, See Figure 1	25°C		0.8			0.8			0.8		V/µs
V_n	Equivalent input noise voltage (see Figure 2)	f = 10 Hz	25°C		21			21			21		nV/√Hz
		f = 1 kHz	25°C		17			17			17		
$V_{N(PP)}$	Peak-to-peak equivalent input noise voltage	f = 0.1 to 1 Hz	25°C		0.16			0.16			0.16		µV
		f = 0.1 to 10 Hz			0.47			0.47			0.47		
I_n	Equivalent input noise current		25°C		0.1			0.1			0.1		pA/√Hz
B_1	Unity-gain bandwidth	See Figure 3	25°C		1.7			1.7			1.7		MHz
ϕ_m	Phase margin at unity gain	See Figure 3	25°C		47°			47°			47°		

electrical characteristics at specified free-air temperature, $V_{CC}\pm = \pm 15$ V (unless otherwise noted)

PARAMETER		TEST CONDITIONS	T_A†	TLE2024 MIN	TYP	MAX	TLE2024A MIN	TYP	MAX	TLE2024B MIN	TYP	MAX	UNIT
V_{IO}	Input offset voltage		25°C			1000			750			500	μV
			Full range			1200			950			700	
α_{VIO}	Temperature coefficient of input offset voltage		Full range		2			2			2		μV/°C
	Input offset voltage long-term drift (see Note 4)	$V_{IC}=0,\ R_S=50\ \Omega$	25°C		0.006			0.006			0.006		μV/mo
I_{IO}	Input offset current		25°C		0.6	6		0.5	5		0.4	4	nA
			Full range			7			6			5	
I_{IB}	Input bias current		25°C		50	60		45	55		40	50	nA
			Full range			70			65			60	
V_{ICR}	Common-mode input voltage range	$R_S=50\ \Omega$	25°C	−15 to 13.5	−15.3 to 14		−15 to 13.5	−15.3 to 14		−15 to 13.5	−15.3 to 14		V
			Full range	−15 to 13.2			−15 to 13.2			−15 to 13.2			
V_{OM+}	Maximum positive peak output voltage swing	$R_L=10\ k\Omega$	25°C	13.8	14.		13.9	14.2		14	14.3		V
			Full range	13.7			13.7			13.8			
V_{OM-}	Maximum negative peak output voltage swing		25°C	−13.7	−14.		−13.7	−14.1		−13.7	−14.1		V
			Full range	−13.5			−13.6			−13.6			
A_{VD}	Large-signal differential voltage amplification	$V_O=\pm 10$ V, $R_L=10\ k\Omega$	25°C	0.4			0.8	4		1	7		V/μV
			Full range	0.4			0.8			1			
CMRR	Common-mode rejection ratio	$V_{IC}=V_{ICR}$ min, $R_S=50\ \Omega$	25°C	92	106		94	105		97	108		dB
			Full range	88			90			93			
k_{SVR}	Supply-voltage rejection ratio ($\Delta V_{CC}\pm/\Delta V_{IO}$)	$V_{CC}\pm=\pm 2.5$ V to ± 15 V	25°C	98	112		100	115		103	117		dB
			Full range	93			95			98			
I_{CC}	Supply current	$V_O=0$, No load	25°C		940	1200		940	1200		940	1200	μA
			Full range			1200			1200			1200	
ΔI_{CC}	Supply current change over operating temperature range		Full range		50			50			50		μA

†Full range is −40°C to 85°C

NOTE 4: Typical values are based on the input offset voltage shift observed through 168 hours of operating life test at $T_A = 150$°C extrapolated to $T_A = 25$°C using the Arrhenius equation and assuming an activation energy of 0.96 eV.

TLE2024I, TLE2024AI, TLE2024BI
EXCALIBUR HIGH-SPEED LOW-POWER PRECISION
QUAD OPERATIONAL AMPLIFIERS

electrical characteristics at specified free-air temperature, V_{CC} = 5 V (unless otherwise noted)

PARAMETER	TEST CONDITIONS	T_A†	TLE2024 MIN	TYP	MAX	TLE2024A MIN	TYP	MAX	TLE2022B MIN	TYP	MAX	UNIT
V_{IO} Input offset voltage		25°C			1100			850			600	μV
		Full range			1300			1050			800	
α_{VIO} Temperature coefficient of input offset voltage		Full range		2			2			2		μV/°C
Input offset voltage long-term drift (see Note 4)	V_{IC} = 0, R_S = 50 Ω	25°C		0.005			0.005			0.005		μV/mo
I_{IO} Input offset current		25°C		0.6	6		0.5	5		0.4	4	nA
		Full range			7			6			5	
I_{IB} Input bias current		25°C		45	60		40	55		35	50	nA
		Full range			70			65			60	
V_{ICR} Common-mode input voltage range	R_S = 50 Ω	25°C	0 to 3.5	−0.3 to 4		0 to 3.5	−0.3 to 4		0 to 3.5	−0.3 to 4		V
		Full range	0 to 3.2			0 to 3.2			0 to 3.2			
V_{OH} High-level output voltage	R_L = 10 kΩ	25°C	3.9	4.2		3.9	4.2		4	4.3		V
		Full range	3.7			3.7			3.8			
V_{OL} Low-level output voltage	V_O = 1.4 V to 4 V, R_L = 10 kΩ	25°C		0.7	0.8		0.7	0.8		0.7	0.8	V
		Full range			0.95			0.95			0.95	
A_{VD} Large-signal differential voltage amplification	V_O = 1.4 V to 4 V, R_L = 10 kΩ	25°C	0.2	1.5		0.3	1.5		0.4	1.5		V/μV
		Full range	0.1			0.1			0.1			
CMRR Common-mode rejection ratio	V_{IC} = V_{ICR} min, R_S = 50 Ω	25°C	80	90		82	92		85	95		dB
		Full range	80			82			85			
k_{SVR} Supply-voltage rejection ratio ($\Delta V_{CC}/\Delta V_{IO}$)	V_{CC} = 5 V to 30 V	25°C	98	112		100	115		103	117		dB
		Full range	93			95			98			
I_{CC} Supply current	V_O = 2.5 V, No load	25°C		700	1000		700	1000		700	1000	μA
		Full range			1000			1000			1000	
ΔI_{CC} Supply current change over operating temperature range		Full range		30			30			30		μA

† Full range is −40°C to 85°C

NOTE 4: Typical values are based on the input offset voltage shift observed through 168 hours of operating life test at T_A = 150°C extrapolated to T_A = 25°C using the Arrhenius equation and assuming an activation energy of 0.96 eV.

534

operating characteristics at specified free-air temperature, $V_{CC}\pm = \pm 15$ V (unless otherwise noted)

	PARAMETER	TEST CONDITIONS	T_A†	TLE2024			TLE2024A			TLE2024B			UNIT
				MIN	TYP	MAX	MIN	TYP	MAX	MIN	TYP	MAX	
SR	Slew rate at unity gain	$V_O = \pm 10$ V, See Figure 1	25°C	0.7	0.9		0.7	0.9		0.7	0.9		V/µs
			Full range	0.63			0.63			0.63			
V_n	Equivalent input noise voltage (see Figure 2)	f = 10 Hz	25°C		19	50		19	50		19	50	nV/√Hz
		f = 1 kHz	25°C		15	30		15	30		15	30	
$V_{N(PP)}$	Peak-to-peak equivalent input noise voltage	f = 0.1 to 1 Hz	25°C		0.16			0.16			0.16		µV
		f = 0.1 to 10 Hz	25°C		0.47			0.47			0.47		
I_n	Equivalent input noise current	See Figure 3	25°C		0.1			0.1			0.1		pA/√Hz
B_1	Unity-gain bandwidth	See Figure 3	25°C		2.8			2.8			2.8		MHz
ϕ_m	Phase margin at unity gain	See Figure 3	25°C		52°			52°			52°		

†Full range is −40°C to 85°C

operating characteristics at specified free-air temperature, $V_{CC}\pm = 5$ V (unless otherwise noted)

	PARAMETER	TEST CONDITIONS	T_A	TLE2024			TLE2024A			TLE2024B			UNIT
				MIN	TYP	MAX	MIN	TYP	MAX	MIN	TYP	MAX	
SR	Slew rate at unity gain	$V_O = 1$ V to 3 V, See Figure 1	25°C		0.8			0.8			0.8		V/µs
V_n	Equivalent input noise voltage (see Figure 2)	f = 10 Hz	25°C		21	50		21	50		21	50	nV/√Hz
		f = 1 kHz	25°C		17	30		17	30		17	30	
$V_{N(PP)}$	Peak-to-peak equivalent input noise voltage	f = 0.1 to 1 Hz	25°C		0.15			0.16			0.16		µV
		f = 0.1 to 10 Hz	25°C		0.47			0.47			0.47		
I_n	Equivalent input noise current	See Figure 3	25°C		0.1			0.1			0.1		pA/√Hz
B_1	Unity-gain bandwidth	See Figure 3	25°C		1.7			1.7			1.7		MHz
ϕ_m	Phase margin at unity gain	See Figure 3	25°C		47°			47°			47°		

TLE2024C, TLE2024AC, TLE2024BC
EXCALIBUR HIGH-SPEED LOW-POWER PRECISION
QUAD OPERATIONAL AMPLIFIERS

electrical characteristics at specified free-air temperature, $V_{CC\pm} = \pm 15$ V (unless otherwise noted)

PARAMETER		TEST CONDITIONS	T_A†	TLE2024 MIN	TYP	MAX	TLE2024A MIN	TYP	MAX	TLE2024B MIN	TYP	MAX	UNIT
V_{IO}	Input offset voltage		25°C			1000			750			500	µV
			Full range			1200			950			700	
α_{VIO}	Temperature coefficient of input offset voltage		Full range		2			2			2		µV/°C
	Input offset voltage long-term drift (see Note 4)	$V_{IC} = 0$, $R_S = 50\ \Omega$	25°C		0.006			0.006			0.006		µV/mo
I_{IO}	Input offset current	$V_{IC} = 0$,	25°C		0.6	6		0.5	5		0.4	4	nA
			Full range			6			5			4	
I_{IB}	Input bias current		25°C		50	60		45	55		40	50	nA
			Full range			65			60			55	
V_{ICR}	Common-mode input voltage range	$R_S = 50\ \Omega$	25°C	−15 to 13.5	−15.3 to 14		−15 to 13.5	−15.3 to 14		−15 to 13.5	−15.3 to 14		V
			Full range	−15 to 13.5			−15 to 13.5			−15 to 13.5			
V_{OM+}	Maximum positive peak output voltage swing	$R_L = 10\ k\Omega$	25°C	13.8	14.1		13.9	14.2		14	14.3		V
			Full range	13.7			13.8			13.9			
V_{OM-}	Maximum negative peak output voltage swing		25°C	−13.7	−14.1		−13.7	−14.1		−13.7	−14.1		V
			Full range	−13.6			−13.6			−13.6			
A_{VD}	Large-signal differential voltage amplification	$V_O = \pm 10$ V, $R_L = 10\ k\Omega$	25°C	0.4	2		0.8	4		1	7		V/µV
			Full range	0.4			0.8			1			
CMRR	Common-mode rejection ratio	$V_{IC} = V_{ICR}$ min, $R_S = 50\ \Omega$	25°C	92	102		94	105		97	108		dB
			Full range	88			90			93			
k_{SVR}	Supply-voltage rejection ratio ($\Delta V_{CC\pm} / \Delta V_{IO}$)	$V_{CC\pm} = \pm 2.5$ V to ± 15 V	25°C	98	112		100	115		103	117		dB
			Full range	93			95			98			
I_{CC}	Supply current	$V_O = 0$, No load	25°C		940	1200		940	1200		940	1200	µA
			Full range			1200			1200			1200	
ΔI_{CC}	Supply current change over operating temperature range		Full range		20			20			20		µA

†Full range is 0°C to 70°C

NOTE 4: Typical values are based on the input offset voltage shift observed through 168 hours of operating life test at T_A = 150°C extrapolated to T_A = 25°C using the Arrhenius equation and assuming an activation energy of 0.96 eV.

electrical characteristics at specified free-air temperature, V_{CC} = 5 V (unless otherwise noted)

PARAMETER		TEST CONDITIONS	T_A†	TLE2024			TLE2024A			TLE2024B			UNIT
				MIN	TYP	MAX	MIN	TYP	MAX	MIN	TYP	MAX	
V_{IO}	Input offset voltage		25°C			1100			850			600	µV
			Full range			1300			1050			800	
α_{VIO}	Temperature coefficient of input offset voltage		Full range		2			2			2		µV/°C
	Input offset voltage long-term drift (see Note 4)	V_{IC} = 0, R_S = 50 Ω	25°C		0.005			0.005			0.005		µV/mo
I_{IO}	Input offset current		25°C		0.6	6		0.5	5		0.4	4	nA
			Full range			6			5			4	
I_{IB}	Input bias current		25°C		45	60		40	55		35	50	nA
			Full range			65			60			55	
V_{ICR}	Common-mode input voltage range	R_S = 50 Ω	25°C	0 to 3.5	-0.3 to 4		0 to 3.5	-0.3 to 4		0 to 3.5	-0.3 to 4		V
			Full range	0 to 3.5			0 to 3.5			0 to 3.5			
V_{OH}	High-level output voltage	R_L = 10 kΩ	25°C	3.9	4.2		3.9	4.2		4	4.3		V
			Full range	3.7			3.7			3.8			
V_{OL}	Low-level output voltage		25°C		0.7	0.8		0.7	0.8		0.7	0.8	V
			Full range			0.95			0.95			0.95	
A_{VD}	Large-signal differential voltage amplification	V_O = 1.4 V to 4 V, R_L = 10 kΩ	25°C	0.2	1.5		0.3	1.5		0.4	1.5		V/µV
			Full range	0.1			0.1			0.1			
CMRR	Common-mode rejection ratio	V_{IC} = V_{ICR} min, R_S = 50 Ω	25°C	80	90		82	92		85	95		dB
			Full range	80			82			85			
k_{SVR}	Supply-voltage rejection ratio (ΔV$_{CC}$ / ΔV$_{IO}$)	V_{CC} = 5 V to 30 V	25°C	98	112		100	115		103	117		dB
			Full range	93			95			98			
I_{CC}	Supply current	V_O = 2.5 V, No load	25°C		700	1000		700	1000		700	1000	µA
			Full range			1000			1000			1000	
ΔI_{CC}	Supply current change over operating temperature range		Full range		15			15			15		µA

† Full range is 0°C to 70°C

NOTE 4: Typical values are based on the input offset voltage shift observed through 168 hours of operating life test at T_A = 150°C extrapolated to T_A = 25°C using the Arrhenius equation and assuming an activation energy of 0.96 eV.

537

operating characteristics at specified free-air temperature, $V_{CC}\pm$ = ± 15 V (unless otherwise noted)

PARAMETER		TEST CONDITIONS	T_A†	TLE2024			TLE2024A			TLE2024B			UNIT
				MIN	TYP	MAX	MIN	TYP	MAX	MIN	TYP	MAX	
SR	Slew rate at unity gain	V_O = ±10 V, See Figure 1	25°C	0.7	0.9		0.7	0.9		0.7	0.9		V/μs
			Full range	0.65			0.65			0.65			
V_n	Equivalent input noise voltage (see Figure 2)	f = 10 Hz	25°C		19	50		19	50		19	50	nV/√Hz
		f = 1 kHz	25°C		15	30		15	30		15	30	
$V_{N(PP)}$	Peak-to-peak equivalent input noise voltage	f = 0.1 to 1 Hz	25°C		0.16			0.16			0.16		μV
		f = 0.1 to 10 Hz	25°C		0.47			0.47			0.47		
I_n	Equivalent input noise current		25°C		0.1			0.1			0.1		pA/√Hz
B_1	Unity-gain bandwidth	See Figure 3	25°C		2.8			2.8			2.8		MHz
ϕ_m	Phase margin at unity gain	See Figure 3	25°C		52°			52°			52°		

†Full range is 0°C to 70°C

operating characteristics at specified free-air temperature, V_{CC} = 5 V (unless otherwise noted)

PARAMETER		TEST CONDITIONS	T_A	TLE2024			TLE2024A			TLE2024B			UNIT
				MIN	TYP	MAX	MIN	TYP	MAX	MIN	TYP	MAX	
SR	Slew rate at unity gain	V_O = 1 V to 3 V, See Figure 1	25°C		0.8			0.8			0.8		V/μs
V_n	Equivalent input noise voltage (see Figure 2)	f = 10 Hz	25°C		21	50		21	50		21	50	nV/√Hz
		f = 1 kHz	25°C		17	30		17	30		17	30	
$V_{N(PP)}$	Peak-to-peak equivalent input noise voltage	f = 0.1 to 1 Hz	25°C		0.16			0.16			0.16		μV
		f = 0.1 to 10 Hz	25°C		0.47			0.47			0.47		
I_n	Equivalent input noise current		25°C		0.1			0.1			0.1		pA/√Hz
B_1	Unity-gain bandwidth	See Figure 3	25°C		1.7			1.7			1.7		MHz
ϕ_m	Phase margin at unity gain	See Figure 3	25°C		47°			47°			47°		

PARAMETER MEASUREMENT INFORMATION

(a) SINGLE-SUPPLY

(b) SPLIT-SUPPLY

NOTE A: C_L includes fixture capacitance.

FIGURE 1. SLEW RATE TEST CIRCUIT

(a) SINGLE-SUPPLY

(b) SPLIT-SUPPLY

FIGURE 2. NOISE VOLTAGE TEST CIRCUIT

NOTE A: C_L includes fixture capacitance.

(a) SINGLE-SUPPLY

(b) SPLIT-SUPPLY

FIGURE 3. UNITY-GAIN BANDWIDTH AND PHASE MARGIN TEST CIRCUIT

539

PARAMETER MEASUREMENT INFORMATION

NOTE A: C_L includes fixture capacitance.

(a) SINGLE-SUPPLY   (b) SPLIT-SUPPLY

FIGURE 4. SMALL-SIGNAL PULSE RESPONSE TEST CIRCUIT

typical values

Typical values as presented in this data sheet represent the median (50% point) of device parametric performance.

TYPICAL CHARACTERISTICS

table of graphs

				FIGURE
V_{IO}	Input offset voltage		Distribution	5
I_{IB}	Input bias current	vs	Common-mode voltage	6
		vs	Temperature	7
I_I	Input current	vs	Differential input voltage	8
V_{OM}	Maximum peak output voltage	vs	Output current	9
		vs	Temperature	10
V_{OH}	High-level output voltage	vs	High-level output current	11
		vs	Temperature	12
V_{OL}	Low-level output voltage	vs	Low-level output current	13
		vs	Temperature	14
$V_{O(PP)}$	Maximum peak-to-peak output voltage swing	vs	Frequency	15, 16
A_{VD}	Differential voltage amplification	vs	Frequency	17
		vs	Temperature	18
I_{OS}	Short-circuit output current	vs	Supply voltage	19, 21
		vs	Temperature	20, 22
I_{CC}	Supply current	vs	Supply voltage	23
		vs	Temperature	24
CMRR	Common-mode rejection ratio	vs	Frequency	25
SR	Slew rate	vs	Temperature	26
	Pulse response		Small-signal	27, 28
			Large-signal	29, 30
$V_{N(PP)}$	Peak-to-peak equivalent input noise voltage		0.1 to 1 Hz	31
			0.1 to 10 Hz	32
V_n	Equivalent input noise voltage	vs	Frequency	33
B_1	Unity-gain bandwidth	vs	Supply voltage	34
		vs	Temperature	35
ϕ_m	Phase margin	vs	Supply voltage	36
		vs	Capacitive load	37
		vs	Temperature	38
	Phase shift	vs	Frequency	17

TYPICAL CHARACTERISTICS

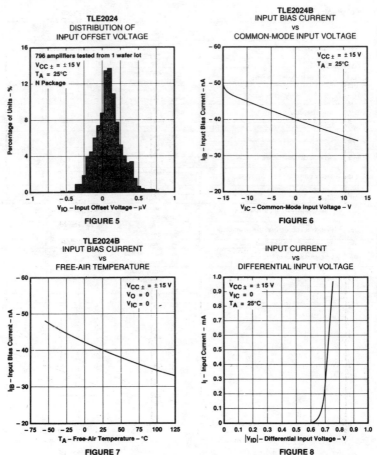

**TLE2024
DISTRIBUTION OF
INPUT OFFSET VOLTAGE**

796 amplifiers tested from 1 wafer lot
$V_{CC\pm} = \pm 15$ V
$T_A = 25°C$
N Package

V_{IO} – Input Offset Voltage – µV

FIGURE 5

**TLE2024B
INPUT BIAS CURRENT
vs
COMMON-MODE INPUT VOLTAGE**

$V_{CC\pm} = \pm 15$ V
$T_A = 25°C$

V_{IC} – Common-Mode Input Voltage – V

FIGURE 6

**TLE2024B
INPUT BIAS CURRENT
vs
FREE-AIR TEMPERATURE**

$V_{CC\pm} = \pm 15$ V
$V_O = 0$
$V_{IC} = 0$

T_A – Free-Air Temperature – °C

FIGURE 7

**INPUT CURRENT
vs
DIFFERENTIAL INPUT VOLTAGE**

$V_{CC\pm} = \pm 15$ V
$V_{IC} = 0$
$T_A = 25°C$

$|V_{ID}|$ – Differential Input Voltage – V

FIGURE 8

†Data at high and low temperatures are applicable only within the rated operating free-air temperature ranges of the various devices.

542

TYPICAL CHARACTERISTICS[†]

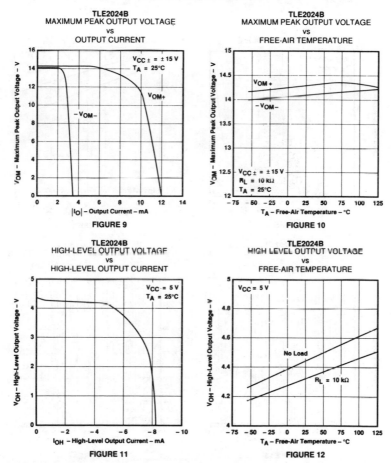

TLE2024B
MAXIMUM PEAK OUTPUT VOLTAGE
vs
OUTPUT CURRENT

FIGURE 9

TLE2024B
MAXIMUM PEAK OUTPUT VOLTAGE
vs
FREE-AIR TEMPERATURE

FIGURE 10

TLE2024B
HIGH-LEVEL OUTPUT VOLTAGE
vs
HIGH-LEVEL OUTPUT CURRENT

FIGURE 11

TLE2024B
HIGH-LEVEL OUTPUT VOLTAGE
vs
FREE-AIR TEMPERATURE

FIGURE 12

[†]Data at high and low temperatures are applicable only within the rated operating free-air temperature ranges of the various devices.

TYPICAL CHARACTERISTICS†

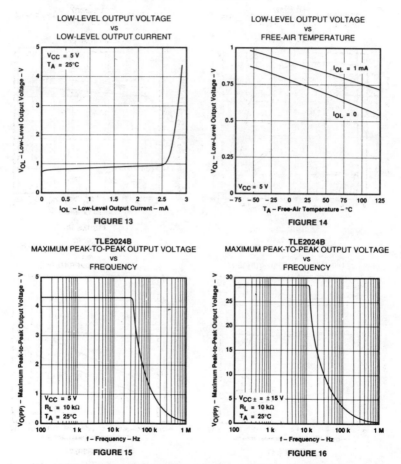

LOW-LEVEL OUTPUT VOLTAGE
vs
LOW-LEVEL OUTPUT CURRENT

FIGURE 13

LOW-LEVEL OUTPUT VOLTAGE
vs
FREE-AIR TEMPERATURE

FIGURE 14

TLE2024B
MAXIMUM PEAK-TO-PEAK OUTPUT VOLTAGE
vs
FREQUENCY

FIGURE 15

TLE2024B
MAXIMUM PEAK-TO-PEAK OUTPUT VOLTAGE
vs
FREQUENCY

FIGURE 16

†Data at high and low temperatures are applicable only within the rated operating free-air temperature ranges of the various devices.

TYPICAL CHARACTERISTICS†

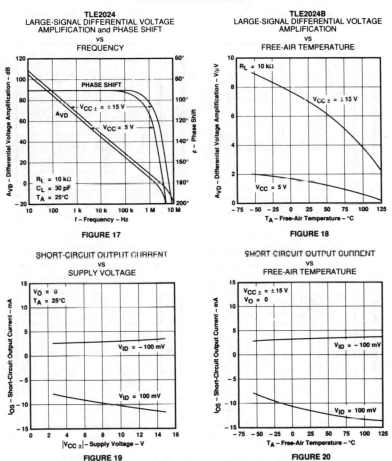

TLE2024
LARGE-SIGNAL DIFFERENTIAL VOLTAGE
AMPLIFICATION and PHASE SHIFT
vs
FREQUENCY

FIGURE 17

TLE2024B
LARGE-SIGNAL DIFFERENTIAL VOLTAGE
AMPLIFICATION
vs
FREE-AIR TEMPERATURE

FIGURE 18

SHORT-CIRCUIT OUTPUT CURRENT
vs
SUPPLY VOLTAGE

FIGURE 19

SHORT-CIRCUIT OUTPUT CURRENT
vs
FREE-AIR TEMPERATURE

FIGURE 20

†Data at high and low temperatures are applicable only within the rated operating free-air temperature ranges of the various devices.

TYPICAL CHARACTERISTICS†

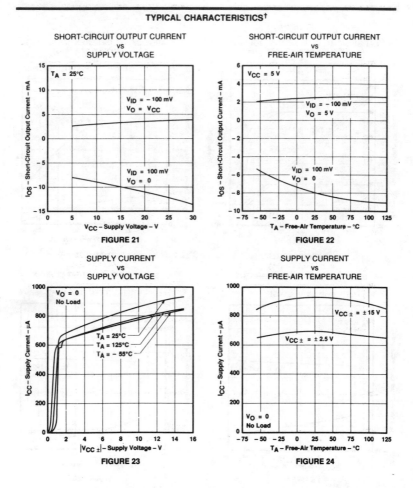

SHORT-CIRCUIT OUTPUT CURRENT
vs
SUPPLY VOLTAGE

FIGURE 21

SHORT-CIRCUIT OUTPUT CURRENT
vs
FREE-AIR TEMPERATURE

FIGURE 22

SUPPLY CURRENT
vs
SUPPLY VOLTAGE

FIGURE 23

SUPPLY CURRENT
vs
FREE-AIR TEMPERATURE

FIGURE 24

†Data at high and low temperatures are applicable only within the rated operating free-air temperature ranges of the various devices.

TYPICAL CHARACTERISTICS[†]

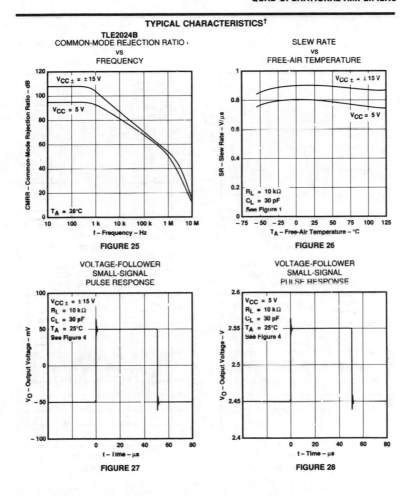

TLE2024B
COMMON-MODE REJECTION RATIO
vs
FREQUENCY

FIGURE 25

SLEW RATE
vs
FREE-AIR TEMPERATURE

FIGURE 26

VOLTAGE-FOLLOWER
SMALL-SIGNAL
PULSE RESPONSE

FIGURE 27

VOLTAGE-FOLLOWER
SMALL-SIGNAL
PULSE RESPONSE

FIGURE 28

[†]Data at high and low temperatures are applicable only within the rated operating free-air temperature ranges of the various devices.

TYPICAL CHARACTERISTICS

VOLTAGE-FOLLOWER
LARGE-SIGNAL
PULSE RESPONSE

$V_{CC\pm} = \pm 15$ V
$R_L = 10$ kΩ
$C_L = 30$ pF
$T_A = 25°C$
See Figure 1

V_O – Output Voltage – V

t – Time – µs

FIGURE 29

VOLTAGE-FOLLOWER
LARGE-SIGNAL
PULSE RESPONSE

$V_{CC} = 5$ V
$R_L = 10$ kΩ
$C_L = 30$ pF
$T_A = 25°C$
See Figure 1

V_O – Output Voltage – V

t – Time – µs

FIGURE 30

PEAK-TO-PEAK EQUIVALENT
INPUT NOISE VOLTAGE
0.1 TO 1 Hz

$V_{CC\pm} = \pm 15$ V
$T_A = 25°C$

$V_{N(PP)}$ – Peak-to-Peak Input Noise Voltage – µV

t – Time – s

FIGURE 31

PEAK-TO-PEAK EQUIVALENT
INPUT NOISE VOLTAGE
0.1 TO 10 Hz

$V_{CC\pm} = \pm 15$ V
$T_A = 25°C$

$V_{N(PP)}$ – Peak-to-Peak Input Noise Voltage – µV

t – Time – s

FIGURE 32

TYPICAL CHARACTERISTICS†

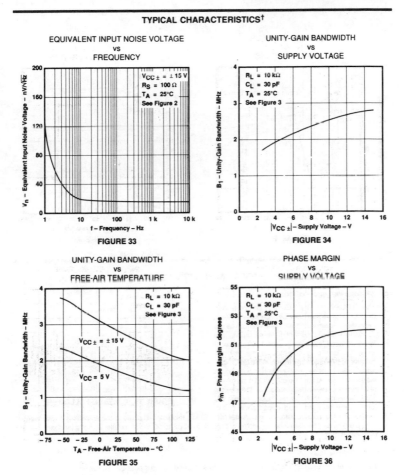

EQUIVALENT INPUT NOISE VOLTAGE
vs
FREQUENCY

FIGURE 33

UNITY-GAIN BANDWIDTH
vs
SUPPLY VOLTAGE

FIGURE 34

UNITY-GAIN BANDWIDTH
vs
FREE-AIR TEMPERATURE

FIGURE 35

PHASE MARGIN
vs
SUPPLY VOLTAGE

FIGURE 36

†Data at high and low temperatures are applicable only within the rated operating free-air temperature ranges of the various devices.

549

TYPICAL CHARACTERISTICS†

PHASE MARGIN
vs
LOAD CAPACITANCE

FIGURE 37

PHASE MARGIN
vs
FREE-AIR TEMPERATURE

FIGURE 38

†Data at high and low temperatures are applicable only within the rated operating free-air temperature ranges of the various devices.

TYPICAL APPLICATION DATA

voltage-follower applications

The TLE2024 circuitry includes input protection diodes to limit the voltage across the input transistors; however, no provision is made in the circuit to limit the current if these diodes are forward-biased. Note that this condition can occur when the device is operated in the voltage-follower configuration and driven with a fast, large-signal pulse. It is recommended that a feedback resistor be used to limit the current to a maximum of 1 mA to prevent degradation of the device. Also, remember that this feedback resistor will form a pole with the input capacitance of the

FIGURE 39. VOLTAGE-FOLLOWER

device. For feedback resistor values greater than 10 kΩ, this pole will degrade the amplifier's phase margin. This problem can be alleviated by adding a capacitor (20 pF to 50 pF) in parallel with the feedback resistor (see Figure 39).

input characteristics

The input of any unused amplifiers should be tied to ground to avoid possible oscillation.

ANSWERS TO EXERCISES

Chapter 1

1.1   (a) Circuit of Figure 1.2(a) with R_1 = 100 kΩ,
       R_2 = 500 kΩ
    (b) Circuit of Figure 1.2(a) with R_1 = 2 kΩ,
       R_2 = 40 kΩ
    (c) Circuit of Figure 1.2(b) with R_2/R_1 = 99
    (d) Circuit of Figure 1.8 with R = 100 kΩ, C = 0.1 μF
    (e) Circuit of Figure 1.5 with R_1 = 400 Ω

1.2   (a) -6 V, (b) $+6$ V, (c) -2 V, (d) -2 V, (e) $+6$ V,
    (f) $+10$ V, (g) -1 V

Chapter 2

2.1   (a) R_{in} = 10 kΩ, R_f = 100 kΩ
    (b) 1%, (c) 0.05%

2.2   5, 10^9 Ω

2.3   1 mA, 0.02%, 50 MΩ

2.4 $\beta = R_s/(R_s + R_f)$, $1/\beta = 1 + R_f/R_s$, where R_s is the
    resistance of the current source. If $R_s \gg \infty$ $\beta \gg 1$,
    $1/\beta \to 1$
    $\beta - R_p/(R_p + R_f)$, $1/\beta = 1 + R_f/R_p$, where
    $R_p = R_1 // R_2 // R_3$
    $\beta = 1$, $1/\beta = 1$
    $\beta = R_1/(R_1 + R_2)$, $1/\beta = 1 + R_2/R_1$
    $\beta = R/(R + 1/[j\omega C])$, $1/\beta = 1 + 1/j\omega CR$

2.5   (a) 0 dB, (b) 6 dB, (c) 10 dB, (d) 20 dB, (e) 40 dB,
    (f) 60 dB, (g) 120 dB

2.6   (a) 16 dB, (b) 24 dB, (c) 10 dB, (d) 50 dB, (e) 20 dB,
    (f) -40 dB, (g) -26 dB, (h) 3 dB, (i) -3 dB

2.7   (a) 10 mV, 84.3° lead
    (b) 70.7 mV, 45° lead
    (c) 100 mV, 5.7° lead

2.8 $\dfrac{1}{1 + j\omega CR}$, $\dfrac{1}{1 + \dfrac{1}{j\omega CR}}$, $\dfrac{1 + j\omega CR_2}{1 + j\omega C(R_1 + R_2)}$

2.10 (a) (i) 500 kHz, (ii) 100 kHz, (iii) 20 kHz
     (b) (i) $e = -(4e_1 + 3e_2 + 2e_3)$, (ii) $1/\beta$ = 10,
      (iii) 100 kHz, (iv) 2%

2.11 (a) (i) 706 kHz, 0.47 μs, (ii) 50 kHz, 6.7 μs
    (b) (i) 65 kHz, 5.2 μs, (ii) 56 kHz, 6 μs

2.12 (a) (i) 5 pF, 8 $\times$ 10^5 Hz, 4 V/μs (ii) 10 pF, 8 $\times$ 10^5 Hz, 2 V/μs
    (b) 400 pF, 0.05 V/μs

2.13 (a) 7.96 kHz, (b) 15.9 kHz

2.14 $\zeta = 0.294$, overshoot 38%
    $C_{f(min)} = 58$ pF, overshoot 4.3%

2.15 5.73 $\times$ 10^6 Hz (equation A.2.6), 35° phase margin (equation A.2.8), 4.44 dB, 5.66 $\times$ 10^6 Hz, 58 μs
    (a) 5, 2.83 $\times$ 10^6 Hz, (b) 10, 1.29 $\times$ 10^6 Hz

2.17 (i) (a) 0.7 V, (b) 0.02 V, (c) 0.02 V, (d) 9.9 kΩ
    (ii) (a) 0.25 V, (b) 0.011 V, (c) 0.011 V

2.18 (a) 0.57 V, (b) 0.11 V

2.19 0.28 V, 0.14 V, 0.07 V
    0.1 V, 0.05 V, 0.025 V

2.20 10^4, 0.01%

2.21 (a) 2.4 μV, (b) 2.4 μV, (c) 4.1 μV

2.22 (a) (i) 310 nV, (ii) 363 nV, (iii) 734 nV, (iv) 2.05 μV
    (b) (i) 20.4 pA, (ii) 20.6 pA, (iii) 26.7 pA, (iv) 41.6 pA
    (c) (i) 0.31 μV, 0.37 μV, 2.05 μV; (ii) 0.36 μV, 0.43 μV, 2.1 μV; (iii) 0.74 μV, 0.87 μV, 3.1 μV; (iv) 2 μV, 2.4 μV, 6.1 μV

2.23 (a) 144 μV, 864 μV (b) 370 μV, 2.2 mV

Chapter 4

4.1 $e_o = -(e_1 + 2e_2 + 10e_3)$, 0.14%

4.2 (a) 101 (40 dB), (b) 202 (46 dB), (c) 1.98 $\times$ 10^4 Hz, (d) 0.41 V

4.3 (a) 66 dB, (b) 10^4 Hz, (c) 0.25 V

4.4 62 dB, 4 $\times$ 10^4 Hz

4.5 10°, 90°, 1.6 $\times$ 10^4 Hz

4.6 (a) 1 μA, (b) 53 nA
    (a) 1.1 μA, (b) 0.4 μA

Chapter 5

5.1 Inputs with $-$ 10 V as reference use $R_{a_1} = 100$ kΩ $R_{b_1} = 1$ MΩ, $R_{a_2} = 100$ kΩ, $R_{b_2} = 125$ kΩ
Inputs with + 10 V as reference are applied through a unity gain inverter, diodes reversed, use $R_{a_3} = 100$ kΩ, $R_{b_3} = 333$ kΩ, $R_{a_4} = 100$ kΩ, $R_{b_4} = 250$ kΩ

5.2   Break points, 0 V, 1 V, 3 V. Slopes, $-1, -1/2, -1/4$
5.3   180 mV, 167 mV
5.5   150 mV
5.6 $R_2 = 15$ MΩ, $R_4 = 54.6$ kΩ
5.7   0.07 V
5.8   (a) 2.5 nA, (b) 0.1 μA
5.9 $R_2 = 10$ kΩ, $R_4 = 54.7$ kΩ
5.10 $e = 1/(10^4 I_o) e_{in}^2$, $e_o = e_{in}^2/10$

Chapter 6

6.1   70 mV/s
6.2   Circuit of Figure 6.8 with $R_1 = 1$ MΩ, $R_2 = 500$ kΩ,
       $R_3 = 100$ kΩ; 2.1 mV/s, 2 mV/s
6.3   0.018 Hz, 0.38%, 10°, 1.5%
6.4 5×10^3
6.5   16 Hz
6.6   4 kHz, 160 Hz, 14 mV

Chapter 7

7.1   See Figure 7.3, make $R_2 = 19R_1$, $E_{ref} = 2.89$ V and
       clamp output levels to $+ 5$ V and $- 1$ V
7.2   See Figure 7.2, input signals applied through resistors
       R, $R/2$, $R/3$ and a reference $- 5$ V applied through a
       resistor $R/6$
7.3 $t_1 = 1.32$ ms, $t_2 = 2.35$ ms
7.4 $R_3 = 49$ kΩ, $R_4 = 196$ kΩ
7.5 $- 4.2$ V to $- 0.95$ V
7.6   Square wave 20 V peak to peak, triangular wave 12 V
       peak to peak; 1 ms, 95 ms, 105 ms

Chapter 8

8.1 10^5 Hz, 16 Hz, 3 mV, noise gain $1/\beta$ increases, upper
       frequency bandwidth limit becomes 8.33×10^4 Hz,
       200 mV
8.2   0° to 176°
8.3 $C_{eff} = 1000$ μF, $R_{eff} = 100$ Ω, 2.05 mV/s
8.4   15 s, 16 s

8.5   10 Hz, 25 V

8.6   11.5 μs, 40 mA, 5 mV/s

Chapter 9

9.4   5.6 mΩ, 2.582 mV

9.5   100 kΩ, 200 kΩ

9.6   (a)   79 μV

      (b)   62 mV

      (c)   95.34 mV

9.7   10 kΩ, 10 kΩ, 0.32 μF, 7.95 nF

9.8 $A_o = 100$, $f_o = 159$ Hz, $Q = 50.5$

      $R_g = 1$ MΩ, $R_7 = 1$ MΩ, $R_8 = 10$ kΩ

      $e_{max} = 0.1$ V, $R_1 = R_2 = 796$ kΩ

      $R_4 = 1.99$ MΩ

Chapter 10

10.1   0.013 pF

10.2   Figure 10.6 $C_p = 5$ pF

10.3 $R_1 = R_2 > 48$ kΩ

10.4 $R_3 = 5$ MΩ

10.5 $R_3 = 99$ kΩ, $R_5 = 690$ kΩ

10.6 -3 V to $+33$ V, 4

INDEX